Rereading the Fossil Record

Rereading the Fossil Record

*The Growth of Paleobiology
as an Evolutionary Discipline*

DAVID SEPKOSKI

THE UNIVERSITY OF CHICAGO PRESS   CHICAGO AND LONDON

The University of Chicago Press, Chicago 60637
The University of Chicago Press, Ltd., London
© 2012 by The University of Chicago
All rights reserved. Published 2012.
Paperback edition 2015
Printed in the United States of America

21 20 19 18 17 16 15   2 3 4 5 6

ISBN-13: 978-0-226-74855-9 (cloth)
ISBN-13: 978-0-226-27294-8 (paper)
ISBN-13: 978-0-226-74858-0 (e-book)
DOI: 10.7208/chicago/9780226748580.001.0001

Library of Congress Cataloging-in-Publication Data

Sepkoski, David, 1972–
    Rereading the fossil record : the growth of paleobiology as an evolutionary discipline /
David Sepkoski.
        p. cm.
    Includes bibliographical references and index.
    ISBN-13: 978-0-226-74855-9 (hardcover : alkaline paper)
    ISBN-10: 0-226-74855-3 (hardcover : alkaline paper)   1. Paleobiology—History.
2. Paleontology—History.   I. Title.
    QE719.8.S47 2012
    560.9—dc23

                                                                          2011035589

♾ This paper meets the requirements of ANSI/NISO z39.48–1992 (Permanence of Paper).

FOR MY FATHER

Contents

Rereading the Fossil Record

When Stephen Jay Gould was five years old, his father took him to the American Museum of Natural History in New York to see the institution's great collection of dinosaurs. Gould later recalled that as he stood in front of the *Tyrannosaurus*, "a man sneezed; I gulped and prepared to utter my *Shema Yisrael*. But the great animal stood immobile in all its bony grandeur, and as we left, I announced that I would be a paleontologist when I grew up" (Gould 1980d, 267). Gould did indeed grow up to be arguably the best known paleontologist in the world, but he never studied dinosaurs. Rather, his doctoral dissertation was on Bermudian land snails, and most of his scientific career was spent investigating the dynamics of evolutionary change using abstract models, computer simulations, and theoretical generalizations. Gould did not just become a paleontologist; he was a "paleobiologist," a designation he helped popularize and install as an important subfield of evolutionary biology. Gould's career was a microcosm of an important change that took place during the second half of the 20th century, that transformed paleontology and influenced the way broad questions about the history of life were incorporated into the developing field of evolutionary biology. This is a book about that transformation.

More specifically, this book is about the development of a subfield of paleontology—"paleobiology"—over the 40 years roughly between 1945 and 1985. It is, essentially, a story about how a small group of scientists solved a problem. That problem was both conceptual and practical. Paradoxically, despite the fact that fossils provide indispensable information about the way evolution has unfolded, paleontology was not considered a serious part of evolutionary biology from the time of Darwin up through the last few decades of the 20th century. There are a variety of reasons for this, which this book will explore in detail, but the main one is quite simple: As Darwin himself recognized, the fossil record is imperfect. The geological record, as he famously put it, is like a book with many missing pages, "a history of the world imperfectly kept, and written in a changing dialect; of this history we possess the last volume alone, relating only to two or three countries. Of this volume, only here and there a short chapter has been preserved; and of each page, only here and there a few lines" (Darwin 1964, 310–11). This is one key component of the problem that faced paleontologists: how to use their major resource in a way that would produce reliable information about the patterns and processes of evolution.

The historical consequence of what I will call "Darwin's dilemma" was that paleontology was marginalized by the rest of the community of evolutionary biology as a "merely descriptive" discipline during much of the 20th century. As the prominent geneticist John Maynard Smith once put it, "The attitude of population geneticists to any palaeontologist rash enough to offer a contribution to evolutionary theory has been to tell him to go away and find another fossil, and not to bother the grownups" (Maynard Smith 1984, 401). This was the second part of the problem: how to convince scientists in other fields that paleontology was a legitimate evolutionary discipline. The dilemma faced by paleontologists interested in establishing paleontology as a legitimate evolutionary discipline, welcomed to the "high table" of evolutionary biology, was thus twofold. It involved both asserting the theoretical value and autonomy of paleontological analysis of the fossil record *and* repositioning paleontology within the larger disciplinary matrix of evolutionary biology. It was a problem that demanded both an intellectual and an institutional response.

The solution reached by a small but disproportionately influential group of paleontologists was, effectively, to reinvent their discipline. Instead of being an "idiographic" field concerned mostly with digging up, describing, and cataloguing individual fossils, paleontology would now focus on large-scale quantitative analyses of patterns in the history of life. This new approach underscored the uniqueness—the essentialness—of paleontology's main resource: the fossil record. Without the perspective of paleontology, there were certain questions in evolutionary biology that simply could not be answered. Central among these were an understanding of macroevolution, or the patterns of evolutionary change visible above the level of the population or species over long periods of time, and the dynamics of mass extinctions, events where substantial portions of the biota (anywhere from 10% to 95%) became extinct in a geological instant. The image of the paleontologist would change, too. Gone was the picture of a dusty fieldworker who spent his life absorbing the minutiae of a single group of extinct organisms. The new model paleontologist was trained in biology as well as geology, was adept at quantitative analysis, was prepared to employ general theoretical models to explain how evolution worked, and might be more comfortable seated at a computer than at a fossil preparation table.

To accompany such a radical change, these paleontologists created a new label for their field: paleobiology. This name emphasized the close relationship between the analyses of life's past and its present. While advocates of paleobiology can be found well back into the early 20th century, paleobiology's most distinctive era was the period between 1970 and 1985, when a "paleobiological revolution" brought sudden visibility and notoriety to the discipline and many of its practitioners.

One purpose of this book is to show how this happened: to document, from the time of Darwin to the recent past, how paleontologists reinvented their discipline by creating a new identity for themselves. This is the first serious attempt to write the history of paleobiology, and my aim here is to create a point of departure for further analysis of paleobiology by historians, sociologists, and philosophers of science. I argue that the guiding principle behind paleobiology was a deliberate manipulation of Darwin's famous "book" metaphor. If the fossil record was widely considered to be an imperfect text, the strategy of paleobiologists was to "reread" that text in a manner that could produce reliable evolutionary insight. This was a consciously adopted metaphor, and I have identified three main approaches to "rereading" the fossil record that were developed by paleobiology's practitioners. In the first approach, paleobiologists attempted a "literal reading" in which the fossil record, with all its notorious gaps and inconsistencies, was taken at face value as a reliable document. There never were, in other words, any missing pages or volumes; the discontinuities in the fossil record existed because the history of life is discontinuous. The major example of literal rereading is the theory of punctuated equilibria advanced in the early 1970s by Gould and Niles Eldredge, and its origins and consequences are traced over the last several chapters of this book.

However, around the same time a radically different strategy emerged, which I have labeled "idealized rereading." Here the physical particulars

of the fossil record were all but ignored, and the history of life—the species, genera, families, etc., that make up the actual record—was modeled as a series of homogeneous data points (or "particles") using very
simple parameters. Crucially, instead of invoking adaptation and selection to explain evolutionary patterns, processes such as speciation and
extinction were assumed to be stochastic (or random). While Gould was
also involved in this approach, its major proponents were Thomas J. M.
Schopf and David Raup, both of whom throughout the 1970s explored
the possibility of a "stochastic paleontology" in which the life histories
of individual organisms or groups were no more important than those
of individual molecules in a volume of gas. This approach was "idealized," because the fossil record itself was not the basis for the models
produced, and it effectively rendered the inadequacies of the fossil record irrelevant.

Finally, I argue that a third strategy emerged in the late 1970s, which I
have called "generalized rereading." This approach combined the other
two, and it ultimately became the dominant methodology in analytical
paleobiology during the 1980s. It sought to resolve Darwin's dilemma by
amassing such an enormous quantity of data that valid statistical generalizations could be made about the patterns in life's history. It accepted
the heuristic value of null hypotheses that modeled the history of life as
an interaction of basically stochastic forces, but also acknowledged the
inescapable pull of history. Generalized models might explain the basic
parameters for how the evolution, diversification, and extinction of life
work, but they must also be based on our best available empirical knowledge of the actual fossil record. While Gould often championed this approach in later popular writings, its main practitioners were Raup and
Jack Sepkoski, who pioneered a generalized interpretation of the history
of Phanerozoic life through statistical analysis of patterns of diversification and extinction in the marine fossil record.

So this book is in part an intellectual history of the paleobiology
movement. But of equal importance is the history of paleobiology as an
institutional endeavor. The strategies used by paleontologists to establish
their science, I argue, illuminate how disciplines change, and specifically
how individual scientists self-consciously manipulate their disciplinary
identities. One of the singular features of the "revolutionary" period
of paleobiology (the 1970s and early 1980s) is how self-aware its proponents were about what they were doing. These paleontologists—Gould,
Schopf, Raup, Eldredge, Sepkoski, and a select few others—intentionally

set about to cause a revolution in their field, and quite self-consciously mapped out the strategy they would pursue. Part of that strategy involved distancing themselves from "traditional" paleontology by producing work that would appeal to the wider community of evolutionary biologists. One task was to create a body of general theory that would draw attention to paleobiology and emphasize the importance of their data and perspective. Another element was engineering institutional support for paleobiology: as David Hull has convincingly shown in the case of systematics, ideas require institutions to support them (Hull 1988). This included establishing academic centers for paleobiology, recruiting students, and developing relationships with scientists in allied disciplines. A crucial element in the success of paleobiology was the founding (and survival) of a new journal—titled simply *Paleobiology*—that served as the mouthpiece for the movement. The journal was conceived, planned, and operated by members of the movement, and its first editor, Tom Schopf, was able to use it as a platform from which to promote a very distinctive vision of what paleobiology was.

Finally, the ultimate establishment of paleobiology required that the rest of the world sit up and take notice. Much of the effort between the 1950s and the 1970s was directed towards internal change—to reinventing the image of paleontology. In the 1980s, however, paleobiologists finally achieved recognition outside of paleontology. Here Gould was the master propagandist, tirelessly articulating the importance of paleobiology to other paleontologists, to biologists, and even to the general public. The reward for this effort was also a broader visibility that saw paleobiology frequently discussed in major general science journals like *Science* and *Nature*, and even in the popular media.

The conceptual and intellectual transformation of paleontology was not accomplished overnight. While the majority of this book examines what I call the "revolutionary" phase of paleobiology in the 1970s and early 1980s, when paleobiologists most actively and successfully crusaded for visibility and acceptance, the first several chapters lay out the historical background to these events. I start by examining the roots of the conceptual and methodological problems surrounding evolutionary interpretation of the fossil record, beginning with Darwin's own view of the reliability of fossil evidence. I then trace the consequences of this view for the conceptual and professional development of paleontology up through the formation of the modern evolutionary synthesis in the 1940s. As I show, despite the historical marginalization of paleontology within evolutionary biology, even during this earlier period a number of paleontologists actively pursued theoretical evolutionary questions using the fossil record as a basis. The establishment of the modern synthesis was a turning point, when paleontology was established as a legitimate (though not necessarily coequal) component of evolutionary biology, and I discuss the contributions of paleontologists like George Gaylord Simpson to this project. The two decades after World War II were important in establishing the theoretical and institutional agenda of paleobiology, and I describe the efforts of Simpson, Norman Newell, and others to carve out disciplinary space for theoretical paleontology. This involved pioneering new methods and approaches to studying evolutionary questions using the fossil record, but also establishing institutional and pedagogical centers for paleobiology that laid the groundwork for the next generation of paleobiologists. Finally, I document the origins of the "revolutionary" phase in paleobiology as an outgrowth of this earlier historical context, and in the second half of the book I examine the formulation, implementation, and promotion of a new "paleobiological agenda" in the 1970s that, by the mid-1980s, came to have a transformative effect on the field.

One final advertisement to the reader: I have a very personal connection to this history. My father, J. John "Jack" Sepkoski Jr., was one of the major contributors to the paleobiological movement I describe during the 1970s and 1980s. At various stages in this project, people have asked me whether this connection might compromise my "objectivity." My answer has been that I believe all historians have investments of one kind or another in their subject matter—we are none of us ever truly objective when writing about subjects that are meaningful to us. While it is somewhat rare for an academic historian to write about such a close family member, I do not think that my family relationship with one of the protagonists in this book makes my analysis any less objective or more biased. In researching this project, I treated my father the same as any other source: I read his published papers, examined letters, notebooks, and manuscripts in his archival collection, and discussed his work with the paleontologists I interviewed. When my father was alive, we often discussed his work in broad terms, but the possibility that I might one day write about it never crossed either of our minds. In fact, when he died in 1999, I was finishing a dissertation on 17th-century mathematics; I have no idea what he would have said about my pursuing this project.

I imagine, though, that his first reaction would have been to be embarrassed by the attention.

This is a work of history, not of advocacy. Historians construct arguments; mine has to do with the characteristics of the intellectual and institutional growth of an important subdiscipline of evolutionary biology. Readers will find that I am just as interested in the failures, conflicts, and dissent that have marked the growth of paleobiology as I am in any success and consensus that was achieved. To put it another way, as a historian I can argue that paleobiology did achieve a remarkable prominence in a fairly short time thanks, directly, to the interventions of a few key players. What those scientists defined as "paleobiology" is now a central interest in the field of paleontology, and the work, institutions, and journals those activists helped to build are thriving today. I will not argue, however, that this was "good" for the profession, or that paleobiology has received more or less attention than it deserves from evolutionary biologists. Those are legitimate questions for a paleontologist or an evolutionary biologist, but as I am neither, I have no stake in the matter. If readers who do belong to those disciplines are moved to reevaluate the legacy of paleobiology—either positively or negatively—from reading this book, then I will be very pleased. But that is not why I wrote it. I wrote this book for several reasons: to tell a story about how scientists and scientific disciplines construct identities and promote agendas; to shed light on a subject in the history of evolutionary biology that has not received much attention from historians; and to encourage other historians, sociologists, and philosophers to take an interest as well. In the end, I also wrote this book to learn something, if very indirectly, about my own history, and to continue a conversation that began many years ago when my father first brought me along on a geology field trip, and which sadly ended much too soon when he died at age 50 just over a decade ago.

CHAPTER ONE

Darwin's Dilemma

*Paleontology, the Fossil Record,
and Evolutionary Theory*

Darwin's Dilemma

It is well documented that paleontological and geological evidence were vitally important to Charles Darwin in establishing his theory of evolution via descent with modification, particularly because the historical evidence of the fossil record enabled him to argue for temporal evolutionary succession of past forms. This first became evident to him during his voyage on HMS *Beagle* in the 1830s, when he observed the succession of a variety of forms of fossil animals like the giant sloth *Megatherium* and the armadillo-like *Glyptodon* along the length of the South American continent. In the first and successive editions of *Origin*, Darwin devoted many pages to discussing the significance of fossil succession, and it is no exaggeration to say that paleontology formed a major pillar of his argument for evolution. Yet in what appears in retrospect a profound irony, even as Darwin elevated the significance of the evidentiary contribution of fossils, he also had a major hand in condemning paleontology—the newly emerging professional discipline devoted to their study—to the status of a second-class discipline. One of his greatest anxieties was that the "incompleteness" of the fossil record would be used to criticize his theory: that the apparent "gaps" in fossil succession could be cited as negative evidence, at the very least, for his proposal that all organisms have descended by minute and gradual modifications from a common ancestor. Darwin worried that at worst, the record's imperfection would be used to argue for the kind of spontaneous, "special" creation of organic forms promoted by theologically oriented naturalists whose theories he hoped to obviate. His strategy in the *Origin*, then, was to scrupulously examine every possible vulnerability in his theory, and as a result he spent a great deal of space apologizing for the sorry state of the fossil record.

Indeed, Darwin devoted an entire chapter to this problem, entitling it "On the Imperfection of the Geological Record." Even as he made the case that fossil data were vital for a true understanding of organic history, he cited the paucity of transitional forms between species as an inherent and potentially intractable problem for geologists and paleontologists. "We have," he wrote, "no right to expect to find in our geological formations, an infinite number of those fine transitional forms, which on my theory assuredly have connected all the past and present species of the same group into one long and branching chain of life" (Darwin 1964 301). The metaphor Darwin chose in his apology for the fossil evidence was that of a great series of books from which individual pages had been lost and were likely unrecoverable. "I look at the natural geological record," he continued, "as a history of the world imperfectly kept, and written in a changing dialect; of this history we possess the last volume alone, relating only to two or three countries. Of this volume, only here and there a short chapter has been preserved; and of each page, only here and there a few lines" (Darwin 1964 310–11).

This metaphor was not Darwin's own invention; he first encountered it while reading Charles Lyell's *Principles of Geology*, where Lyell wrote:

> Let the reader suppose himself acquainted with just one-tenth part of the words of some living language, and that he is presented with several books purported to be written in the same tongue ten centuries ago. If he finds that he comprehends a tenth part of the terms in the ancient volumes, and that he cannot divine the meaning of the other nine-tenths . . . . He must feel at once convinced that, in the interval of ten centuries, a great revolution in the language had taken place. . . . So if a student of Nature, who, when he first examines the monuments of former changes upon our globe, is acquainted with only one-tenth part of the processes now going on upon or far below the surface, or in the depths of the sea, should still find that he comprehends at once the import of the signs of all, or even half the changes that went on in the same regions some hundred or thousand centuries ago, he might declare without hesitation that the ancient laws of nature have been subverted. . . . In truth, there is no part of the evidence in favour of the uniformity of the sys-

tem, more cogent than the fact, that with much that is intelligible, there is still more which is yet novel, mysterious and inexplicable in the monuments of ancient mutation in the earth's crust. (Lyell 1830, vol. 1, 461–62)

Darwin recorded his approval of this metaphor in his "Notebook D" of 1838: "Lyell's excellent view of geology, of each formation being merely a page torn out of a history, & the geologist being obliged to fill up the gaps, is possibly the same with the philosopher, who has [to] trace the structure of animals & plants—he get[s] merely a few pages" (Darwin 1987, 352–53). The metaphor continued to dominate Darwin's thinking about the evidence of transmutation in the fossil record: in "Notebook E," begun in 1839 but not completed until 1856, he endorsed Adam Sedgwick and Roderick Murchison's view of gradational organic change and asked whether "we give up the whole system of transmut[ation], or believe that time has been much greater, & that systems, are only leaves out of whole *volumes*" (Darwin 1987, 433).

Since the metaphor of the incomplete book clearly had currency in the middle part of the 19th century, the "blame" for its corresponding (and discouraging) message to future paleontologists cannot be laid entirely at Darwin's door. But it is important to note that in an *evolutionary* context, the incompleteness of the fossil record takes on enhanced significance. While Lyell eventually accepted transmutation, his *Principles* assumed that organic form was static, and his geology adhered to the strict uniformitarian view that the conditions and processes of the earth and its inhabitants did not vary greatly over time. Transitional forms were not expected, and if organisms were missing from particular localities or strata where they were expected to be found, Lyell assumed that they were simply waiting to be discovered in some other place. Sedgwick's case was even easier: as a follower of Cuvier's "catastrophist" geology, he actually *expected* gaps to be present in the geologic record, which corresponded to Cuverian "revolutions" or cataclysmic, transformative events.

It was thus only after transmutation came into the picture that the paucity of the fossil record became a significant issue. Darwin's theory revolutionized paleontology, since the fossil record became a vital source of evidence that evolution had occurred and for interpreting the history of organic change. Darwin's "dilemma," however, was that he both needed paleontology and was embarrassed by it. Even as he celebrated the contributions of paleontologists, he simultaneously undercut any claims their emerging discipline might have had for autonomy within evolutionary theory. Without evolution, paleontology made interesting, descriptive observations about the form and distribution of once-living creatures; without paleontology, there was far less evidence that evolution had happened. But on its own, paleontology could offer no independent contribution to evolutionary theory, since that theory depended on evidence from biology, breeding, biogeography, geology, heredity, and other fields in order to make the paleontological data meaningful. In other words, paleontology without the support of evolutionary theory could not decisively settle any questions about the nature of organic history—it *required* Darwinian evolutionary theory to contextualize its contributions, and at the same time to excuse its flaws. At least, this is how Darwin and many of his immediate supporters consciously or unconsciously framed the situation—and, as we will see, this had a significant impact on the next hundred years of paleontological theory.

Paleontology after Darwin

Of course, Darwin himself had no reason to feel any special guilt about the unforeseen consequences of his attitude towards paleontology. When he was developing his theory of evolution, biology and paleontology had not yet become firmly established as independent disciplines, and as a "naturalist" he simply marshaled and interpreted the available evidence from all fields as they best supported his argument. But the aftermath of the publication of *Origin* was a period that saw significant disciplinary reorganization, and one result was that scientists became increasingly aware of distinct disciplinary identities. A number of historians have written about the emergence of the experimental tradition in biology during the second half of the 19th century, which contributed greatly to the direction evolutionary study took after 1859.[1] In mimicking some of the laboratory practices and methods of established disciplines like physics and chemistry, biologists greatly enhanced the prestige and autonomy of their field. The emphasis in post-*Origin* biology was on identifying cell structures responsible for heredity (e.g., chromosomes) and

1. See, for example, Gasking 1970, Allen 1975, Pauly 1987, Nyhart 1996, and Nyhart 2009.

studying the physiological processes of biological development (such as patterns in ontogeny; Bowler 1989).

This turn towards biology as a laboratory science indirectly contributed to the formation of a disciplinary identity for paleontology. Darwin had stated, more or less, that paleontology would make limited contributions towards understanding evolution, so for his supporters there was no great urgency to scrutinize the fossil record. In fact, Darwin's supporters were more likely to want to push paleontology into the background: as William Coleman argues, "To the biologist that [fossil] record posed more problems than it resolved . . . . the incompleteness of the recovered fossil record, in which a relatively full historical record for any major group was still lacking, was the very curse of the transmutationist" (Coleman 1971, 66). As a result, there were really only three alternatives available to paleontologists with regard to evolutionary theory: (1) to ignore any special theoretical relevance of paleontological data and focus purely on descriptive studies of morphology and stratigraphy, (2) to accept the Darwinian position but nonetheless try to improve the quality of the record of isolated fossil lineages to support Darwin's theory, or (3) to reject Darwinian evolution and seek some other theoretical explanation of evolution in which fossil evidence could be brought more directly to bear.

Over the next hundred years, and perhaps even longer, the majority of working paleontologists tended to take option 1, which was essentially agnostic towards evolutionary theory. This did not mean rejecting Darwin or evolution; it simply meant not attempting to make any direct contribution to illuminating evolutionary patterns and processes. In the early 20th century this attitude became even more prevalent as the burgeoning petroleum industry's demand for paleontological expertise swelled the ranks of paleontology with scientists whose interest in the field was "economic" (Rainger 2001). Between 1859 and 1900, option 2 probably described the smallest number of actual paleontologists, although a number of naturalists (such as T. H. Huxley) with significant paleontological expertise did contribute paleontological apologia to Darwinism. In the 19th century, paleontology was dominated by vertebrate paleontologists, and paleontological research examining the morphology and anatomy of larger mammals, fish, birds, and dinosaurs was most conspicuous. Certain lineages, such as the early horses, proved to have well-preserved records and provided modest contributions towards

validating natural selection. However, among paleontologists with ambition to contribute to evolutionary theory, the most popular option
was 3—to explore non-Darwinian evolutionary mechanisms. Lamarck's
theory of directional evolution, in which acquired characteristics could
be passed from parents to offspring, remained popular, as did a number of similar theories that attributed directionality to the fossil record.
American paleontologists, in particular, were drawn to Lamarckism
and to orthogenesis, which tended to assume that "directional" evolutionary patterns—such as parallelism and convergence—reflected an internal evolutionary "guiding force." Late-19th-century paleontologists'
subscription to these non-Darwinian evolutionary beliefs had the short-
term consequence of contributing to what some scholars have termed
the "eclipse of Darwin," but the longer-term and more significant effect was that paleontology isolated itself from what would be the mainstream, Darwinian attitude of evolutionary biologists in the first half of
the 20th century (Bowler 1983, chs. 4, 6, 7).[2] As Coleman notes, during
the late 19th century, "in no discipline did expectations appear so great
but frustrations prove so common as in paleontology," which "as a consequence long maintained most ambiguous relations with orthodox Darwinism" (Coleman 1971, 80).

   Of course, not every paleontologist in Darwin's day accepted Darwin's dismal conclusion about the fossil record or its predictions for
the future contributions of paleontology. For example, Darwin's coun-
tryman John Phillips—a paleontologist who conducted the first thorough accounting of the accumulated fossil record and interpretation
of its results—strongly disagreed with Darwin's position. Phillips is remembered today primarily as the inventor of the three great stages in
the history of life: the Paleozoic, Mesozoic, and Cenozoic eras, which
corresponded to discontinuities he observed in the proportions of various major taxa present (or absent) in the succession of the earth's strata.
Phillips depicted the history of life as a series of three, overlapping diversity curves, the first two of which terminate as a new curve begins
its ascent (Phillips 1860, 66; fig.1.1). Phillips's model was ambitious for
its time, but he defended its legitimacy in part by criticizing Darwin's
opinion that "we possess . . . merely fragments of the record, which indeed never was complete. . . . Thus we must not expect to be able to arrange the fossil remains in a really however broken series, since the true

2. See also Rainger 1982 and Rainger 1991.

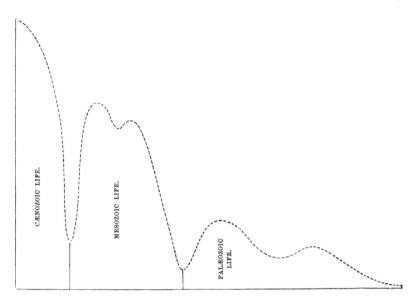

FIGURE I.I. John Phillips's depiction of the history of life as three successive diversity curves. John Phillips, *Life on the Earth; Its Origin and Succession* (Cambridge and London: Macmillan and Co., 1860), 66.

order and descent may be, and for the most part is, irrecoverably lost." Rather, he countered, "surely this imperfection of the geological record is overrated. With the exceptions of the two great breaks at the close of the Paleozoic and Mesozoic periods, the series of strata is nearly if not quite complete, the series of life almost equally so" (Phillips 1860, 206–7). Phillips's approach to the history of life was simultaneously progressive and conservative: it was progressive both in the sense that it argued for the epistemological significance of the fossil record and because it promoted a view of the history of life that was directional—*pace* Lyell, time's arrow moved steadily forward, and the fossil record demonstrated irrecoverable changes in the pattern of life's history (Gould 1987; Rudwick 2005a; Rudwick 2005b). On the other hand, as Peter Bowler has observed, Phillips's view of progressive change exhibits traces of the "conservative, idealistic philosophies that stood opposed to the liberal image of gradual transformation built upon by evolutionists such as Darwin and Herbert Spencer" (Bowler 1996, 353).

Among Darwin's supporters, the most outspoken apologist for the paleontological record was probably Darwin's "bulldog," Thomas Henry Huxley. While Huxley differed with Darwin about the origin of new types (e.g., higher taxa), he nonetheless defended Darwinian descent with modification using evidence from paleontology (Lyons 1999; Desmond 1982, ch. 3). In his 1870 address to the Geological Society, "Paleontology and the Doctrine of Evolution," Huxley argued that "it is generally, if not universally, agreed that the succession of life has been the result of a slow and gradual replacement of species by species; and that all appearances of abruptness of change are due to breaks in the series of deposits, or other changes in physical conditions" (Huxley 1894, 343). Huxley emphasized the uniformitarian assumption in this view that "the continuity of living forms has been unbroken from the earliest times to the present day," and concluded with the Lyellian-sounding proposal that "the hypothesis I have put before you requires no supposition that the rate of change in organic life has been either greater or less in ancient times than it is now; nor any assumption, either physical or biological, which has not its justification in analogous phenomena of existing nature" (Huxley 1893, 343 and 388). In private, however, Huxley was unconvinced that the evolution of higher taxa did not involve some form of saltation, or evolutionary "leaps." In an 1859 letter he wrote to Lyell that "the fixity and definite limitation of species, genera, and larger groups appear to me to be perfectly consistent with the theory of transmutation. In other words, I think *transmutation* may take place without transition." This did not mean that he rejected natural selection, or even that he doubted gradualism, but at least in the year *Origin* was published he felt the paleontological evidence "lead[s] me to believe more and more in the absence of any real transitions between natural groups, great and small" (Huxley to Lyell, June 25, 1859, quoted in Herbert 2005).

If Darwin's reading of the fossil record had difficulty gaining traction even among his staunchest allies, it is easy to understand why many paleontologists less ideologically committed to Darwinism felt compelled to pursue an entirely different theoretical approach. There is some truth to the perception that paleontology offered few contributions to evolutionary theory between the publication of *Origin* and the period of the modern evolutionary synthesis in the 1930s and 1940s, but this interpretation hardly tells the whole story. Many paleontologists did in fact pursue descriptive work in morphology and stratigraphy with very little interest in evolutionary theory, but there was a sizable minority who had genuine theoretical and evolutionary ambitions. This group included a number of prominent late-19th- and early-20th-century paleontologists, including the American vertebrate specialists O. C. Marsh, E. D. Cope, Alpheus

Hyatt, H. F. Osborn, W. D. Matthew, and William Gregory, and such European paleontologists as George Mivart, Alexandr Kovalevskii, Othenio Abel, Louis Dollo, Wilhelm Waagen, Karl Alfred von Zittel, and Otto Schindewolf. When in 1944 George Gaylord Simpson published his landmark *Tempo and Mode in Evolution*, it certainly marked a turning point in terms of paleontology's reception among evolutionary biologists, but Simpson was hardly sui generis. There were, in fact, a great many paleontologists between 1860 and 1940 who pursued evolutionary theory. The problem was—at least from the perspective of the eventual framers of the modern synthesis—these paleontologists pursued the wrong *kind* of evolutionary theory.

It is worth taking a moment to consider what, to a paleontologist, the fossil record seems to indicate about evolutionary processes and patterns, and how that interpretation might differ from the perspective of a biologist. As Bowler points out, 19th-century research into systematics was concerned especially with reconstructing fossil phylogenies. This involved arranging fossils into likely sequences based on "structural resemblances" and attempting to extrapolate evolutionary development across morphological and stratigraphic gaps. For this reason, "morphology thus became the first center of evolutionary biology" (Bowler 1996, 41). The problem, however, is that the sequence of morphology in the fossil record does not always clearly indicate the steps in evolutionary sequence taken by a particular lineage. Fossils are often too rare, too poorly preserved, or just plain missing, and it is left to the paleontologist to connect the dots in as likely a fashion as possible. A feature that appears to stand out in the fossil record is the appearance of fairly linear trends, even among distantly or unrelated groups, towards similar morphological features. One example is the phenomenon of parallelism, or the tendency for multiple lines of descent to follow "a more or less identical sequence of morphological stages" after divergence from a common ancestor (Bowler 1996, 70). Another apparent trend is convergence, in which wildly different groups independently settle on the same adaptive response to an environment without the presence of a recent common ancestor (as in the evolution of wings in both birds and bats). The question posed to paleontologists was whether these phenomena could be explained by simple Darwinian natural selection, or whether some other force or mechanism was required.

In other words, many late 19th-century paleontologists were sufficiently impressed by the apparent *linearity* of evolution that they felt compelled to reject Darwinian natural selection as its primary cause (Bowler 1983, 118ff). "Linearity" should be understood as the appearance of trends in the fossil record that cannot be explained simply as the random product of accumulated adaptive responses to environment—trends, such as increasing body size, that appear to be "preprogrammed" into evolutionary development. The common name for such a belief is "orthogenesis," which holds that evolution proceeds in an upwardly linear (orthogonal) path along a predetermined trajectory. The essential assumption of orthogenesis is that trends are produced independently of adaptive fitness, even to the ultimate detriment of a lineage, as in the case of so-called senescence or "racial senility" of a group, which leads to eventual extinction.

Orthogenesis (and Lamarckism) particularly caught on in America, where Swiss émigré Louis Agassiz appears to have played a central role by promoting an "idealist philosophy of nature" from his position as curator of the Museum of Comparative Zoology at Harvard (Bowler 1983, 120). This theory appealed particularly to paleontologists—and to vertebrate paleontologists especially—who were attempting to reconstruct lineages across millions of years based on scant, often incomplete fossil evidence. E. D. Cope was perhaps foremost in the 19th-century American school of orthogenetic paleontology, and his views are preserved in the so-called Cope's law of increasing body size over evolutionary history. As for the question of why paleontologists rather than biologists were drawn to orthogenetic theories, the answer is quite straightforward: trends become apparent only when time is made a dimension of evolutionary study, and paleontology is the branch of evolutionary biology that deals with the temporal evidence of evolution. Trends have had a persistent interest for paleontologists, and indeed—as later chapters will discuss—much of 20th-century evolutionary paleobiology has been concerned with identifying and evaluating evidence for apparent trends in the fossil record. Where 19th-century orthogenetic paleontologists ran afoul of Darwinism was in asserting that internal forces guided these trends independently of adaptation; a major task of 20th-century paleobiology has been to explain the appearance and existence of trends without invoking non-Darwinian, speculative mechanisms.

In any event, the late 19th century saw the development of a framework for a genuinely paleontological evolutionary theory. There is no question, however, that paleontology was also pushed outside of the developing institutional framework of biology during this period; as Ronald Rainger argues, the subject was increasingly shut out of university departments and relegated to museums, where it "remained largely peripheral to the developments occurring in American biology." Where paleontology was given a foothold in academic science faculties, it was most often made a subdivision of geology, where biological, evolutionary work was discouraged (Rainger 1988, 219). This pattern is seen in the careers of two of the leading American proponents of paleontological evolutionary theory, the vertebrate specialists Marsh and Cope. While Marsh had institutional support at Yale, both he and Cope were essentially free agents who relied on personal wealth and independent funding to pursue collection and research on North American dinosaur and mammal fossils.[3] Marsh was fortunate to have the support of his wealthy uncle George Peabody, who endowed Yale's Peabody Museum, as well as a lucrative position with the US Geological Survey, and therefore enjoyed a large support staff to assist in his work. Cope was less fortunate, and spent much of his career in Philadelphia scrambling for funds in an effort to keep up with Marsh (Rainger 1991, 12–13).

Of the two, Cope made the more serious effort to develop a fully articulated, non-Darwinian evolutionary theory. Marsh's interests were primarily descriptive and systematic, but while "he rarely examined questions concerning the process or pattern of evolution," his systematic studies of dinosaurs, birds, and fossil horses nonetheless "substantiated evolution" (Rainger 1991, 14). Darwin, in particular, admired Marsh's reconstruction of birds with teeth, and Huxley was quite impressed with his work on fossil horses during a visit to the United States. It was Cope's Lamarckian evolutionary theory, however, that had the most influence on contemporary paleontology, even though it seems to have been motivated more by theological and idealistic, rather than strictly empirical, concerns (Bowler 1983, 123; Rainger 1991, 13). Essentially, Cope's theory combined adaptive response to environment (via Lamarckian inheritance of acquired characteristics) with an internal directing force that accounted for the nonadaptive, directional trends he believed he observed in the fossil record. He applied this theory to a variety of evolutionary trends: the evolution of mammalian teeth, the progression of horse feet and hooves, and of course the evolution of increasing size and complexity (Bowler 1983, 124–25). Cope's work had a significant impact

3. The literature on Marsh and Cope—and particularly their infamous 'feud'—is voluminous. See Jaffe 2000, Wallace 1999, and Brinkman 2010.

on the development of non-Darwinian evolutionary ideas, but perhaps his more lasting influence on the development of paleontological evolutionary theory was in imparting a sense of theoretical mission to his protégé H. F. Osborn, who would be instrumental in establishing paleontology as an institutional mainstay at the American Museum of Natural History after the turn of the century.

In any case, it is fair to say that the promise of paleontology as a central evolutionary discipline, which had appeared so bright in the mid-19th century, had faded considerably by 1900. The most spectacular advances in the field had been in the collection of large vertebrate fossils, and broad, empirical studies of evolutionary pattern and process—such as Phillips's accounting of the fossil record—were not actively pursued. Rightly or wrongly it was also perceived that paleontologists had abandoned Darwinism and natural selection, which alienated those evolutionary biologists who were still committed to Darwinian orthodoxy, and paleontology would pay heavily for that perception when Darwinism emerged triumphant in the mid-20th century. Finally, from an institutional perspective, paleontology was in danger of losing all contact with biology: isolated in geology and museum collections departments, paleontologists had little regular interaction with experimental biologists. This led to mutual mistrust and incomprehension between the two fields that was only exacerbated after the "genetic turn" in biology following the rediscovery of Mendel. Darwin may have considered paleontology, geology, and biology to be equal partners in the enterprise of evolutionary natural history, but as the 20th century began, these fields were separated by a fairly wide gulf.

Twentieth-Century Paleontology before the Modern Synthesis

Paleontology and the Emerging Synthesis

In the revisionist history of biology offered by supporters of the modern synthesis, there is a tendency to view the pioneering American vertebrate paleontologist George Gaylord Simpson as the savior of modern paleontology—the person who rescued the field from descriptive oblivion and enabled its reintegration with evolutionary biology (Gould 1980a, 154; Laporte 2000, 2). For all of its genuine originality, however, Simpson's contribution was magnified by a particular confluence of developments inside and outside paleontology. And while Simpson's paleontological theory provided much of the inspiration for modern paleo-biology, it also contributed to the perpetuation of prejudices towards paleontology established back in Darwin's day.

As Betty Smokovitis has argued, the modern synthesis was in its essence a project of unification. As such, its goals were institutional and political as much as theoretical and empirical. Its major architects— Theodosius Dobzhansky, Ernst Mayr, Sewell Wright, Julian Huxley, and Simpson—attempted to construct a particular view of evolutionary theory that enshrined certain practices (experimental genetics, population biology), endorsed particular theoretical alternatives (natural selection, genetic drift), and excluded others (orthogenesis, saltationism). According to Smokovitis, this tradition involved "preserving the whole of the disciplinary ordering of knowledge and an Enlightened worldview," one that was "liberal, humanistic, and secular" (Smocovitis 1996, 99–100). Paleontological theories of evolution were not singled out for exclusion by the architects of the synthesis—biologists such as Hugo de Vries and Richard Goldschmidt, who promoted macromutations or saltations, attracted much greater hostility—but the idealistic orthogenetic and La-marckian theories endorsed by Cope and others clearly had no place in the synthetic view.

Nor, as it turned out, was the methodology adopted in the 19th century by most vertebrate paleontologists adequate to meet the demands of the synthetic conception of rigorous quantitative science: vertebrate paleontology was first and foremost descriptive, and quantitative paleontological analysis was limited to the most cursory kinds of anatomical measurements and tabulations. Biology, on the other hand, underwent a quantitative revolution in the first several decades of the 20th century, where "the attachment of numbers to 'nature'—and the growing measurability and testability of natural selection within a populational framework" helped produce "a mechanistic and materialistic science of evolution that could rival Newtonian physics" (Smocovitis 1996, 122 and 127). The impetus for this transformation was the discovery of quantitative laws of heredity, such as the Hardy-Weinberg principle of stable genetic equilibrium, which established a mathematical basis for confirming the expectations of natural selection in populations (Provine 1971). Paleontologists simply had no way of translating their data into terms that population biologists and geneticists appreciated and could make use of— and until Simpson stepped to the fore, they remained mostly invisible to the synthesists.

A sampling of statements regarding paleontology and its potential contributions by important biologists between 1900 and 1945 reveals the extent to which the discipline had sunk in the eyes of the larger evolutionary community. For example, Thomas Hunt Morgan, whose study of the genetics in populations of fruit flies produced the modern field of genetics, offered the following sneering evaluation of paleontology in 1916:

> Paleontologists have sometimes gone beyond this descriptive phase of the subject and have attempted to formulate the 'causes,' 'laws' and 'principles' that have led to the development of their series. . . . The geneticist says to the paleontologist, since you do not know, and from the nature of your case you can never know, whether your differences are due to one change or to a thousand, you can not with certainty tell us anything about hereditary units which have made the process of evolution possible. And without this knowledge there can be no understanding of the causes of evolution (Morgan 1916, 25–27).

An echo of this attitude is found in Julian Huxley's *Evolution: The Modern Synthesis* (1942), which served as a kind of manifesto for the synthetic movement. Huxley opined that many paleontologists had been misdirected towards orthogenesis and Lamarckism because "the paleontologist, confronted with his continuous and long-range trends, is prone to misunderstand the implications of a discontinuous theory of change" (Huxley 1942, 31). Paleontological data is inherently suspect because the fossil record is incomplete: it is often poorly preserved, and the material which *is* preserved is insufficient to inform theoretical conclusions. Or, as Huxley put it bluntly, "paleontology is of such a nature that its data by themselves cannot throw any light on genetics or selection. . . . All that paleontology can do . . . is to assert that, as regards the type of organisms which it studies, the evolutionary methods suggested by geneticists and evolutionists shall not contradict its data" (Huxley 1942, 38).

This attitude is visible even among the more moderate and accommodating biologists of the synthesis era. Despite more openly courting paleontological participation in the synthetic project, Mayr, Dobzhansky, and other biologists were very specific about the kinds of information paleontology could supply, and were careful to circumscribe that contribution within the bounds of what was emerging as the neo-Darwinian framework of the modern synthesis. In particular, Mayr and others noted that the field of macroevolution, the study of the larger patterns of evolution observable at taxonomic levels above the species, was a general source of confusion. In *Systematics and the Origin of Species* (1942), Mayr wrote that in the past "the field of macroevolution [had been] left more or less to the paleontologist and anatomist," leading to "difficulties and misunderstandings, since paleontologists, taxonomists, and geneticists talk three different languages" (Mayr 1942, 291). Nonetheless, Mayr cautioned that "even if some of the generalizations and interpretations of the paleontologist and the taxonomist are wrong or expressed in unfortunate (for example Lamarckian) terms, this does not invalidate the facts on which these interpretations are based, and nothing is gained by ignoring them." He added, "We must be grateful to the paleontologists and the anatomists who have attempted to establish some general laws" (Mayr 1942, 292). Gratitude did not, however, necessarily extend to *accepting* those theories, and Mayr was quick to point to the incompleteness of the fossil record as an impediment to establishing paleontological generalizations.

Macroevolution, "Paleobiology," and the Synthesis

Although the term "macroevolution" does not appear in the literature before 1927, it can be argued that the concept is as old as evolutionary theory itself. Darwin's theory was inspired in large part by his dissatisfaction with previous crypto-mystical evolutionary concepts like "power of life," and with contemporary explanations of organic form and taxonomic classification that appealed to divine intervention. For Darwin, natural selection is the ubiquitous process that gradually adds up small variations to produce large changes: famously, "It may be said that natural selection is daily and hourly scrutinizing, throughout the world, every variation, even the slightest; rejecting that which is bad, preserving and adding up all that is good" (Darwin 1964, 84). Darwin explained what we would now call macroevolutionary trends through the phenomenon of divergence, or the tendency for offshoots of a particular lineage to differentiate from one another over time. In answering the question of "how it is that varieties . . . become ultimately converted into good and distinct species," he felt he needed no answer other than natural selection (Darwin 1964, 61). This helps explain Darwin's later supporters' reflexive insistence on natural selection as the sole evolutionary mechanism. That reflexivity was part of Darwin's defense against creationism; in considering the evolution of complex organs, he explained that "if it could be demonstrated that any complex organ existed, which could not possibly have been formed by numerous, successive, slight modifications, my theory would absolutely break down" (Darwin 1964, 189). Darwin championed what has been called the "variational" theory of evolution, which emphasized the nonprogressive nature of the evolutionary process. In order to avoid giving ground to the argument from design, Darwin was unable to consider macroevolution as a process independent from natural selection on individuals.

The term "macroevolution" itself first appeared in the Russian biologist Iurii Filipchenko's 1927 work *Variabilität und Variation* (which was published in German). Filipchenko argued that in the contemporary, genetics-dominated field of biology, "an evolution of the higher systematic groups (a kind of macroevolution)" remained to be explained, and he noted "the absence of an internal connection between genetics and a theory of descent, the latter of which is chiefly the concern of macroevolution" (Filipchenko 1927, 93–94).[4] Filipchenko's work had only limited direct influence in the West, but his ideas helped shape the views of his student, Theodosius Dobzhansky, whose book *Genetics and the Origin of Species* (1937) became a pillar of the modern synthesis. Dobzhansky adopted his mentor's term for large-scale patterns of evolution, but he rejected Filipchenko's suggested causal separation of the mechanisms of macroevolution from microevolution. In a statement that became paradigmatic of the synthetic approach to macroevolution, Dobzhansky argued, "There is no way towards understanding of the mechanisms of macroevolutionary changes, which require time on geologic scales, other than through an understanding of microevolutionary processes observable within a span of a human lifetime, often controlled by man's will, and sometimes observable in laboratory experiments." He confidently concluded that "the words 'microevolution' and 'macroevolution' are relative terms, and have only descriptive meaning; they imply no difference in the underlying causal agencies" (Dobzhansky 1951, 16–17).

This position essentially became the official party line for synthetic biology. Mayr's view of macroevolution mirrored Dobzhansky's, arguing that "there is only a difference of degree, not one of kind" between micro- and macroevolution, and concluding that "all the processes and phenomena of macroevolution and the origin of higher categories can be traced back to intraspecific variation" (Mayr 1942, 291–98). As the synthesis began to "harden" in the later 1940s and 1950s, additional support

---

4. All translations are mine unless otherwise noted.

for this "extrapolationist" approach came from systematics, comparative morphology, and paleontology (Gould 1983 and (Provine 1986). In general, any deviance from extrapolationism met with a frosty reception from the synthesists. As we will see, Simpson, who became an important architect of the synthesis in his own right, felt compelled to modify his initial and apparently heterodox views about macroevolution to stay in the good graces of Mayr, Dobzhansky, and Wright. The biologist Richard Goldschmidt, on the other hand, was cast into oblivion for directly challenging the synthetic account. Goldschmidt is infamous for proposing that macromutations, or "hopeful monsters," account for all major taxonomic novelties, and his name has been banished to the fringes of biology ever since. In his 1940 book *The Material Basis of Evolution*, he argued for a complete decoupling of macroevolution from microevolution, the latter of which, he maintained, "does not lead beyond the confines of the species." Goldschmidt advanced the radical proposal that "species and the higher categories originate in single macroevolutionary steps as completely new genetic systems" as the result of sudden, drastic mutations (Goldschmidt 1940, 396).

It should be stressed that any animus shown towards paleontology by synthetic biologists was directed primarily at stamping out orthogenesis and Lamarckism, both of which were antithetical to the synthetic theory. Biologists at the time may have been under the impression that most paleontologists were anti-Darwinian, but, as Bowler notes, "by the 1920s . . . the excessive claims for the evolution of nonadaptive characters were being greeted with increasing skepticism by English-speaking paleontologists," indicating a trend that "predates the emergence of the modern synthetic theory of evolution in the 1940s, showing that paleontologists were playing an important role in challenging the old non-Darwinian theories" (Bowler 1996, 360). Interestingly, it was also during this time that the term "paleobiology" came into wider usage. The earliest record of the term is an 1893 paper in the *Quarterly Journal of the Geological Society of London* by S. S. Buckman, who commented on the usefulness of a term that "I may call 'palæo-biology'" (Buckman 1893, 127). However, the source for the eventual widespread usage of the label was the Austrian vertebrate paleontologist Othenio Abel, who began using the term *päleobiologie* to describe biologically-informed paleontology as early as 1912. In that year he published *Grundzüge der Paläobiologie der Wirbeltiere* (Fundamentals of vertebrate paleobiology), followed by *Paläobiologie der Cephalopoden* in 1916; his most widely read work among English-speaking paleontologists was *Paläobiologie und Stammesgeschichte* (Paleobiology and phylogeny), published (but never translated) in 1929.

Abel is an interesting case: as a distinguished professor of paleontology at the universities of Vienna and Göttingen, he had a significant influence on German paleontology before World War II, and was responsible for founding the journal *Palaeobiologica* in 1928 as the official organ of the Viennese *Paläobiologischen Gesellschaft* (Paleobiological Society) (Reif 1986; Reif 1999). Theoretically, he was sympathetic to the idealist tradition of directional evolution and he supported a version of orthogenesis, but as Bowler notes, he also "made at least a pretense of conforming to a mechanistic language" in presenting his theory (Bowler 1996, 359). For example, in *Paläobiologie und Stammesgeschichte* Abel wrote that "we need assume neither a supernatural principle of perfection, nor a principle of progression, nor a vital principle," but that nonetheless "the phenomenon of orthogenesis, which has often been disputed but now can no longer be denied, is transmitted by the mechanical law of inertia into the organic world" (Abel 1980, 399). However, he was also a strong proponent of the biological basis for paleontological theory, and his orthogenetic beliefs were not cultured in isolation from biology, nor did he reject all of the adaptationist tenets of Darwinism. In fact, he argued that "research on adaptation had originally cultured the nucleus of paleobiology" in Darwin's day (Abel 1980, v), and he lamented the subsequent exclusion of paleontology from biology:

> One should think that through the appearance of this work [*Origin of Species*], which produced such an enormous revolution in biology, paleontological research would all at once be steered onto a new path, and that the basis for these sorts of [paleobiological] investigations were prepared here. It is so much the more astounding that paleontology held itself in the background for so long, and can scarcely take its place in that eternally lively discussion, in any case not to the extent as the depth of the available knowledge of fossils allowed at that time. The cause of the delayed entrance of paleontology into the path of biological research on phylogeny lay less in its not yet very considerable breadth of principles of observation and comparison, but rather first of all, because at that time Darwin's theory of selection produced such a disturbance on the field of biology that the scientific understanding of fossil remains lay nearly exclusively in the hands of geology (Abel 1980, 5).

In other words, Abel argued that paleontology had been prevented from taking its place at the evolutionary "high table" in part by its subordination to geology, a complaint that would become more and more common among paleontologists over the next several decades. Abel concluded, however, that paleobiology had a decisive role to play in evolutionary theory: "Among all phylogenetic research disciplines paleobiology stands alone in being able to demonstrate historical documentation, and to make readable and to draw conclusions from these facts" (Abel 1980, vi).

Abel's work was read fairly widely by American paleontologists, and was cited repeatedly by Simpson. Simpson did not approve of Abel's reliance on orthogenesis to explain the evolution of horses (at one point calling Abel's belief "naïve"), but Abel's general message about the ambitions of paleontology was more warmly received (Simpson 1944, 149). For example, in 1926 Simpson published a paper on the evolution of Mesozoic mammals, which he described as "a study in paleobiology, an attempt to consider a very ancient and long extinct group of mammals not as bits of broken bone but as flesh and blood beings" (Simpson 1926, 228). This was Simpson's first use of the term "paleobiology," and he prominently cited Abel's *Grundzüge der Paläobiologie der Wirbeltiere*. Simpson also recalled many years later, "While still in graduate school I found Othenio Abel's books particularly interesting and useful" (Simpson, quoted in Mayr and Provine 1980, 456; see also Abel 1912). Simpson was a committed reader of German scientific literature (he later reviewed German paleontological publications in the journal *Evolution* for his linguistically challenged colleagues), and at the very least his and others' familiarity with Abel's work probably accounts for the origin of the term "paleobiology" in its modern context (Kutschera 2007).

Abel's case is partial evidence of broader interest among paleontologists in extending the biological relevance of their research well before the period of the synthesis. In the United States the Department of Vertebrate Paleontology (DVP) at the American Museum of Natural History, where Henry Fairfield Osborn set up shop in the late 19th century, was a center of biologically inspired paleontology. Osborn inherited a suspicion of Darwinian natural selection from Cope, and his pet theory was a variant of orthogenesis he eventually called "aristogenesis," or evolution towards improvement. Osborn strongly endorsed greater theoretical independence for paleontology, but he also did much to alien-

ate the biologists with whom he interacted. For example, he asserted that
the solution to "the twenty-five-century problem of the origin of biomechanical adaptations" is derivable "solely from paleontology, while
wholly beyond the ken of zoölogy or experimentalism," and rather in-
temperately claimed that "the larger number of modern zoölogists are
committing suicide by adapting a Darwinian creed . . . . [and are] travers-
ing a swamp of useless inquiry let by the will-of-the-wisp of expectation"
(Osborn 1933, 160 and 163). Naturally this did not endear Osborn to his
Columbia University colleague T. H. Morgan, and it probably explains
some of Morgan's hostility towards paleontologists. Rainger argues that
Osborn's feud with Morgan actually set paleontology back in the eyes of
biologists, and that "to an extent, Osborn's ideas exacerbated the scientific and institutional isolation of vertebrate paleontology in the United
States" (Rainger 1991, 246).

  In the first several decades of the 20th century, paleontologists often reflected the attitudes projected onto their discipline by biologists.
Rainger concludes that despite the biological interests of people like Osborn and Abel, "interest in such biological questions did not transform
the discipline of paleontology. In the 1920s, just as in the 1880s, many
students of the fossil record remained preoccupied with descriptive, taxonomic questions, and vertebrate paleontology was still primarily a museum science" (Rainger 1988, 244). According to Rainger, this state of
affairs persisted until Simpson offered his radical reevaluation of paleontological goals and methods. This assessment is probably accurate
for the bulk of paleontological practice in the first part of the 20th century, but it is important not to diminish the continuity between Simpson
and his predecessors, nor to overstate the discontinuity between Simpson's approach and prior paleontological theory. Paleontologists up to
and during the synthesis elaborated a theoretical agenda for their discipline, and Simpson's voice was perhaps just the loudest and most persua-
sive among those of his contemporaries.

The Paleontological Society and Evolutionary Theory

The establishment of a professional society for American paleontologists
provides a useful perspective on the development of paleontology during
the first half of the 20th century. In 1908, a committee established by the
Geological Society of America met to discuss forming a separate profes-

sional organization for paleontologists. The committee was chaired by Yale invertebrate specialist Charles Schuchert, and was composed of paleontologists F. B. Loomis, David White, T. W. Stanton, S. W. Williston, and Osborn. The organization's first president was John M. Clarke, who in 1910 established the tradition of giving an annual address, which has continued to the present day.[5] While no official mandate determines the subject matter of these addresses, over the years many exiting presidents have taken the opportunity to consider major directions or challenges in the field, and have often explicitly offered a "state of the discipline" assessment. As such, these presidential addresses provide a window on emerging attitudes among professional paleontologists in the period leading up to the modern synthesis. Thomas J. M. Schopf, who has analyzed these addresses in some detail, calculates that between 1909 and 1979, questions about the history and "philosophy of paleontology and predictions for the future" led all other topics in popularity, in a total of 20 addresses. Additionally, a further eight addresses considered topics involving evolution and extinction, so that out of 52 papers whose subjects are known, more than half were concerned with the "big picture" of paleontology (Schopf 1980). While presidents of the society—having been elected by the general membership for extraordinary contributions to the field—may not accurately represent the attitudes of rank-and-file paleontologists, their addresses do suggest that paleontology was hardly theoretically moribund before 1950.

In 1914 Osborn presented a short address comparing "vertical changes in morphology with comparable changes over the geographic range of a species" (Schopf 1980). The central question he asked was: "In what respects do the characters observed in a genus ascending and developing in geologic time resemble or differ from the characters observed in a genus distributed in geographic space?" (Osborn 1914, 411) In essence, Osborn proposed that time and space are equally important axes in plotting morphological evolution—or, as he put it, "To institute a true comparison between a geographic series and a geologic series precisely the *same methods of observation* should be employed" (Osborn, 1914, 415). Interestingly, Osborn did not mention either orthogenesis or aristogenesis, but argued that the pattern of evolution "marks the steps from

5. For a history of the Paleontological Society, see Schopf 1980. See also Rainger 2001. On early 20th-century American paleontology, see Rainger 1997, Rainger 1993, Rainger 1988, Rainger 1986, and Rainger 1982.

'species' to 'species' and the minute transition stages between species" without speculating on the mechanism that produced this change (Osborn, 1914, 411–12). One of the synthesists' central objections to orthogenesis was that it potentially cast aside the Darwinian assertion of gradual, incremental change through selection. At least on this occasion, however, Osborn professed agnosticism with respect to mechanism, stating, "Whether the causes of these changes are to be sought in heredity or ontogeny or environment or selection, or in the interactions of these four coefficients of evolution, is a problem which remains obscure" (Osborn, 1914, 416).

The first presidential address to strongly advocate the adoption of methods and conclusions from genetics was delivered in 1922 by William Diller Matthew, Osborn's longtime assistant in the DVP at the American Museum. According to Rainger, Matthew developed a strong personal dislike for Osborn, and he came to reject Osborn's aristogenetic evolutionary theory in favor of fairly orthodox Darwinism. Osborn relied heavily on Matthew's abilities in biostratigraphy and paleobiogeography, and as a result Matthew developed a much greater sensitivity for the accuracy of the fossil record than did his patron (Rainger 1991, 183–84). However, this knowledge led Matthew to downplay the completeness of the fossil record, and it encouraged his skepticism about larger, theoretical claims made solely on the basis of paleontological evidence.

Matthew's address to the Paleontological Society, "Recent Progress and Trends in Vertebrate Paleontology," was presented as a critical review of the "adequacy" of the "foundations" (i.e., the fossil data) on which paleontology is built (Matthew 1923, 401–2). Matthew's argument was decidedly ambivalent: on the one hand, he defended the legitimacy of paleontological contributions to evolutionary theory, but on the other, he cautioned paleontologists about overreaching their evidence. Concerning the attitude of many geneticists towards paleontology, Matthew argued that he did "not altogether agree with a distinguished Columbia professor [Morgan] who declared not long ago that paleontologists had no business to reason on or draw conclusions from their specimens, but should content themselves with describing and illustrating them. . . . Nevertheless, I do think we should distinguish far more sharply between provisional and tentative conclusions based on scanty and fragmentary data and those which are really proven by adequate evidence" (Matthew 1923, 415). As an example, he reminded his audience that "the paleontologist . . . is dealing with true genetic sequences, exact or approximate; with the evolution of species and genera of animals, not merely with illustrations of how certain structures may have evolved" (Matthew 1923, 417). Nonetheless, he did hold out hope for paleontology's independent contribution to evolutionary theory; ironically, he felt this would only be assured by diligent descriptive work documenting specimens in stratigraphic sequences in order to document "the evolution of species and not merely of structures" (Matthew 1923, 417). Then, Matthew argued, "Professor Morgan's strictures on paleontological evolution, which are aimed really at the old methods, not our modern standards, will be no longer justified" (Matthew 1923, 418).

Matthew's message undeniably made an impression on Simpson, with whom he worked in the field in 1924, the year after the address was published. Simpson replaced Matthew in the DVP in 1927, and in his autobiography he recalls being "awed, and instructed" by his senior colleague "at least as much as by my major formal professor" (Simpson 1978, 33–34). Matthew is also credited by historians for having an instrumental role in turning paleontology towards genetics and, eventually, its place in the modern synthesis, and he helped turn the tide against paleontological support for directional evolution (Rainger 1991, 214; Bowler 1996, 417). From another perspective, though, it appears that Matthew— and, later, Simpson—helped deprive paleontologists of their major arguments for theoretical independence from biology. The not-so-subtle message Matthew preached in his presidential address and elsewhere was that paleontology had theoretical value *if* it could tell geneticists what they wanted to hear. Unlike Osborn, Matthew accepted Darwin's disclaimer about the imperfection of the fossil record and consequently, intentionally or not, prepared the path for paleontology toward an inferior position at the "high table" of evolutionary theory. Several generations of later paleontologists would chafe under the constraints he and Simpson helped impose on their field.

Whether or not as a result of Matthew's influence, the next several presidential addresses to the Paleontological Society—at least the ones that asked broad questions about the field—showed an increasing accommodation of the Darwinian attitude towards the fossil record. Joseph Cushman's address "The Future of Paleontology" in 1937 returned explicitly to Darwin's metaphor of the book in its assessment of paleontology's limitations. He wrote that

to the paleontologist is given a rare privilege, for it is his stewardship of the
oldest book of records that we possess, the book which contains all that we
know of the progress of life on this planet of ours. True, the book has had
very hard usage and is imperfect. Many of its pages are badly torn and blot-
ted; others have later records written across a page of earlier ones, so that it
is difficult to trace the written lines. Many of its pages are entirely missing
(Cushman 1938, 359).

Cushman did not entirely dismiss the theoretical ambitions of paleontologists, but like Matthew he did charge paleontologists with the mostly
descriptive task of filling in links in fossil succession. In 1942, Lloyd William Stephenson's "Paleontology: An Appraisal" offered similar language and a similar message. Stephenson's variation on Darwin's metaphor proclaimed variously that "in Nature's great history book much
of the story is recorded," that "the book is in sections, scattered here
and there over the face of the earth" with "many pages [that] have not
been turned," and that "hints of the contents of the book are gained,
but only by turning the pages all the way back is the full story revealed"
(Stephenson 1942, 376). Interestingly, Stephenson did not comment on
"missing pages," perhaps suggesting greater optimism than his predecessors, but he advocated what can only be described as a handmaid role
for paleontology among the other sciences: "We must look to the organic
chemist, the physiologist, the biologist, and the experimental breeder for
discoveries that reveal the actual processes by which the evolution of organisms is accomplished. . . . it [then] becomes the duty and privilege of
paleontologists to seek out confirmatory evidence of evolution offered
by fossils" (Stephenson 1942, 377).

In two final addresses, both from the mid- to late 1940s (and therefore
in the wake of the synthesis), arguments were presented for the independence of paleontology, but here the complaints were directed at geology
and not biology. In considering "Paleontology in the Post-War World,"
B. F. Howell described his fellow paleontologists as being "all too prone
to look upon themselves as mere hand-maids to geology and to think
of paleontology as nothing more than the tail on the geological dog."
Howell optimistically encouraged his colleagues to rather think of their
discipline as "a sister science to biology that both draws from, and contributes to, geology," and to assert that paleontology was "an independent science, worthy of recognition as such" (Howell 1945, 375). How-

ever, his positive message was somewhat dampened by his reminder that "since paleontology cannot, because of its nature, be a very exact science paleontologists cannot think or write in such mathematically definite terms as can other scientists as the mathematicians, the physicists, and the chemists, or even as nearly exactly as can the biologists" (Howell 1945, 375).

J. Brookes Knight's 1945 address was titled "Paleontologist or Geologist," and he addressed the same topic as Howell. Knight made an even more strident argument for the independence of paleontology from geology, and even proposed that universities should establish separate departments for each discipline (Knight 1947, 282). He also argued that "because paleontology is the study of the life of the past it is a biological science," and that "its attitudes and techniques are—or should be far more than they are among invertebrate paleontologists—dominantly biological" (Knight 1947, 282–83). Knight did not, however, comment on whether paleontology had much to offer biological evolutionary theory; in fact, the impression his address gave was that the benefits from a détente between the two disciplines would accrue mostly for paleontology.

All in all, the picture of paleontology's contributions to evolutionary theory and relationship to biology is fairly heterogeneous and complex during the decades before the synthesis. While many paleontologists did accept Darwin's description of the subordinate role for paleontology, others explored important ways by which paleontology could produce biologically sensitive theoretical work. Some advocated non-Darwinian mechanisms, while others, including Matthew, promoted an agenda whereby paleontologists could develop theoretical autonomy without distancing themselves from colleagues in biology. Especially notable was emerging use of the term "paleobiology" to describe this endeavor. This term would become the rallying cry for theoretical paleontology during the 1970s, but it would first gain popularity among pathbreaking paleontologists like Simpson and Norman Newell during the 1950s and 1960s. Overall, then, there are strong reasons for interpreting the history of theoretical paleontology in the 20th century as, in large part, a continuous development. Before the "revolution" of the 1970s came an important "renaissance" in the preceding decades that, in many ways, made the later events possible. First, however, we must examine the impact of the modern synthesis on paleontology, and in particular the influence of one of its architects and staunchest defenders, G. G. Simpson.

Paleontology and the Modern Synthesis

The "modern synthesis" of evolutionary biology has been fairly consistently defined by historians as the sum total of theoretical development,
roughly between 1937 and 1950, whereby the genetic principles of Mendelian heredity were accommodated to Darwin's theory of natural selection (Provine 1971; Smocovitis 1996; Mayr and Provine 1980; Mayr
1982a; Cain 1993; Cain 2002; Allen 1975). In other words, biologists applied the knowledge of heredity accumulated by geneticists in the first
decades of the 20th century to the principles of gene flow determined
by ecologists and biologists studying adaptation and selection in populations. The resulting synthesis defined evolutionary biology as a study
of the movement, via inheritance and mutation, of *genes* within *populations*. One of the most important aspects of the synthetic approach was
the development of a quantitative understanding of how genes move in
populations, which allowed biologists to confirm that Darwin's qualitative assessment of the sufficiency of natural selection to produce evolution agreed with the modern understanding of genetics. The doctrine
produced by the end of the synthesis period became commonly known
as "neo-Darwinism."

Historians have also emphasized that the synthetic project was, in
large part, an institutional project. The architects of the synthesis called
for dialogue between a variety of evolutionary disciplines, including genetics, ecology, systematics, zoology, anatomy, and paleontology, which
necessitated the orchestration of what Joe Cain has called "institution-
alized cooperation." According to Cain, this allowed the architects to
cross "disciplinary boundaries in pursuit of common problems . . . to ensure inclusion and to elevate the status of fields and practices otherwise
deemed marginal within biology" (Cain 1993, 2). This is part of the process of "unification" that Smokovitis describes as involving several discrete steps, beginning with Dobzhansky's successful translation of mathematical population genetics into terms comprehensible to the average
naturalist, and concluding with the establishment of the Society for the
Study of Evolution and its journal, *Evolution* (Smocovitis 1996, 99–127).

Dobzhansky's *Genetics and the Origin of Species* provided the
wake-up call to biologists, but it was Julian Huxley and Mayr who took
the lead in advancing the institutional agenda of the synthesis. Shortly
after the publication of Dobzhansky's monograph in 1937, Huxley be-

gan organizing a movement in Britain to redefine the field of systematics in light of advances in genetics. This activity culminated in the book *The New Systematics* (1940), which Huxley edited, whose project "was to integrate the various studies of divergence and isolation and relate them to taxonomic groups and evolutionary mechanisms" (Cain 1993, 4–5). Meanwhile, Huxley pressed for a similar reform project in America, and was able to generate sufficient interest among a number of important biologists, including Alfred Emerson, Dobzhansky, and Mayr, to launch a working group in 1940 called the Society for the Study of Speciation (Cain 1993, 7; Smocovitis 1994, 1–2). This group was short-lived, as the intervention of World War II and the arrival of Dobzhansky to the faculty at Columbia University shifted the center of the synthesis to New York, where the American Museum also played a prominent role. There the Columbia biologist L. C. Dunn oversaw publication of the Columbia Biological Series of monographs, whose titles included Dobzhansky's *Genetics and the Origin of Species*, Mayr's *Systematics and the Origin of Species* (1942), Simpson's *Tempo and Mode in Evolution* (1944), Bernhard Rensch's *Evolution above the Species Level* (1959), and other seminal works of evolutionary biology. This series was an enormously effective tool for promoting the agenda of the synthesis, and it was centered around Dobzhansky's influential interpretation of population genetics. It also provided a vehicle for members of disciplines outside genetics to promote their own theoretical legitimacy: as Smokovitis argues, these monographs "were written by individuals who, engaging in dialogue with Dobzhansky, in turn legitimated as they grounded *their* disciplines with Dobzhansky's evolutionary genetics" (Smocovitis 1996, 134).

By 1942 a new group, the Committee on Common Problems in Genetics and Paleontology (CCP), was established through the auspices of the National Research Council to consider a wider array of evolutionary questions. During its first two years Mayr took the lead in its organization, publishing regular newsletters that informed its scattered membership of current issues and topics of debate (Cain 1993, 12). When Simpson returned to the American Museum from military service in 1944, he assumed the chair of the CCP and immediately began planning its transformation into a larger, more permanent organization that would appeal to scientists in all evolutionary disciplines. At the 1946 meeting of the American Association for the Advancement of Science, the Society for the Study of Evolution was established, finally giving the synthesis a legitimate professional organization and a journal, *Evolution*, that was sponsored by the American Philosophical Society (Cain 1993, 13; Smocovitis 1994, 5–6).

From one perspective, Simpson's role was a legitimate triumph for paleontologists, in that their discipline was recognized so prominently in the institutionalization of the synthetic theory. Without question this was largely due to Simpson's efforts, which were undeniably heroic. It would be a mistake, however, to conclude that paleontology was, either before or after this unification project, a fully equal and respected partner in the community of neo-Darwinian evolutionary biology. In fact, biologists exerted considerable pressure to ensure friendly paleontologists' cooperation in adhering to the synthetic party line. In 1944, the Princeton vertebrate paleontologist Glenn Jepsen admitted to Mayr that "paleontology presents good evidence, as you know, that evolution proceeds by microgenetic rather than macrogenetic alterations and that this evidence is in harmony with experimental genetics." Jepsen appealed to Mayr's colleague, the invertebrate paleontologist Kenneth Caster: "I hope you will be willing to make a statement on this subject" (Jepsen, quoted in Cain 1993, 12). The concern here, of course, was that many paleontologists had in the past been seduced by the idea of macromutations and saltations as explanatory mechanisms for major evolutionary change. Indeed, Mayr recalled many years later that "most paleontologists were either saltationists or orthogenesists, while those we believe to have been neo-Darwinists failed to write general papers or books" (Mayr 1980b, 28). As a response, biologists "effectively controlled the identity of biology through the synthesis period," Cain concludes, and "'synthesis,' from this perspective, meant the expansion of laboratory work together with the subsumption of descriptive studies by field and museum workers who were 'brought into line'" (Cain 1993, 17–18).

The trend in paleontological statements about evolution, then—as evidenced, for example, in the presidential addresses of the Paleontological Society—reflects paleontology's move closer to the mainstream of biological evolutionary theory and away from saltationism and orthogenesis, but it also indicates the extent to which the "disciplining" efforts of synthetic biologists were successful in engineering agreement with neo-Darwinian principles. Paleontologists would certainly benefit from greater participation in the community of evolutionary biologists: more secure institutional positions, greater respect for their data, better access to mainstream publications and conferences, and a larger stake in theoretical discussions all followed over the next few decades. But there was a cost as well: as Patricia Princehouse argues, "In large part the modern synthesis served to sideline major research traditions in paleontology" (Princehouse 2003, 21). One of those traditions involved approaching macroevolutionary analysis of the fossil record with confidence that paleontologists had unique access to patterns and processes of evolution undetectable by their "neontologist" colleagues.

Simpson's Place in the Synthesis

Simpson's role in this aspect of the story is complex. He was perhaps the most influential paleontologist of the 20th century, and his masterpiece, *Tempo and Mode in Evolution*, has been read by generations of paleontologists and biologists. As chair of the seminal CCP, he had an active role in defining the agenda of the modern synthesis. He was also, however, in many respects an iconoclast, and even as he appealed to the biological understanding of evolution promoted by the other major architects of the synthesis, his vision of evolution was often at odds with the dominant interpretation of evolution promoted by his biologist colleagues. Simpson appears to have been aware of how radical his views were to both biologists and paleontologists. Echoing the sentiments of many contemporary biologists, he later recalled that "at the time when I began to consider this subject [of evolution] I believe that the majority of paleontologists were opposed to Darwinism and neo-Darwinism, and most were still opposed in the early years of the synthetic theory" (Simpson, quoted in Mayr 1980a, 455).

There is little in Simpson's early career that would appear to have marked him as a revolutionizer of paleontology, but a few facts from his intellectual biography are noteworthy. As a graduate student at Yale he trained under the paleontologist Richard Swann Lull, whom Léo Laporte described as promoting a "paleobiological approach," and in addition to his reading of Abel, Simpson avidly read D'Arcy Thompson's *On Growth and Form*, a highly quantitative study of growth and structural constraints in biology (Laporte 2000, 18). But Simpson's biological training was mostly adventitious, and it appears that he solidified his strong commitment to biologically oriented paleontology only after his appointment at the AMNH in 1927. Here, Simpson seems to have consciously modeled himself after his predecessors Osborn and Matthew, and he benefited from interacting with colleagues like W. K. Gregory, a former Osborn student at Columbia. And while Simpson's *Tempo and Mode* would help put the final nails in the coffin of orthogenesis, Laporte argues that Osborn's legacy was nonetheless profound for having given "warrant to paleontologists to pursue theoretical questions," and for providing "a forceful example to Simpson of what the important questions were in paleontology" (Laporte 2000, 24).

Before *Tempo and Mode*, Simpson had already announced his intention to revitalize paleontology by increasing the discipline's analytical rigor. Simpson's second marriage, to Anne Roe, had a profound effect on his scientific career. With Roe, who had a doctorate in psychology, Simpson began preparing a book on biological statistics in the mid-1930s. The product of this collaboration was a textbook titled *Quantitative Zoology* (1939), which was a primer in mathematical and statistical analysis for zoologists and paleontologists. In the preface, Simpson and Roe noted that while it is "proper" for zoologists to avoid relying on an a priori mathematical framework, the behaviors and characteristics of actual organisms nonetheless can be profitably translated into a symbolic language (Simpson and Roe 1939, vii). The central problem the authors hoped to correct was the fact that "whether from inertia, from ignorance, or from natural mistrust . . . most zoologists and paleontologists distrusted the overt use of any but the very simplest and most obvious numerical methods" (Simpson and Roe 1939, viii). Over the course of the book, then, Simpson and Roe introduced readers to concepts and methods such as producing frequency distributions, measuring dispersion, plotting and fitting data to curves, sampling, correlation and regression, and plotting graphs. All of these techniques, they claimed, could be learned with no advanced knowledge of statistics or calculus.[6]

Simpson began writing *Tempo and Mode* while he and Roe were still finishing *Quantitative Zoology*, but its publication was delayed until 1944, after Simpson returned to the United States from active military duty. In the preface he noted that "the final revision was made under conditions of stress," and that because of those circumstances several "important studies" relevant to his subject had been omitted (Simpson 1944, vi). Simpson was obliquely referring here to Huxley's *Evolution: The Modern Synthesis* and Mayr's *Systematics and the Origin of Species*, both of which had been published in 1942 after he had begun his mili-

---

6. In 1960 this book was released in an updated edition with the addition of the biologist Richard Lewontin as coauthor.

tary service. Simpson had, however, read Dobzhansky's *Genetics and the Origin of Species*, and that work had a profound influence on his vision of evolutionary paleontology. In later years he stressed the importance of this encounter: "The book profoundly changed my whole outlook and started me thinking more definitively along the lines of an explanatory (causal) synthesis and less exclusively along lines more nearly traditional in paleontology" (Simpson, quoted in Mayr 1980a, 456). It also "opened a whole new vista to me of really explaining the things that one could see going on in the fossil record and also by study of recent animals," and it allowed him to relate his own paleontological research to the exciting new work in genetics (Simpson, quoted in Laporte 2000, 25).

Probably the single most important influence Dobzhansky had on Simpson was to push him to think about the history of life, and the evidence of the fossil record, in terms of the genetics of once-living populations. The major argument of *Tempo and Mode* is that what happens on the Darwinian population level explains transformations in the fossil record, and that those transformations can be explained using models of population genetics. Paleontology, Simpson stressed, could be useful for uncovering the mechanisms that drive evolution, and not just for documenting the physical historical record. As he wrote in the introduction, "Like the geneticist, the paleontologist is learning to think in terms of populations rather than of individuals and is beginning to work on the meaning of changes in populations" (Simpson 1944, xvi). Simpson's great insight was that paleontology could be modeled after population biology with the additional dimension of *time*; he described *Tempo and Mode* as a work in "four-dimensional" biology, where the distribution and transformation of organisms could be tracked in a temporal "geography" analogous to physical geography. He emphasized that the temporal or historical element of paleontology offered a unique and critical perspective to evolutionary theory, and the importance of this message cannot be overstated: after *Tempo and Mode*, temporal biogeography became the center of paleontological evolutionary theory (as later chapters in this book will show).

Tempo and Mode was indeed revolutionary in its insights, and its impact is all the more remarkable because in just 217 pages of text, Simpson hardly had the opportunity to document many of the claims he made. His book is a theoretical manifesto, and one of the rare instances in the history of science where a highly original work almost instantly managed to win nearly universal approval from colleagues. In part, the success of

Simpson's book lay in the way it adapted paleontological evidence to the growing consensus in the biological community that neo-Darwinian selection and genetics offered the most accurate model of evolution. As Stephen Jay Gould notes, *Tempo and Mode* expressed reasonable pluralism about the mechanisms of selection, but Simpson explicitly "regarded natural selection as the primary controller of rate and direction, and adaptations to local environments as its result" (Gould 1980a, 161). In other words, Simpson followed the tradition of Darwin and his successors in viewing evolution as the continual transformation of lineages via slight modifications, the agents of which were hereditary patterns and genetic drift as described by population biology and genetics.

Simpson's most influential innovation to the synthetic program was his elaboration of Sewall Wright's concept of the "adaptive grid" or "landscape." Wright had proposed that possible adaptations (understood as potential morphologies) could be modeled as a topographical "map," showing concentrations of convergent adaptations as peaks, with the lower topography representing inadaptive or uncommon characters. Following Dobzhansky, Simpson treated the peaks in the topography as species, or as points of convergent fitness or ecological niches (Simpson 1944, 89–93; fig. 1.2). What makes paleontology unique, he argued, was that these "selection landscapes" could be correlated with geological time and stratigraphy, thus providing insight into evolutionary trends (e.g., the appearance of directionality in the fossil record) that did not show up at the resolution of population genetics. Simpson's argument that these trends could all be explained via neo-Darwinian microevolution provided some of the most damning evidence against orthogenetic and Lamarckian evolution yet presented.

One of *Tempo and Mode*'s broadest and most lasting contributions to evolutionary theory was its suggestion that paleontology and the fossil record have something unique to say about macroevolution. Recall that the synthetic view of macroevolution—endorsed by Dobzhansky, Mayr, Huxley, and many others—held that major evolutionary patterns at the higher taxonomic levels are simply extrapolated effects of microevolution. As a paleontologist, however, Simpson had a keener eye than his colleagues Mayr and Dobzhansky for the apparent discontinuities in the fossil record, and his approach to macroevolution reflected this. In *Tempo and Mode* he argued that the justification for extrapolating macroevolution from microevolution was not a settled matter. His theory broke evolution into three tiers. The first tier, microevolution, basically followed

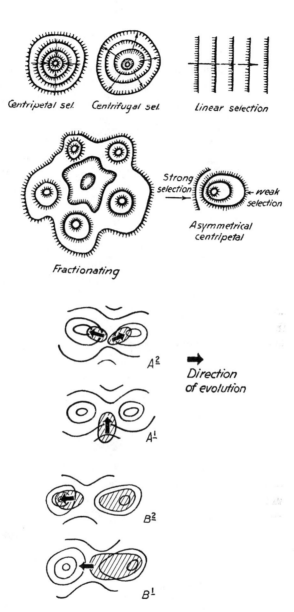

FIGURE 1.2. G. G. Simpson's topographical illustration of "selection landscapes." *Top*, the hachures are on the "downhill" side, and the direction of selection is "uphill." *Bottom*, two patterns of phyletic splitting using topographic imagery. George Gaylord Simpson, *Tempo and Mode in Evolution* (New York: Columbia University Press, 1944), 90 and 92.

the synthetic account. Macroevolution, the second tier, accounted for broader patterns by the accumulation of microevolutionary processes—again, much as the synthesis proposed. However, it was a third process, which Simpson labeled "mega-evolution," that brought about major taxonomic changes. In order to account for the seemingly abrupt transitions in the fossil record, Simpson introduced the concept of "quantum evolution," which described accelerated evolutionary change among small, geographically isolated populations that had come into disequilibrium (Simpson 1944, 206). Simpson suggested that quantum evolution probably used basic microevolutionary mechanisms of random mutation and natural selection (and not saltations), but he emphasized that such accelerated change might constitute an independent process. While his theory was not necessarily opposed to the broader synthetic view, Simpson provocatively urged that "if the two [macro- and microevolution] proved to be basically different, the innumerable studies of micro-evolution would become relatively unimportant and would have minor value in the study of evolution as a whole" (Simpson 1944, 97).

We will return to the subject of quantum evolution in a later chapter when we consider Niles Eldredge and Gould's theory of punctuated equilibria, which was strongly influenced by Simpson. It is sufficient to note here that, while not explicitly in conflict with the synthetic theory of evolution, Simpson's approach to macroevolution and the fossil record was somewhat idiosyncratic. Gould notes that while Simpson's approach to the fossil record was "consistent with genetic models devised by neontologists" in that it held "adaptation as the primary cause and result of evolutionary change," and maintained that "continuous transformation of populations" explains directional patterns in evolutionary history, Simpson nonetheless left the door open for other explanations (Gould 1980a, 161). In particular, Simpson argued that sequential discontinuities in the fossil record (especially across taxonomic categories) might not always be artifacts of imperfect preservation: "The development of discontinuities between species and genera, and sometimes between still higher categories, so regularly follows one sort of pattern that it is only reasonable to infer that this is normal and that sequences missing from the record would tend to follow much the same pattern" (Simpson 1944, 98). In fact, he continued, "the face of the fossil record really does suggest normal discontinuity at all levels"—an observation whose significance for evolutionary theory was ambiguous (Simpson 1944, 99). While Simpson noted that many observed gaps in the record were likely "a taxonomic artifact," he felt that this did not adequately explain "the systematic occurrence of the gaps between larger units" (Simpson 1944, 107).

Based on this observation, Simpson divided paleontological interpreters of the fossil record into two categories: those who had dismissed gaps as "a phenomenon of the record" and thus of no special evolutionary significance, and those who argued "that transitional forms never existed." Simpson's conclusion was "that neither extreme [was] likely to be correct," although he did reckon that the true answer lay "nearer to the first" (Simpson 1944, 115). He dismissed "the paleontological evidence for the saltation theory of mega-evolution," which was "solely the systematic nature of the breaks in the record." While he admitted that "the evidence against this theory is largely indirect," he maintained that "it is cumulative and in sum is conclusive." If these gaps corresponded to genuine saltations, we would expect a total absence of intermediate forms, but this is not the case. Furthermore, Simpson argued, our current understanding of genetics provides no mechanism for such decisive and sudden transformations (Simpson 1944, 115–16). Discontinuities in the fossil record can, however, be tied to genuine evolutionary patterns, such as variability in rates of evolution. Simpson noted that while most paleontologists had assumed that evolutionary rates remain constant over time, this assumption was "quite unjustified on the basis of what is recorded of the facts and theories of population genetics or of what can be inferred as to the gaps in the record" (Simpson 1944, 118). While some groups, like the Equidae, demonstrate mostly uniform rates, such groups nearly always "represent large populations living under relatively stable environmental and ecological conditions." Smaller groups existing in unstable conditions should be expected to experience abnormally high rates of evolution and also great fluctuation of those rates, which in some cases may account for discontinuities in the record (Simpson 1944, 118–19). While Simpson did not claim, as paleobiologists in the 1970s often would, that the fossil record is a mostly complete document, neither did he endorse Darwin and many fellow paleontologists' characterization of that record as devoid of theoretical interest. Perhaps his most important message on this topic was that "incompleteness is an essential datum and . . . can be studied with profit" (Simpson 1944, 105).

Simpson's examination of rates of evolution raises a final important feature of *Tempo and Mode*: its commitment to a quantitative, analytical method. As suggested earlier, many biologists and even some paleontologists had dismissed all hope of paleontology ever reaching the quantitative sophistication of most other scientific disciplines, including biology. This pessimism contributed greatly to paleontology's theoretical subordination as a discipline, and quite likely discouraged many bright, analytical students from pursuing the profession. Simpson, however, was determined to apply the methodology he promoted in *Quantitative Zoology* to paleontological data, and in this regard his effort was genuinely revolutionary. Gould later called Simpson's "use of quantitative information . . . [his] second greatest departure from traditional paleontological practices," and characterized it as "a novel style . . . [of] drawing models (often by analogy) from demography and population genetics and applying them to large-scale patterns of diversity in the history of life" (Gould 1980a, 158–59).

Simpson's most significant use of quantification was his treatment of taxonomic survivorship, or the measure of the longevity of a particular taxon or group. Simpson's approach was to gather taxonomic data from fossil catalogues like K. A. von Zittel's *Grundzüge der Palaeontologie*, from paleontological monographs, or from other systematics literature, and then to tabulate the group's longevity based on its first and last appearances in the record (as will be described in chapter 8, this method would later be labeled "taxic paleobiology"). Next, following a method established by the demographer Raymond Pearl for statistical analysis of human populations, he plotted curves representing survivorship over time as a percentage of the initial population (Pearl 1920; fig. 1.3). By modifying this data with a number of straightforward statistical devices, Simpson was able to draw out several very interesting conclusions: general patterns of survivorship appear on the whole to follow the same, diminishing parabolic curves, although different groups (he compared bivalve mollusks to carnivorous placental mammals) have widely differing rates (Simpson 1944, 24–26). These curves can also be correlated with extant fauna, comparisons can be made within groups over different periods of time, and generalizations about major fauna can be produced, all of which, Simpson noted, shed important light on evolutionary patterns. In succeeding chapters, we will again and again confront examples of "Simpsonian" analytical techniques applied to paleobiology; Simpson's presentation of paleontology's amenability to theoretical modeling and statistical techniques had the greatest possible influence on later generations of paleontologists, as what I will describe as the "generalized" approach to reading the fossil record.

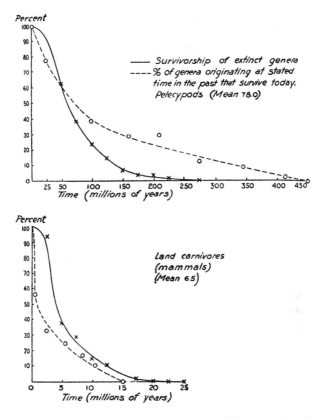

FIGURE 1.3. Survivorship curves for genera of pelecypods (bivalves) and land carnivores. The crosses and circles are data points along which the curves have been fitted. George Gaylord Simpson, *Tempo and Mode in Evolution* (New York: Columbia University Press, 1944), 25.

Interestingly, Léo Laporte notes that paleontologists were not Simpson's most vocal initial supporters (Laporte 2000, 32). Rather, it was biologists and geneticists who first reacted most positively to *Tempo and Mode*, perhaps because most paleontologists were still entrenched in the old nonbiological way of thinking. Simpson himself seems to have anticipated this: he wrote that while "phylogeny and morphogenesis continue to be chief aims of paleontological research," his goal was "to discuss the 'how' and—as nearly as the mystery can be approached—the 'why' of evolution, not the 'what.'" He saw his new agenda as "one more suggestive of new lines of study," adding that "it is more immediately interesting to the nonpaleontological biologist" (Simpson 1944, xviii). Biologists tended to agree. In his review in *The American Naturalist*, eminent marine biologist Carl L. Hubbs called the book "a first rank treatise . . . brilliant, but not flashy," and endorsed Simpson's tripartite division of modes of evolution (Hubbs 1945, 271–72). Ecologist G. Evelyn Hutchinson wrote in *American Journal of Science* that the book "must be studied thoroughly by everyone concerned with the mechanism of evolution" (Hutchinson 1944, 356–58), while Huxley's review in *Nature* proclaimed it "a stimulating book, undoubtedly destined to be the parent of much new work and many valuable conclusions" (Huxley 1945, 3–4). Reviewers also applauded Simpson's contribution to the growing synthesis: Dobzhansky's review in *The Journal of Heredity* called it "the long awaited synthesis of paleontology and genetics" (Dobzhansky 1945, 113–15), while the geneticist Mark Graubard wrote in *The Scientific Monthly* that "there can be no question that the appearance of this book will help greatly eliminate the mutual distrust which insulated" the two disciplines (Graubard 1945). Finally, the geneticist H. Bentley Glass rated Simpson's contribution along with Dobzhansky's and Mayr's treatises, commenting that *Tempo and Mode*'s "genius lies, in the first place, in the application of genetic concept [*sic*] to paleontological data; in the second place, to the grasp and originality of Simpson's thought" (Glass 1945, 261).

Simpson was understandably gratified by this outpouring of support from biologists, noting in the preface to his next major work, 1953's *Major Features of Evolution*, the "uniformly kind" reception *Tempo and Mode* received" (Simpson 1953, xi). One response that seems to have given him pause, however, was Sewall Wright's lengthy "Critical Review" in the journal *Ecology*. Simpson had relied heavily on Wright's concept of adaptive topography and his "shifting balance" view of selection, and Wright used his review as much as an opportunity to promote his own theories as to evaluate Simpson's. Wright was far more measured and critical in his tone than other reviewers, noting, for example, that while he mostly approved of Simpson's conclusions, he found "remarkably little difference between Dr. Simpson's interpretation and that which had seemed to him . . . the natural deduction from genetic principles" (Wright 1945, 415). The review took issue with Simpson's treatment of variability, commenting that Simpson "underestimates the amount of evolution that could occur by mere shifts in gene frequency," and it criticized Simpson's dismissal of the importance of intergroup selection

(Wright 1945, 416). Wright did endorse Simpson's distinction between the mechanisms of micro- and mega-evolution, but took much of the credit himself for having independently established the principle (Wright 1945, 417–18). Despite his "enthusiasm" for Simpson's book, with which he concluded his review, Wright's assessment of the divide between paleontology and genetics was also considerably less rosy than that of other reviewers: while "a contact has been established that did not exist a few years ago," it was nonetheless "too much to say that paleontologists and geneticists as groups have reached agreement" (Wright 1945, 415). Surprisingly, the notoriously thin-skinned Simpson seems to have been genuinely pleased with Wright's review, and wrote to him to thank him "for both your favorable remarks and your excellent criticisms," adding that he "quite agree[d] both with your corrections and with the expansion you have made" (Simpson to Wright, 8 January 1946, quoted in Provine 1986, 417).

However, it appears that when Simpson sat down to update *Tempo and Mode* in the early 1950s, a major shift in his thinking had taken place. Specifically, he now significantly downplayed the importance of what he had called "perhaps the most important outcome" of *Tempo and Mode*: "The attempted establishment of the existence and characteristics of quantum evolution" and its concomitant endorsement of Wright's mechanism of genetic drift (Simpson 1944, 206). Exactly why Simpson did this is the subject of debate. In the introduction to *Major Features of Evolution*, as the revised book was rechristened, Simpson wrote, "When a revision of *Tempo and Mode in Evolution* was called for, my feeling was that the book had served its purpose and should be allowed to fossilize quietly." In particular, he noted that while the idea of extending population genetics to paleontology had been "new and exciting" in the early 1940s, "that idea is now a commonplace" (Simpson 1953, ix). So instead of simply revising *Tempo and Mode*—which "would have been like installing air conditioning in a pioneer's log cabin"—Simpson wrote what was effectively a new book (Simpson 1953, x–xi).

Major Features follows the same general plan as *Tempo and Mode*, but it is significantly longer (by more than a hundred pages) and contains much more detailed evidence to support its arguments. It is also more explicitly in line with the synthetic interpretation of evolution, a fact not lost on one reviewer who commented, "Simpson is a neo-Darwinian, and throughout his book, he has ably championed the position that combinations of small genetic changes, and their selection under natural conditions, are the real forces in evolution in the development of new forms" (Durrant 1954, 601). As Gould put it less charitably several decades later, *Major Features* "moves toward a more rigid selectionism . . . [which] marks an important change both in Simpson's thought and in the history of the modern synthesis" (Gould 1980a, 161). Specifically, *Major Features* now stressed that phyletic evolution is the overwhelming cause of evolutionary change and speciation: "Phyletic splitting of lineages, including those from which higher categories up to the highest later develop, thus occurs by speciation at their bases. . . . The paleontological evidence cannot exclude the possibility of exceptions, but it confirms the conclusion in particular examples, and there is nothing in the record that requires or suggests exceptions" (Simpson 1953, 383–84). Furthermore, Simpson was now committed to the adaptationist program of the synthesis, and he maintained that phyletic evolution is "cumulative and [those features] that have persistence of direction are evidently nonrandom and apparently always adaptive" (Simpson 1953, 385). Quantum evolution, with which Simpson concluded *Tempo and Mode*, describing it as "the dominant and most essential process in the origin of taxonomic units" (Simpson 1944, 206), now received only a few pages and was effectively dismissed as merely a "limiting case of phyletic evolution" (Simpson 1953, 389).

Why did Simpson change his mind so completely about what, at one time, he had presented as the decisive paleontological contribution to evolutionary theory? According to Gould, this transition reflects the "hardening" of the modern synthesis, a period when Dobzhansky, Mayr, and others abandoned the more inclusive and pluralistic tone of the 1940s and began to push for a rigid, adaptationist interpretation of Darwinism that excluded potentially stochastic processes (such as Wright's work on genetic drift) as major evolutionary mechanisms. As Provine (who concurs with Gould) puts it, "The consequence was that biologists viewed the evolutionary process as almost exclusively adaptive rather than nonadaptive, even at the lowest taxonomic levels" (Provine 1986, 404).[7] Gould's less than charitable reading of Simpson's revision was that he capitulated to pressure from the other major architects of the synthesis and dropped his Wrightian, potentially inadaptive interpretation of quantum evolution in favor of mainstream neo-Darwinian microevolution (Gould

7. On the "hardening" or "constriction" of the modern synthesis, see also Gould 1983, Gould 2002, and Provine 1989.

2002, 529). It is difficult to know exactly what triggered Simpson's shift, but Gould has suggested that "tides of fashion, stirred by a few strong individuals, persuaded others and forged a consensus rooted more in sociology than in logic or experience," that created a "with us or against us" mentality in the postwar evolutionary community (Gould 1981, 133). In this reading, Simpson either consciously or unconsciously acted to preserve his (and paleontology's) standing with biologists by dropping elements of his theory that made his colleagues uncomfortable.

Of course, Gould had an agenda of his own when he formed this account of the synthesis, which will be examined in later chapters. His conclusion was that Simpson's message in *Major Features* "contained a disturbing dilemma for its later development as a ruling paradigm in paleontology": it "unified paleontology with evolutionary theory, but at a high price indeed—at the price of admitting that no fundamental theory can arise from the study of major events and patterns in the history of life" (Gould 1980a, 169–70). Most importantly, Simpson's argument for quantum evolution was founded on the assumption that the apparent regularity of abrupt transitions in fossil lineages was too compelling to be dismissed merely as the artifact of imperfect preservation—in other words, that the fossil record was more than simply a book with "missing pages," and consequently that paleontology had a unique contribution to make to evolutionary theory. However, if, as *Major Features* seems to imply, gradual phyletic evolution is the dominant mode in the history of life, then paleontology was potentially cast once again in a role subordinate to genetics and population biology, since—as Simpson and other paleontologists readily admitted—their field could not experiment with populations or examine genetic relationships.

Even putting Gould's agenda aside, we can reasonably conclude that by the early 1950s Simpson had ceded some of the territory *Tempo and Mode* had claimed for paleontology a decade earlier. Ironically, it was only after the publication of *Major Features* that Simpson's work became required reading for the next generation of paleontologists, many of whom only discovered *Tempo and Mode* (which had gone out of print) years later. Nonetheless, Simpson's work had a profound influence on the development of paleobiology over the next several decades, and while he may not have pushed as hard as he might have for paleontology's place in the synthesis, it is also undeniable that in his absence the paleobiological movement might never have started at all. When young paleontologists rediscovered his earlier work in the late 1960s and early 1970s, they

found compelling motivation for a revitalized paleontological evolutionary theory. *Tempo and Mode* made two major arguments for the advantages of paleontology in the study of evolution. First, it offered insight
into "evolutionary rates under natural conditions, the measurement and
interpretation of rates, their acceleration and deceleration, the conditions of exceptionally slow or rapid evolution, and phenomena suggestive
of inertia or momentum"—that is to say, the *tempo* of evolution. Second,
it argued that paleontology can examine "the way, or manner, or pattern of evolution. . . . to derive how populations become genetically and
morphologically differentiated, to see how they passed from one way of
living to another or failed to do so, to examine the figurative outline of
the stream of life and the circumstances surrounding each characteristic
event in that pattern"—or evolution's *mode* (Simpson 1944, xvii–xviii).
These two arguments became central to the theoretical resurgence during the "paleobiological revolution" of the 1970s, and paleobiology from
the late 1960s to the present has continued to mine these insights.

But in the early 1950s the work of synthesizing paleontology with biology had only just begun. With Simpson's help, paleontology was granted
a provisional seat at the evolutionary table, and (as the next chapter will
describe) paleontologists were beginning to find a wider audience via
publications like the new journal *Evolution*. However, as Laporte observes, "Full-scale incorporation of the results of the evolutionary synthesis into a "paleontological tradition" required the subsequent education of a new generation of paleontologists in microevolutionary theory"
before the discipline of "paleobiology" would have a meaningful stake
in evolutionary theory (Laporte 2000, 34). The next generation would
find new and innovative ways of further incorporating biological concepts and methods into paleontology; in doing so, it would also uncover
new arguments for the theoretical autonomy of paleontology that would
sow seeds of eventual conflict with neo-Darwinian synthetic evolutionary theory.

CHAPTER TWO

The Growth of
Theoretical Paleontology

A s discussed in the last chapter, the publication of G. G. Simpson's *Tempo and Mode in Evolution* in 1944 did not create the immediate stir among paleontologists one might have expected, but it did generate interest that Simpson capitalized on with the release of *The Major Features of Evolution* in 1953. By the early 1950s a major transformation was quietly taking place in paleontological approaches to evolutionary theory and the fossil record—one that Simpson certainly played a role in starting. This shift involved several distinct but related aspects. First, paleontologists began to actively assess the institutional status of their discipline—asking whether paleontology "belonged," for example, with geology, or with biology, or rather constituted an independent discipline on its own. Second, paleontologists began more and more to explicitly connect their work with the evolutionary agenda of the modern synthesis, and to publish in outlets (such as the journal *Evolution*) that were read by biologists and geneticists. Even papers in paleontology-specific publications like *Journal of Paleontology* took on a more theoretical cast during this period. Third, and perhaps most significantly, paleontology became quantitative. This is not to say that quantitative methods (measurements and statistical analysis) had been absent from the work of paleontology in the past, but the period between 1950 and 1969 saw a burgeoning interest in attacking broad, synthetic questions about the fossil record (e.g., morphology, evolution, extinction) with a quantitative rigor and sophistication not previously seen in paleontological literature.

Paleontology's Identity Crisis

As early as the late 19th century, paleontologists had recognized the
awkward position of their discipline in relation to geology and biology.
In an 1889 review essay in *Nature*, an anonymous author (known only as
"E. R. C.") put the attitude of many nonpaleontological observers suc-
cinctly: "The palaeontologist has been defined as a variety of naturalist
who poses among geologists as one learned in zoology, and among zoologists as one learned in geology, whilst in reality his skill in both sciences is diminutive" ("E. R. C." 1889, 364). Aside from its rather negative opinion of paleontology (the author also called the study of fossils
"a definite hobby"), this short essay is significant because it marks the
first appearance in the literature of the term "neontology" to refer to
the study of living organisms. Paleontology, by extension, was defined as
the study of ancient life, suggesting that both approaches together make
up the discipline of biology. While the term "neontology" did not gain
immediate traction in the literature during the late 19th and early 20th
centuries, by the mid-20th century many paleontologists and even some
biologists had begun to regularly use this bipartite distinction, marking
the beginnings of a shift away from thinking of paleontology as a subordinate field of geology.

Nonetheless, in 1889 "E. R. C." expressed a view that would become
common among biologists, geologists, and even some paleontologists:
that subordinating paleontology to geology would provide "a better
chance for the cultivation of true geology, which now, to some extent, has
its professional positions, its museums, and its publications invaded by
these specialists [i.e., by paleontologists]" ("E. R. C." 1889, 364). Clearly,
there were more than intellectual issues at stake here. With limited resources available for geology in universities and museums, it was natural for scientists in established departments to want to preserve what
they viewed as the traditional core of their discipline. For geologists, this
meant the study of sediments, minerals, and stratigraphy. In the United
States, paleontology did get a lift with the establishment of the Paleontological Society (PS) in 1908. In its early years the PS was dominated
by invertebrate paleontologists, and this trend continued as demand for
invertebrate specialists increased due to their value in the burgeoning
oil industry. While the PS was formed mainly to attract academic paleontologists, a parallel society, the American Association of Petroleum

Geologists, founded in 1906, and its daughter organization the Society of Economic Paleontologists and Mineralogists (SEPM) represented the interests of geologists and paleontologists in the petroleum industry, and thus initially enjoyed larger membership (Schopf 1980). When the first US paleontology journal, *Journal of Paleontology* (*JP*), was launched in 1927, it was owned not by the PS but rather by SEPM, and its initial mandate was to provide an outlet "primarily for the description of newly discovered microfossils from the oil fields" (Dunbar 1959, 909). Academic paleontologists therefore had practical as well as intellectual reasons to feel somewhat disenfranchised.

However, by the late 1940s paleontologists had sufficient confidence to question the status quo. Recall from the last chapter that in his 1946 presidential address to the PS, J. Brookes Knight made a forceful call to arms for paleontologists to throw off the shackles chaining them institutionally to geology departments (Knight 1947, 282–83). These comments touched off a minor controversy in the paleontological community. The first to respond was J. Marvin Weller of the Walker Museum at the University of Chicago, who rejected Knight's call entirely. Arguing that paleontological stratigraphy was "the heart of geology" and its "single great unifying agency," Weller urged paleontologists to stick close to their geological roots (Weller 1947, 570; see also Rainger 2001). "Invertebrate paleontology is much more closely related to geology than biology," he reasoned, and the two fields were mutually interdependent, whereas biology and paleontology could each "get along" without the other. Weller had little time for vertebrate paleontologists, whom he considered hardly even geologists, and even less interest in the kind of paleontological-biological synergy preached by his many of his peers: "Any student of fossils who does not have a strong, abiding, and well-founded interest in geology . . . is not a paleontologist. He is simply a paleobiologist" (Weller 1947, 572).

While Weller's use of the term "paleobiologist" as an epithet was unusual, his sentiments were connected to quite reasonable concerns about the training and employment of future paleontologists, whom he worried would have inadequate experience with stratigraphy without significant geological education. He predicted that removing geological courses from the curriculum would divert students from "the really important problems" in paleontology "under the delusion" that paleontological work could answer biological questions (Weller 1947, 574). While Weller recognized that a paleontologist's training required "a good foundation in modern biology," he urged that this "should not come at the expense of geology courses." More fundamentally, he appears to have been concerned about the potential that (a) geology would lose its ability to appeal to bright young students, and (b) paleontology would sacrifice its distinctiveness and relevance among the sciences. Weller noted that "invertebrate paleontology has lagged far behind modern biology in both breadth of vision and accomplishment," and he warned that "if few geologists are to be interested in invertebrate paleontology this source of recruits would soon dry up and there would be grave danger that invertebrate paleontology would stagnate for lack of active workers" (Weller 1947, 575).

Norman Newell and the Rise of Paleobiology

Not all paleontologists accepted Weller's dismissive view of "paleobiology." One of the chief advocates of the new disciplinary orientation was Norman D. Newell, who in 1947 had recently arrived as curator of invertebrate paleontology at the American Museum of Natural History. As this chapter will argue, nobody did more to promote the agenda of paleobiology in the 1950s and 1960s than Newell, and his influence, measured both directly through his work and indirectly through his mentoring of students and younger paleontologists, was profound. Newell's hand touched nearly every major aspect of paleobiology during his career, and he was directly responsible for encouraging such fundamental paleobiological topics as the investigation of broad patterns in the fossil record, the development of quantitative approaches to fossil databases, the study of the evolutionary significance of mass extinctions, and the creation of the subdiscipline of "paleoecology." Throughout his career Newell also tirelessly promoted the institutional agenda of paleobiology, and he trained many of the leaders of the movement's next generations. In 1978 he was awarded the Gold Medal for career achievement by the AMNH, and a number of prominent scientists were asked to comment on his work. Stephen Jay Gould reflected:

> When virtually all paleontologists were trained as geologists and had no biological knowledge beyond the basics of invertebrate morphology, Norman Newell saw, virtually alone, that the most exciting future direction in paleontology lay in its relationship to evolutionary theory and to biological thought

in general. I think that only a few very old-fashioned paleontologists would deny today that this prediction has been fulfilled and that American invertebrate paleontology is now in its most exciting phase since the era immediately following Darwin's *Origin of Species*. With his early monographs, and his persistent encouragement of biological thinking, Norman Newell was the godfather of this movement (Stephen Jay Gould to Niles Eldredge, 9 March 1978: AMNH Inv.).

This sentiment was shared by biologists as well: Ernst Mayr commented that "Norman has served as an important bridge between specialized paleontology and evolutionary biology as a whole. . . . [and was] quite instrumental in introducing the evolutionary synthesis into invertebrate paleontology." Mayr also noted, "It should not be left unmentioned what an important impact Newell had on the development of the younger generation of invertebrate paleontologists. Many of the best known people in this field were his students or students of his students" (Ernst Mayr to Niles Eldredge, 1 March 1978: AMNH Inv.). And Simpson, who was often reluctant to praise other scientists, stated, "In my opinion he is a truly great scientist, fully worthy of this honor, and I will value my own medal from that museum even more highly if I have the added honor of sharing it with Norman" (George Gaylord Simpson to Roger Batten, 22 February 1978: AMNH Inv.).

Newell and Invertebrate Paleontology at the AMNH

Newell's formal training came first as an MA student under Raymond Moore and J. Brookes Knight at the University of Kansas, and later for the PhD with Carl Dunbar at Yale. His first appointment, in 1937, was as assistant professor of geology at the University of Wisconsin, where his own first doctoral student was Bernhard Kummel. He was recruited by Simpson to the AMNH in 1946, while the museum was in a period of reorganizing and strengthening its paleontology programs. In taking the position, Newell remarked on his "prevalent impression that invertebrate paleontology has not prospered at the museum . . . and it has been termed the 'step-child' of the monumental work in vertebrate paleontology." The most conspicuous issue was the understaffing of the invertebrate division at the museum, and although Newell regarded "the opportunity to change the situation for the better as one of the challenging opportunities of the position," he also expressed the hope that an additional assistant curator might be hired to broaden the coverage in invertebrate paleontology (Norman Newell to G. G. Simpson, 7 April 1945: AMNH DVP 67, 21). Simpson did not disagree with Newell's assessment, but wrote optimistically that "this is a main reason why we wanted to have you here, and our inviting you to come is evidence of the fact that we do want to infuse new life into your subject." On the issue of additional staffing, Simpson was guardedly optimistic: "Progress in this direction will surely be possible. . . . And we will all want to work together toward a general paleontological program, or in fact geological in the broadest sense" (G. G. Simpson to Norman Newell, 19 April 1945: AMNH DVP 67, 21).

Eventually, additional resources for invertebrate paleontology were allocated, first in the hiring of a new junior curator, John Imbrie, and later in the establishment of a serious program to aggressively build on the museum's invertebrate collections. All in all, through a combination of judicious hiring, new institutional connections, and aggressive expansion of collection and research in invertebrate paleontology, Simpson and others at the AMNH positioned the institution to build on its traditional strength in vertebrate paleontology with an equally strong program in invertebrates. This prepared the institution to take a leading role in the growth of paleontology over the next several decades—a role Newell was instrumental in expanding.

In 1948, shortly after arriving at the museum, Newell published a response in *JP*, coauthored with his Columbia colleague Edwin H. Colbert, to Marvin Weller's critical essay. Newell and Colbert clearly sympathized with Knight's view that paleontology was a "biological science," and they praised Knight for "real service to paleontology and geology in calling attention to serious traditional deficiencies in the training of invertebrate paleontologists" (Newell and Colbert 1948, 264). While they were aware of the practical obstacles to enacting Knight's vision—particularly the institutional limitation that "it is not likely that many universities could be persuaded to erect separate paleontology departments"—they respectfully opined that "Professor Weller's point of view admirably expresses the traditional (and 'narrow') attitude of the geologist toward paleontology," which "is being modified only too gradually." Paleontology was only considered a branch of geology, Newell and Colbert reasoned, "because paleontologists, through lack of adequate training in biology, have made it so" (Newell and Colbert 1948, 265). They proposed a division of paleontology into two categories—stratigraphic paleontology and "paleobiology"—and emphasized that even this dichotomy obscured significant areas of overlap between the two approaches. Many of the goals of paleontology transcended stratigraphy, they stressed, such as phylogeny reconstruction and the improvement of the fossil record, but were also beyond the ken of biologists who lacked paleontological training. And, turning the tables on Weller, Newell and Colbert argued that close traditional association with geology had, "as much as anything . . . [caused] the lack of mature growth of this branch of [invertebrate] paleontology." In their conclusion, Newell and Colbert centered the issue on paleontology's engagement with evolution: "The invertebrate paleontologist in North America has suffered because of his lack of an *evolutionary* viewpoint, the result of a lack of training in biology" (Newell and Colbert 1948, 267).

Weller's response, published in *JP* later that year, indignantly claimed that Newell and Colbert had "seriously misrepresented" his views (Weller 1948, 268). Weller defended his assertions but clarified his view of the importance of biological training, noting: "I am as fully aware as they of the gross neglect of fundamental biological problems by invertebrate paleontologists." However, he refused to budge from his negative assessment of the nascent paleobiological movement, remarking that the disagreement appeared to be more semantic than substantive. "They use 'paleobiologist' with what seems to be exactly the same meaning as 'paleontologist,'" he commented, while in his view "a paleobiologist is a student of fossils without an interest in geology" (Weller 1948, 268). But, semantic quibbles aside, Weller's response also shows that both sides of the disagreement saw the important institutional stakes in the debate. Weller's comment that "the desire to set up paleontology as a separate science suggests a professional inferiority complex" recognized a professional marginalization of paleontology with which Newell was also concerned, but there was clearly deep disagreement about how best to shore up credibility and status for paleontology (Weller 1948, 269). For "traditionalists" like Weller, the solution was to reinforce paleontology's association with geology, even if it meant conceding certain topics to biologists. On the other hand, Newell would continue to stress closer alignment with biology as the route to greater prominence for his discipline, and he tied this agenda to what he was beginning to call "paleobiology."

In addition to pursuing his agenda publicly in forums like *JP*, Newell also worked to change the mentality at his home institution. In 1948 or 1949 he sent his colleagues in the geology department at Columbia a memo entitled "Instruction in Paleobiology," which he described in a handwritten note to Simpson as "part of an unavoidable campaign of missionary work." In the memo he outlined his systematic plan for revising the way paleontology was taught and ultimately practiced:

> The period between the two world wars was characterized by development in invertebrate paleontology chiefly along utilitarian lines, seemingly at the expense of fundamental progress in the science. . . . Because of the traditional union between invertebrate paleontology and geology it has come to be forgotten that the roots of paleontology are in biology, just as geophysics rests on physics. It is a tragedy that paleontology has at last become a "handmaiden to geology." Yet the techniques and mass of data of paleontology are now so distinct from geology and biology that the majority of biologists and geologists do not even know what constitutes urgent problems in paleontology. Although it is seldom accorded the status of a separate science, paleontology is just that.

Newell drew particular attention to the problems in the current pedagogical climate: with "the majority of teachers of paleontology" being "stratigraphers or petroleum geologists, concerned entirely with the application of paleontology to geology. . . . Little progress is being made toward an understanding and interpretation of fossils and their life environment." However, Newell saw an opportunity to change this at Columbia, drawing on the rich resources at the AMNH to develop "a program of instruction in invertebrate paleontology, or paleobiology, at a professional level, adequate for the development of research specialists" (Norman Newell, "Instruction in Paleobiology": AMNH DVP, 67, 21).

By the end of the 1950s, more and more paleontologists were thinking like Newell and Simpson. G. Arthur Cooper's presidential address in 1958, "The Science of Paleontology," addressed this problem head-on, and argued for a biologically oriented paleontology that would nonetheless maintain its independence from its sister field. Cooper was particularly alarmed by a trend towards exclusively stratigraphic paleontology since the 1920s that had seen a simultaneous decline in "descriptive, taxonomic, and morphological" work, the latter of which was a "pure or 'old fashiond' [sic] Paleontology" to which he wanted the profession to return. He emphasized that paleontology was "a great field still in the qualitative stage with basic work for several generations," but agreed that effort would be required to recruit students to carry on the task (Cooper

1958, 1014). He proposed a straightforward approach to this problem by changing pedagogy to "emphasize the zoological and biological sides of the subject rather than the stratigraphical," and by "reorienting the values" of paleontology accordingly (Cooper 1958, 1015). But although he firmly cast his lot with those who "accept the fact that all life is related and that paleontology is in reality paleobiology, an independent science," Cooper's first concern was that paleontologists continue the tradition of morphological description before attacking the larger theoretical problems of evolution (Cooper 1958, 1012). Thus we can see that at this point, the question that most frequently divided paleontologists was whether paleontology belonged with biology or geology, and not whether their discipline had a role in evolutionary theory. Nonetheless, Cooper concluded that with sufficient collection of data, while "biology shows the mechanics of evolution . . . the pattern can only be learned through fossils" (Cooper 1958, 1016).

Newell on Interpreting the Fossil Record

One of the fundamental goals of evolutionary paleobiology was, as Cooper put it, to elucidate the pattern of evolution, and between 1950 and the late 1960s a number of innovative approaches to assessing patterns in the fossil record were developed. A major effort was applied to studying trends in large sample populations, which naturally required a deeper and more thorough understanding of the fossil record. Since invertebrate fossils are far more numerous than vertebrate remains, paleobiology gradually became oriented towards invertebrate paleontology. Simpson may have helped inaugurate the paleobiological movement with his studies of the evolutionary sequence of fossil horses, but workers recognized that well documented evolutionary sequences and reliable representative sample populations were far less likely to be found in vertebrate lineages. Paleobiology is not, in its essence, an exclusively invertebrate field, but practical and technical limitations and requirements directed the research interests of paleobiologists over the next several decades more and more towards invertebrates.

No interpretation of evolutionary patterns could begin, however, without a reasonable degree of confidence in the fossil record itself. It is worth noting that from the beginning of organized, professional paleontology in the United States, workers had been concerned with addressing this problem. The first meeting of the PS, held in Boston in 1909, was organized as a "Conference on the Aspects of Paleontology," and among the topics chosen for discussion was the "Adequacy of the Paleontologic Record." Given the general disrepute that had dogged the fossil record since Darwin's day, it is remarkable that one of the speakers commented that "the faithfulness with which the paleontological record has been kept since the beginning of the Cambrian is a matter of constant surprise" (Bassler 1910, 589), while another boldly predicted that it was "probably only a matter of time before the complete faunal succession can be established" (Calvin 1910, 582–83).

In fact, confidence in the fossil record did steadily improve over the next decades. In 1959, proceedings from a special "Symposium on Fifty Years of Paleontology" were published in the *Journal of Paleontology*, including several papers on the state of the fossil record. Newell's contribution to the symposium offered a theoretical overview of developments in the appreciation of the fossil record, and it explicitly connected those changes to the growth of paleobiology. Newell began by noting that "from the very beginnings of our science there have been two schools, those who study fossils in order to understand stratigraphy, and those who study fossils in order to learn about past life," and he applauded Cooper's presidential address of the previous year calling for greater biological orientation in paleontology (Newell 1959, 489). Newell was pleased to report that "the fossil record is much richer than we formerly supposed," but cautioned that paleontology needed to produce more biologically sensitive workers to meet the demands of the changing profession. He also cited five "truly revolutionary developments of the past three decades": (1) improved collection and preparation of fossils, (2) "recognition of the special importance of populations in taxonomy and evolution," (3) more attention to ecological context, (4) "the application of statistical methods . . . [to] all sorts of paleontological problems," and (5) greater understanding of the geochemistry of fossils (Newell 1959, 490). Newell contrasted the "gradual increase in appreciation of the positive merits of the fossil record" with Darwin's earlier "preoccupation with the deficiencies in the record," and while he noted a continued "lively debate" over interpretations of the record, he cited a "general agreement . . . that many striking patterns of fossil distributions have been confirmed hundreds of times" (Newell 1959, 490–91). With regard to the sheer quantity of paleontological data, Newell pointed to the dramatic improvement in knowledge of the record: whereas in 1910 Charles Schuchert had estimated some one hundred thousand known fossil species, Curt Teichert calculated in 1956 that as many as ten million species existed to be discovered (Newell 1959, 492; Schuchert 1910, 591–92; Teichert 1956). Overall, Newell predicted that "the future prospects for paleontology are, indeed, very bright" (Newell 1959, 499).

Indeed, Newell had played a major role in preparing that bright future. Not only did he take over leadership in paleontology at the AMNH when Simpson departed for Harvard in 1959, but he was consistently the most active promoter of the new paleobiology's agenda for quantification and biological synergy during the 1950s and 1960s. He was also one of the more prolific paleontological contributors during the first years of the journal *Evolution*, the organ of the Society for the Study of Evolution, and his work was vital to bringing the work of paleontologists to the attention of biologists and geneticists—and vice versa. Between 1947 and 1957 Newell published four pieces in *Evolution*, only one fewer than the five he published in *JP*. In fact, one of the *JP* articles was a reprint of an essay, "Infraspecific Categories in Invertebrate Paleontology," that had first appeared in *Evolution* in the fall of 1947. In justifying the unusual step of reprinting a paper only a year after its initial publication, the editor of *JP* commented, "This subject is of such importance that it should be brought to the attention of all paleontologists" (Newell 1947, 225).

Newell's 1947 paper on infraspecific categories was clearly a call to action for both paleontologists and biologists. The opening line proclaimed that "evolution as a modern philosophy requires the synthesis of paleontology, genetics, and neontology," and Newell proceeded to diagnose just what he thought paleontology could and should contribute to that synthesis (Newell 1947, 163). It is noteworthy that Newell favored the term "neontology" here and in later publications to underline his belief that paleontology and biology were be sister disciplines. His target in this essay was the gap between paleontologic and neontologic understanding of "species," which, he argued, prevented greater synthesis between paleontology and biology. The problem involved the common taxonomic practice of basing taxa on single type specimens, which, in Newell's mind, failed to consider "the variability of organisms in taxonomy." Mayr's 1942 *Systematics and the Origin of Species* had helped discourage this kind of "typological thinking" in biology by introducing the "biological species concept." But paleontologists had been slow to adopt that practice, and they understood species as populations of organisms exhibiting graded variability, tending to assign specimens that differed only slightly from one another to separate taxa.

The solution, according to Newell, was for paleontologists to adopt the biological concept of "subspecies" and to develop greater sophistication in establishing methods for discerning the true relationships between related organisms. His definition of subspecies as "entire populations, or races, which have become differentiated through some degree of isolation" was drawn from Mayr, and he explicitly conceived of these populations as geographic and genetic units (Newell 1947, 164). Of course paleontologists are unable to establish genetic relationships between fossil organisms, but Newell argued that paleontology would benefit simply from paying closer attention to the populational and biogeographical vocabulary of biology. For instance, he urged paleontologists to abandon the imprecise concept "variety," which does not necessarily connote a distinct population, in favor of the biological concept "subspecies," which usually does.

Newell's argument was more than simply a semantic one: if paleontologists were indeed to participate fully in the modern synthesis, they must accept the principle that "in all probability, evolution invariably has been accompanied by gradual morphological change" (Newell 1947, 167). Since the time of Darwin, paleontologists had been aware of the likelihood that the fossil record omitted transitional sequences, but before the synthesis paleontologists did not feel particularly constrained by a theoretical necessity to extrapolate a continuous gradation of forms between taxa. Theories like orthogenesis and saltationism allowed alternatives to perfectly graded sequences. However, accepting the synthetic definition of evolution meant accepting Mayr's and Dobzhansky's populational understanding of taxa, in which species are inherently unstable, variable entities, and divisions between taxonomic groups are often very subtle. So Newell's argument was important not only because it drew attention to the isomorphism between paleontological and biological taxonomic definitions, but also because it issued a challenge to paleontologists: if they wanted to sit at the table, they would have to find a way to incorporate their data within the conceptual vocabulary of biology and genetics.

This was no easy task, and paleontologists would spend the next several decades arguing about the problem. Newell's initial suggestions involved quantitative strategies that looked back towards Simpson's earlier work and forward to the approaching quantitative revolution in taxonomy (to be discussed shortly). One strategy involved establishing empirical criteria for determining a population's inherent variability, which required adapting the "normal curve" of variation for particular kinds of organisms (fig. 2.1). Another approach used stratigraphy to estimate the effects of time and geography on speciation—in effect treating evolution not just as a vertical sequence of forms, but as one that involves significant horizontal branching due to geography (barriers and migrations) and infraspecific variation (Newell 1947, 167–69).

Newell's discussion of the paleontological species concept played into a larger discussion and debate that began to unfold on the pages of *Evolution* and carried on throughout the 1950s. This concluded with his participation in a meeting of the British Systematics Association in London in 1954, appropriately titled "The Species Concept in Palaeontology."

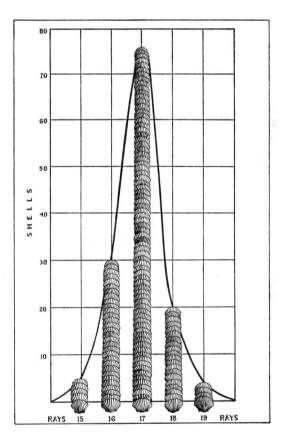

FIGURE 2.1a. Norman Newell's quantitative approach to estimating normal variability in living and fossil populations. The normal curve of variation is illustrated for a random sample of living scallops. Norman D. Newell, "Infraspecific Categories in Invertebrate Paleontology," *Evolution* 1 (1947): 166. © Society for the Study of Evolution.

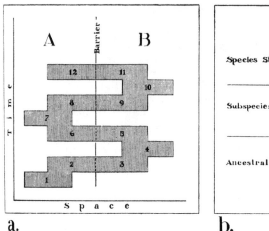
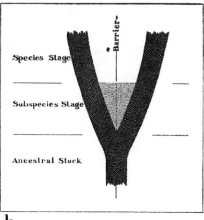

a.   b.

FIGURE 2.1b. Two diagrams represent the fragmentation and splitting of populations. Norman D. Newell, "Infraspecific Categories in Invertebrate Paleontology," *Evolution* I (1947): 169. © Society for the Study of Evolution.

This symposium and its eventual published proceedings were billed as a kind of follow-up to Julian Huxley's 1940 volume *The New Systematics*, also sponsored and published by the Systematics Association. Newell's contribution was an effort to navigate some of the difficulties in applying the biological, populational idea of species to paleontology. He sidestepped the difficulties involved in identifying species from fossil data by declaring that "the genus, which, in practice, is the smallest consistently recognizable unit, has become the working unit of palaeontology" (Newell 1956b, 63). He also reiterated some of his arguments about the direction of the profession, noting particularly "an ideological difference between those palaeontologists whose main, or sole, concern with fossils is the correlation of strata, and those who would learn about ancient life through studies of fossils" (Newell 1956b, 64). Nonetheless he concluded that paleontology was best served by a combination of these approaches, where the "empirical data of stratigraphy" is applied to a biological and ecological "interpretation of ancient life in the broadest sense."

The main target of Newell's analysis, however, was the unique set of problems paleontology faced in applying taxonomic divisions to fossil populations. Here his major concern was preservational bias: while "the fossil record is in fact astonishingly rich and meaningful," the "time dimension" in paleontology complicates matters, since "the selection of species limits in a vertical series might be arbitrary" (Newell 1956b, 67). In other words, the added dimension of time is both a boon and a hindrance to paleontology: within a given "horizontal" sample (i.e., a group of organisms taken from the exact same stratum or "moment" in geological time) it might certainly be possible to distinguish taxa, including species and perhaps even subspecies or varieties. But paleontology also has a vertical dimension, and as the taxa identified from horizontal samples continue forward in time it is extremely difficult to discern where taxonomic limits or divisions should be placed (fig. 2.2). This situation is further complicated by the fact that vertical sequences are almost always interrupted, and the paleontologist is not guaranteed to fill in these gaps by further collection. Finally, as Newell noted, horizontal and vertical perspectives must be combined to get an accurate picture of the influence of geography on phyletic evolution: "It may be doubted . . . that appreciable portions of the evolution of a lineage are completed at one

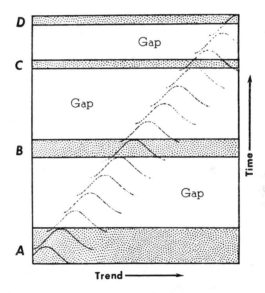

FIGURE 2.2. Newell's illustration of the effect of differential preservation on the fossil record. The succession of curves represents average population variation in a lineage evolving through time. Shaded areas are stratigraphic intervals favorable to preservation. Populations at horizons A, B, and C would be classified as distinct species, while unpreserved intermediate populations would be absent from the taxonomic record. Norman D. Newell, "Fossil Populations," in *The Species Concept in Paleontology: A Symposium.* (London: Systematics Association, 1956), 69.

place without extensive migration accompanied by repeated segregation and reunion of local populations (geographic speciation)" (Newell 1956b, 70). Nonetheless, in the face of such apparently insoluble difficulty, Newell remained confident that "properly conceived and diagnosed, palaeontological species and subspecies can be consistently recognized and studied by the same methods as those employed in neontology" (Newell 1956b, 70–71).

Newell determined that the way to accomplish this was to apply quantitative analysis to the confusing array of fossil data—to let statistics do what the paleontologist was unable to accomplish using traditional, descriptive techniques. In the past, paleontologists had relied on the typological basis for identifying species and higher taxa, but ecological and evolutionary study required paleontology to reorient itself to the neontological population understanding. According to Newell, the "crude procedure" of typology "does not measure up to modern requirements in studies of stratigraphic and evolutionary palaeontology" (Newell 1956b, 71). This was mainly because the typological species concept ignored population variability, which should in each instance follow a normal population curve. A type specimen was normally chosen (i.e., sampled) arbitrarily, and the paleontologist had no guarantee that it "represent[ed] the most frequent condition of populations" (i.e., that it would fall in the middle of a normal variability curve). Instead, the procedure should be to select, ideally as randomly as possible, a group of examples from a population and to estimate, using "biometrical analysis," the range of variation for that population. The trick, according to Newell, was "to summarise in a reasonably accurate way the characteristics of a vast assemblage of individuals, perhaps numbering billions, by means of data provided by a few specimens" (Newell 1956b, 74).

Such a drastic extrapolation could be justified only if one had confidence that the few specimens chosen gave a reasonable indication of the limits of variability in their parent population. Surprisingly, Newell argued, most populations *could* be estimated in such a way, and individual samples were in fact reliable indicators of average variability—provided that they were sampled *randomly*. The mistaken belief that only large and well-documented collections could be analyzed this way had meant "very little headway has been made toward the establishment of uniform practice in quantitative palaeontology"; what we see in Newell's proposal is the solidification of a major argument that statistical analysis could correct for the inadequacies of fossil preservation. This would be perhaps the single most important future direction in paleobiology, but it ultimately depended on a serendipitous convergence of paleontological thinking and technology. As Newell noted several years later, "The recent application of electronic IBM computers in the solution of paleontologic problems" was "more than just another statistical technique"; rather, as he went on to predict, the advent of inexpensive, readily available digital computing meant that "in the near future, we may have at our disposal the means for more or less routine quantitative solutions of all sorts of paleontological problems involving complex interrelationships of many variables" (Newell 1959, 490). In other words, evolutionary paleontology was about to become a quantitative discipline.

Tempo and Mode in Evolution Revisited

Thus far we have been considering the development of evolutionary paleontology as an extension of the program of the modern synthesis.[1] Indeed, paleontologists active in evolutionary theory accepted much more of the synthetic view than they rejected, and it would be false to conclude that at the end of the 1960s paleontology was not overwhelmingly neo-Darwinian in its perspective. Nonetheless, there were elements in the emerging paleobiological movement that offered resistance to—or at least polite disagreement with—some tenets of synthetic orthodoxy. First and foremost was the insistence by many paleontologists that the fossil record was, *pace* Darwin, a source of reliable data. One might argue, however, that this was not a genuine point of divergence from neo-Darwinism, since that theory made no necessary claims about the adequacy of the record (biologists' prejudices aside). More significant were

---

1. It is worth noting that while from the 1940s onward, evolutionary biology was largely centered in the United States and Britain, a significant school of evolutionary thought existed in central Europe—primarily in Germany—that developed independently of the Anglo-American modern synthesis. This tradition harkened back to the nineteenth-century idealistic morphological school of the German Romantic movement, whose members included Johannes Müller, Ernst Haeckel, Adolf Naef, Othenio Abel, Karl Beurlen, and others. Because that earlier tradition centered on morphology and zoology, paleontology grew to have considerably more importance in German evolutionary thought than it did in Anglo-American evolutionary circles. I will pass over much of this fascinating history, but interested readers should consult works by Wolf-Ernst Reif and Patricia Princehouse (Princehouse 2003; Reif 2000; Reif 1986).

the challenges, seen in proposals such as Simpson's "quantum evolution," to the synthetic assumption of a gradual evolutionary tempo. Finally, paleobiological study of the significance of extinction—and particularly mass extinction—as a factor in evolutionary change would also play an important role in the "revolutionary" phase of paleobiology during the 1970s, and here there were clear roots in the 1950s as well. Unsurprisingly, Newell was again at the center of these developments.

One of the questions frequently addressed by paleobiologists from the 1960s onward was whether the fossil record shows evidence of cyclical, discontinuous, or punctuated periods of evolution, radiation, and extinction. Beginning in the 1970s, paleontologists tended to answer this question affirmatively (Gould and Eldredge's theory of punctuated equilibria and David Raup and Jack Sepkoski's argument for periodic mass extinctions being only two of the most notable examples), but even in the 1950s and 1960s there was considerable interest in documenting evolutionary discontinuities. In 1951, for example, the Paleontological Society's annual meeting featured a symposium on the subject of "evolutionary explosions," which was itself a follow-up to a similar symposium at the 1938 PS meeting (Golding 1952, 298). A central topic at the 1951 conference was the influence and legitimacy of "diastrophism," or the theory that the earth's history has been punctuated by periods of geographical upheaval causing dislocations in the earth's crust. Diastrophism had long been the favored explanation for discontinuities apparent in the stratigraphic record, and when it was correlated with stratigraphic evidence for the sudden appearance of new taxa, it became the basis for the standard divisions in the geological timescale. Diastrophic processes included earthquakes, volcanism, flooding, and other violent natural processes; it should be noted, however, that diastrophism did not necessarily imply "catastrophism," since these processes could still fit within a general uniformitarian view of the history of the earth. Likewise, it should also be remembered that while Alfred Wegener in 1912 proposed the theory of continental drift (which now accounts for most of the "diastrophic" phenomena formerly attributed to other mechanisms and is explicitly uniformitarian), that theory did not gain wide acceptance until the 1960s.[2]

Lloyd Henbest summarized the major assertions of diastrophism discussed at the PS meeting in 1951: "The diastrophic theory predicates

2. For a history of the debate concerning continental drift, see Oreskes 1999.

that 1) diastrophism is periodic and synchronous on a world-wide scale; 2) diastrophism is a major control, if not the principal stimulus, of organic evolution; and therefore 3) diastrophism is the ultimate basis for correlating the events of earth history" (Henbest 1952, 299). While he neither rejected nor endorsed a grand diastrophic theory, Henbest did propose that establishing an "independent time-scale" for geology might provide a useful test of the theory, and he encouraged continued correction of the existing scale via techniques like radioisotope dating and stratigraphic correlation of fossils.

The most notable papers in the symposium were Simpson's "Periodicity in Vertebrate Evolution" and Newell's analysis of the same topic for the invertebrate fossil record (Simpson 1952; Newell 1952). The two papers were explicitly presented as part of a joint project which, as Simpson explained, sought "to survey some numerical aspects of the whole fossil record of animals" (Simpson 1952, 360). Simpson's paper reported that the appearance of patterns of explosive evolutionary "climates" "seems to be a biological, evolutionary sequence . . . [that] naturally occurs against the background and within the limitations of the physical environment," although it also concluded that it was "impossible even to make a start at realistic correlation of so regular and continuous a process with intermittent tectonic episodes" (Simpson 1952, 365). In other words, the patterns of explosion in the fossil record, if genuine, do not easily correlate with diastrophic events, nor do they fall conveniently at stratigraphic boundaries. While Simpson granted that certain events in the history of life, like mass extinctions, were sometimes located at such boundaries (in particular the Cretaceous-Tertiary extinction), he thought they were most likely part of a longer, more continuous process and did not "account for the origins of important new types or structural grades of animals" (Simpson 1952, 369). However, Simpson did give credence to "a well-marked periodicity in vertebrate history," although he interpreted it as "suggestive of continuity of evolutionary forces acting regardless of periodic physical events." Furthermore, he concluded that these explosions appeared "to be merely regular, shorter episodes or manifestations in a more basic process which is really secular and not periodic" (Simpson 1952, 371).

While Newell's contribution on the invertebrate record more enthusiastically endorsed genuine patterns of discontinuities, Newell shared Simpson's belief that (a) these discontinuities did not correlate with diastrophic activity, and (b) physical changes in the environment did not play a major role in evolution or extinction. Newell's paper examined the invertebrate fossil record in an attempt to detect "whether or not there have been times characterized by prevalently high or low evolutionary rates among invertebrates," and reported that while patterns at the species level and below were difficult to detect, genus-level data did "reflect evolutionary fluctuations at lower levels" (Newell 1952, 371 and 373). As examples of the kinds of fluctuations he had detected, Newell cited "1) marked increase in total diversity, 2) increase in mean rate of differentiation, and 3) a rise in percentage of new to surviving genera early in the eruption" (Newell 1952, 374). This led him to conclude that many groups showed strongly correlated patterns of evolutionary peaks and valleys "that can hardly be a chance correspondence" (Newell 1952, 380). Specifically, Newell synthesized the data from 26 classes and orders of marine invertebrates, and presented a distinctive pattern of eruptions and extinctions in a striking graph (fig. 2.3). However, the pattern appeared

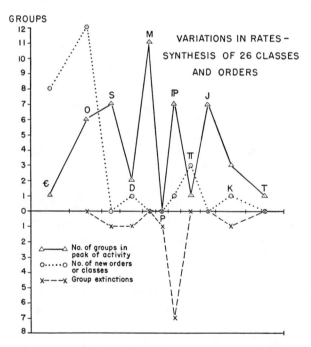

FIGURE 2.3. "Synthetic" graph juxtaposing Newell's analysis of variation of rates in generic differentiation, first appearance, and extinction (last appearance) of marine invertebrates in the fossil record. Norman D. Newell, "Periodicity in Invertebrate Evolution," *Journal of Paleontology* 26 (1952): 383. Courtesy of the Society for Sedimentary Geology (SEPM).

not to correlate with diastrophic periods, and Newell found "no evident reason for supporting that sweeping changes in physical environment can directly affect relative rates of correlation" (Newell 1952, 383). Rather, he suggested that diversification was likely proportionate to the availability of unoccupied niches, and that major periods of extinction were most probably attributable to longer-term, gradual physical events like "the withdrawal of the epicontinental seas from the continents." Thus, while "the rise and fall in apparently evolutionary activity is not at random," patterns in the fossil record are the product of a biological process of selective opportunism involving "the relatively complete occupation of available and accessible habitats" (Newell 1952, 383–84). In other words, at least as far as extinction was concerned, neo-Darwinism had nothing to fear from paleontology.

Mass Extinction and the Fossil Record

Over the next decade, Newell continued to examine patterns in the invertebrate fossil record with a particular eye for relationships between organic and physical histories, and his work directly influenced some of the most important paleobiological theories of the next generation. One of his most lasting legacies was to legitimize the study of mass extinctions as a significant evolutionary process. This legitimization was important: up until the 1960s, the prevailing attitude in the paleontological community was that to seriously discuss the possibility of regular mass extinctions was to invoke the specter of "catastrophism," which was associated either with old, discredited ideas or with the lunatic fringe. In the 20th century, the few scientists who proposed theories of catastrophic mass extinction—including, for example, the German paleontologist Otto Schindewolf, who attributed the extinction of the dinosaurs to sterility caused by bursts of cosmic radiation—were either ignored or ridiculed. It did not help matters that in 1950 Immanuel Velikovsky published the sensationalist *Worlds in Collision*, which argued, among other things, that the planet Venus had begun life as a comet whose perambulations had caused a series of catastrophes on earth during recorded human history, which he extended to a more global theory of geologic catastrophism in his 1955 book *Earth in Upheaval* (Velikovsky 1955).[3]

3. For discussion of Velikovsky's ideas, see Ginenthal et al. 1996.

Of course Velikovsky's ideas were nonsense, but they captured significant public attention, and in their haste to condemn them as such, scientists may have also cast suspicion on more legitimate discussion of catastrophic events in earth history. So in even being willing to discuss catastrophism publicly, Newell was taking a brave stand, and his work may have helped erase some of the taint that surrounded discussion of mass extinctions.

Newell's first important paper on mass extinction, "Catastrophism and the Fossil Record," was published in *Evolution* in 1956. Here Newell addressed Schindewolf's arguments about "the enigmatic, apparently world-wide, major interruptions in the fossil record which mark the boundaries of the eras," and he granted that "abrupt paleontological changes at these stratigraphic levels are real, [and] apparently synchronous" (Newell 1956a, 97). While Newell did not support a "catastrophist" interpretation of the fossil record, he did agree with Schindewolf that "critical events in the history of life evidently were responsible for these world-wide revolutionary changes" (Newell 1956a, 97). Using the data presented in his 1952 paper for the PS symposium on disatrophism, he examined the major faunal disruptions at the Permian-Triassic (P-Tr) and Cretaceous-Tertiary (K-T) boundaries—the latter of which included the disappearance of the dinosaurs. Newell concluded that these genuine mass extinction events "cut a broad swath across diverse and relatively unrelated ecological situations," from which recovery of evolutionary equilibrium was gradual. However, he also questioned Schindewolf's attribution of these changes to "some unknown extra-terrestrial agency," which to him appeared "marshaled to fit the theory" of saltative macromutations, and he questioned whether explanation of mass extinction must make recourse to "some unusual, catastrophic process" (Newell 1956, 99). Since most of the organisms affected by the major mass extinctions were marine, it made more sense to Newell that significant changes in sea level (eustatic changes), would account for extinctions as part of a causal chain that would also include reconfiguration of population dynamics over a considerable period of time. Newell concluded the piece with the observation that "perhaps it is futile to search for a single cause for all of the great mass extinctions," although he certainly did not close the door on further study (Newell 1956a, 101).

In 1962 Newell used his Paleontological Society presidential address, "Paleontological Gaps and Geochronology," to revisit the subject of catastrophic mass extinctions. Interestingly, in this paper Newell revived the "book" analogy discussed in chapter 1, stating that "the fossil record of any area resembles a book in which pages or whole chapters have been removed." Unlike earlier commentators, however, he did not argue that all of these gaps were the result of poor preservation (Newell 1962, 593). He suggested that breaks in stratigraphic sequences could have been caused by physical processes, and pointed to the correlation between two such breaks and the largest known mass extinctions, again citing the P-Tr and K-T boundaries. In his address Newell seemed to be more concerned with bringing attention to the study of these episodes than with announcing an explanation; noting that extinction is a complex and difficult process to understand, he proposed both biological factors (competition and disease) and physical agencies (atmospheric variation, radiation, sea level changes) as potential causes.

The very next year, Newell made a more definitive study, entitled "Revolutions and the History of Life."[4] This was presented at another special GSA symposium, "Uniformity and Simplicity." Newell's paper, along with a popularized version published in *Scientific American* as "Crises in the History of Life," provides a fairly definitive statement of his view of the role of mass extinctions in evolution (Newell 1967; Newell 1963).[5] The symposium paper opened with the bold claim that "the purpose of this essay is to demonstrate that the history of life . . . has been episodic rather than uniform, and to show that modern paleontology must incorporate certain aspects of both catastrophism and uniformitarianism while rejecting others" (Newell 1967, 64). Noting that most geologists thought "change" was "uniform and predictable rather than variable and stochastic," he called for greater openness towards discontinuity and unpredictability, and opined that "catastrophism rightly emphasized the episodic character of geologic history, the rapidity of some changes, and the difficulty of drawing exact analogies between past and present" (Newell 1967, 65). This statement was a remarkable repudiation of uniformitarianism, which was a pillar of both Darwin's theory and the neo-Darwinian interpretation of the modern synthesis, and Newell made it clear that he intended it as such. A major assumption of uniformitarianism is that gaps in the fossil record are the result of biases in deposi-

4. While the symposium took place at the 1963 GSA meeting, the proceedings were not published until 1967.

5. Because the two pieces were composed at the same time and cover substantially similar topics, reference here will be made only to the more scholarly presentation from the GSA symposium.

tion, preservation, or collection, but here Newell endorsed Schindewolf's argument that when such "abrupt changes occur in relatively complete sequences over a large part of the earth, they indicate episodes of greatly increased rate of extinction and evolution" (Newell 1967, 74). He also pointed to other factors, such as the stratigraphic correlation of extinctions of totally unrelated groups, and the tendency for episodes of apparent extinction to be followed by evidence of "episodes of exceptional radiation." This latter point contributed to a model of how extinction and evolution functioned hand-in-glove: Newell proposed that major extinction events cleared the adaptive landscape and opened new niches for surviving organisms to exploit, leading to massive and relatively sudden migrations and the production of new forms. In Newell's words, "a revolution characteristically consisted of two phases: an epoch of extinction and rapid decline by natural selection, accompanied or followed by an epoch of more gradual adaptive radiation" (Newell 1967, 82). Newell supplemented this argument with a graphical juxtaposition of periodic mass extinctions with the pattern of the growth in organismal diversity over geologic time. These patterns together persuasively illustrated the parallel processes of extinction and diversification (fig. 2.4).

Newell's paper also included a much lengthier consideration of causal factors than had his earlier studies, and here Newell more urgently pressed the need to develop explanations for the regular extinction of unrelated groups. After first dismissing proposed causes such as cosmic radiation, oxygen fluctuations, and changes in ocean salinity, he presented a tentative hypothesis of selective elimination by environmental change as the major cause of mass extinction. According to Newell, "This hypothesis postulates widespread, approximately synchronous, environmental disturbances and greatly increased selection pressure," for which he suggested three possible causes (Newell 1967, 84). The first of these causes was migration "involving better adapted immigrants and less adapted natives," which might become more frequent during times of environmental stress. This he posed as a direct challenge to Darwin's assertion that migrations were "selective and continuous," although he noted that it was the least likely source of very sharp discontinuities in the fossil record. The second possible cause was "severe climate changes," such as global ice ages, but Newell discounted its importance since (a) evidence of major climate shifts did not correspond with extinction events, and (b) plants—which we would expect to be especially responsive to climate fluctuations—had not been affected during major mass extinctions

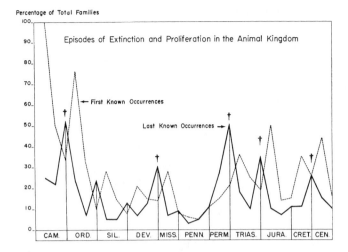

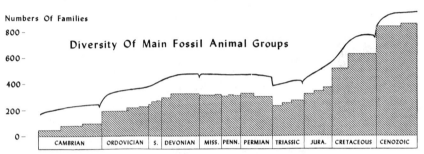

FIGURE 2.4. *Top*, percentage of first (dotted line) and last (solid line) appearances of animal families in the fossil record. The major extinction events at the ends of the Cambrian, Devonian, Permian, Triassic, and Cretaceous periods are clearly visible. *Bottom*, Newell's estimate of variation in familial diversity over geologic time. Norman D. Newell, "Revolutions in the History of Life," in *Uniformity and Simplicity, Geological Society of America Special Paper 89* (Boulder, CO: Geological Society of America, 1967), 79–80.

of animals (Newell 1967, 85). Finally, Newell addressed paleogeographic factors such as changes in sea level, which to him seemed the most likely culprit: "It seems clear that rapid emergence of the continents would result in catastrophic changes in both terrestrial and marine habitats and such changes might well trigger mass extinctions among the most fragile species" (Newell 1967, 88).

Overall, Newell's contributions to the study of mass extinction were significant primarily because of their legitimizing effect within the discipline. By virtue of his standing in the profession and through his

mentoring of young paleontologists, Newell lent considerable respecta-
bility to this area of study. By challenging some of the tenets of uniformitarianism, he also opened the door to more radical critiques of neo-
Darwinism that would be presented by paleobiologists over the next
two decades. Indeed, two of the more active proponents of such revisions were directly influenced by Newell: Gould was Newell's doctoral
student at Columbia between 1963 and 1967, and Niles Eldredge studied
with Newell throughout the 1960s as both an undergraduate and a graduate student. As Eldredge recalls, it was not lost on either him or Gould
that "Newell was the only person in 20th century paleontology who was
talking about the importance of extinction," and this led directly to Eldredge's own interest in patterns of evolution and extinction (Eldredge
interview). To be sure, however, Newell did not associate himself with
the German and Russian "neocatastrophist" movement of the same period, and he stressed that "there is nothing in the record to give support
to catastrophism, as Cuvier understood it."

On the other hand, Newell also argued that neither did the evidence
support "the literal uniformity of Lyell which emphasized slow and uniform instead of episodic changes" (Newell 1967, 89). According to Richard Huggett, "It is plain from these statements that Newell accepted catastrophism and non-actualism," although "these revolutions were not
produced by the grand catastrophes envisioned by Cuvier." While it is
doubtful that Newell would ever have accepted the label "catastrophist," it is undeniable that professionally and pedagogically he provided
much of the inspiration for the coming paleobiological critique of the
neo-Darwinian Synthesis (Huggett 1997, 165–66). Perhaps even more
importantly, Newell stressed that characterizing evolution and extinction as episodic, discontinuous, and stochastic did not mean abandoning a quest for general regularities, and did not necessitate abandoning a
systematic, quantitative study of the fossil record. As he put it, "The record of past revolutions in the animal kingdom is understandable by application of basic principles of modern science. In this sense, the present
is the key to the past" (Newell 1967, 89). As we will soon see, one of the
central themes in the modern paleobiological movement would be the
explanability—and even predictability—of complex, dynamic phenomena such as evolution and extinction. And in this regard most of the paleontologists at the forefront of this research over the next two decades
were, either directly or indirectly, Newell's students.

CHAPTER THREE

The Rise of Quantitative Paleobiology

O ne of the major trends in mid-20th century paleontology that we have already touched on briefly was a movement towards greater quantitative and analytical rigor. A number of scholars have argued that it is a mark of modern scientific disciplines that they increasingly tend towards greater quantification (Porter 1986; Porter 1995). While this may or may not be the case generally, it is certainly true that as paleobiology developed from the 1950s through the 1970s, quantification became closely associated with theoretical work on evolution. It is impossible to avoid the suspicion that paleontologists' desire for greater analytical rigor was motivated partially by the perception that biologists (and geneticists in particular) would only take paleontology seriously when its data could be expressed numerically. After all, since the pioneering work of Ronald Fisher and Sewall Wright, evolutionary biology (particularly in the hands of biologists like J. B. S. Haldane) had become a mathematical discipline, and paleontologists were not above feeling something akin to "physics envy." But it is also true that the kinds of analysis paleontologists like G. G. Simpson and Norman Newell wanted paleontologists to pursue—broad studies of the fossil record—required mathematical treatment. From the 1950s to the 1980s, then, theoretical questions pushed paleontology in a quantitative direction; at the same time, advances in statistical analysis and particularly in computational technology allowed paleontologists to examine their data in new ways. Rereading the fossil record involved, in large part, reading it *mathematically.*

This trend was undoubtedly also sparked by movements in other fields, such as systematics, which had its own quantitative "revolution" during this time and from which paleontologists drew methods and inspiration. But quantitative thinking found unique outlets in paleobiology, particularly in addressing problems involving large data sets and requiring complex multivariate analysis. Unquestionably, this approach would not have reached fruition before digital computers with large data storage capacities and reasonably accessible user interfaces became available, as they were by the mid-1960s. All in all, by the end of the 1960s, what had been a fairly informal collection of interests two decades earlier had now coalesced as central themes of "paleobiology," and one of the movement's essential characteristics was quantitative, theoretical analysis of the fossil record.

The Beginnings of Quantitative Paleobiology

As with many other aspects of its development, paleobiology drew much of its inspiration towards becoming genuinely quantitative from parallel movements in other, related fields. In the case of statistical analysis, an enormous influence on paleontology was the rise of the numerical taxonomy movement in the 1950s and 1960s, much of which played out on the pages of the journal *Systematic Zoology*, a publication frequently read and contributed to by paleontologists. This is not to say that paleontologists contributed no original ideas towards their own quantitative development. A number of them, including Simpson, Newell, and their AMNH colleague John Imbrie, were capable mathematicians who developed insightful and sophisticated statistical methods. However, debates within the field of numerical taxonomy helped raise awareness of statistics within the paleontological profession more broadly, and several of the texts its practitioners produced became the major introductions to quantitative analysis for a generation of paleontologists. Paleobiologists would not have their own, exclusively paleobiological primers on the subject until the late 1970s, by which time the field had already undergone a transformation. The sources of that transformation must therefore be located, at least in part, outside the discipline itself.

The Statistical Frame of Mind

Joel Hagen has thoroughly documented the rise of a "statistical frame of mind" in systematics and taxonomy during the decades after World

War II that took advantage of postwar advances in computing technology (Hagen 2003). In particular, two important books promoted quantitative techniques in this field: Simpson and Anne Roe's *Quantitative Zoology*, which was first published in 1939 and later revised and updated by Richard Lewontin in 1960, and Robert Sokal and F. James Rohlf's *Biometry*, which appeared in 1969 (Simpson 1960; Sokal and Rohlf 1969). While Simpson and Roe's original volume was relatively simplistic by later standards, both works were extremely important for pointing taxonomy in a more statistical direction and for demonstrating how quantitative analysis could help solve biological problems—such as those involving population dynamics—that were complex and multivariate. At the same time, journals like *Systematic Zoology* began to publish papers that were quantitative in orientation, making quantitative ideas in systematics accessible to a wider audience. Rohlf and Sokal first published a description of taxonomy using factor analysis in the journal in 1962, initiating a discussion concerning the use of computers and multivariate statistics that played out in the journal for the next ten years (Rohlf and Sokal 1962).

Statistical thinking was not new to biology in the 1950s, but the greater number of biologists and zoologists did not wholeheartedly accept its importance until at least the mid-1960s. In part, this involved a generational divide: even Simpson, who was quite mathematically adept, belonged to the generation that had struggled over R. A. Fisher's "esoteric" and difficult mathematical evolutionary theory, and Simpson and Roe's 1939 primer was explicitly designed as an introduction to the kinds of complex problems that frightened average biologists (Hagen 2003, 355–59). Another undeniable ingredient in the generational gap was the rapid growth of computing technology; in 1939, Simpson and Roe expected their readers to perform calculations by hand or, at most, with a mechanical or electric calculator, but by 1960 computers were widely available at most universities. However, this shift was not "merely" technological: as Hagen has convincingly shown, the adoption of the "statistical frame of mind" involved a significant methodological and even philosophical reorientation among biologists. Whereas in the past statistics had been regarded as a useful tool for aiding some aspects of data analysis in taxonomy and systematics, the new view, promoted especially by Sokal and Rohlf, argued that computer-assisted statistics "could play both a heuristic and an analytical role in biological research" (Hagen 2003, 368). While this emphasis was somewhat controversial in the general systematics community (as evidenced by the fierce debates over numerical taxonomy in the 1960s and 1970s), the heuristic value of statistical analysis would become an extremely important aspect of quantitative paleobiology in the 1970s, when it inspired the kinds of simulation and model building promoted by Stephen Jay Gould, Thomas Schopf, David Raup, and others (Hull 1988).

The Development of Multivariate Analysis in Paleobiology

In 1948 and 1949, paleontologist Benjamin Burma argued for greater quantitative orientation in paleontology in two linked papers in *Journal of Paleontology*, "Studies in Quantitative Paleontology: I. Some Aspects of the Theory and Practice of Quantitative Invertebrate Paleontology," and "Studies in Quantitative Paleontology II. Multivariate Analysis: A New Analytical Tool for Paleontology and Geology." The first paper was presented as a theoretical prolegomenon to a statistical approach, where Burma argued that the time had come for paleontology to shed "the qualitative outlook [that] has been the mark of its youthful stages" in favor of a quantitative approach which would characterize the discipline's "mature" phase (Burma 1948, 725). Beginning with the view that "paleontological collections are samples," Burma presented a variety of reasons why statistics would aid the study of invertebrate fossil populations, beginning with the need for paleontologists to develop tools that allow reliable extrapolations to be made based on the small samples paleontologists are able to work with. This is the same argument Newell would make several years later, and like Newell, Burma was especially concerned with the problem of determining the limits of a population's variability in order to properly assign specific and subspecific classifications (Burma 1948, 729–31).

Burma also recognized that traditional univariate statistical methods (such as those promoted by Simpson and Roe) were hampered by their inability to simultaneously consider multiple characters. For this reason he declared multivariate analysis to be "the method of preference," although he noted that the complicated and time-consuming nature of multivariate calculations might put off many paleontologists (Burma 1948, 732). Burma thus saved a technical discussion of multivariate statistics for the second part of the paper, which was published the following year. The essential difference between univariate and multivariate techniques is that whereas statistics involving one or two variables plots values on a two-dimensional coordinate system (where values for each specimen are represented as a line or scatter), multivariate analysis plots the dispersion of *n* variables in an *n*-dimensional cluster. The only added complication is the number of calculations required in the latter approach. As Burma pointed out, the mathematics is effectively the same in both cases. He stressed, however, that "an electrically driven" calculator was a basic necessity, and that investigators studying problems with many variables faced diminishing returns as they taxed the limits of available technology (Burma 1949, 96).

Burma's paper highlighted a general problem facing paleontologists and biologists interested in conducting increasingly sophisticated statistical analysis of populations: despite continued refinement of statistical techniques, multivariate analysis required too much computation to be viable using hand computation. Therefore, major advances in quantitative paleontology depended ultimately on the advent of programmable digital computers, which allowed many of the complex computations to be programmed and run automatically, reducing the task of the individual investigator to collecting and entering the raw data. But this advance was not yet available in the early 1950s, meaning that paleontologists were left to imagine the potential of statistics mostly as a future innovation.

Another statistically sophisticated paper from this early period was Everett C. Olson and Robert L. Miller's "A Mathematical Model Applied to a Study of the Evolution of Species," published in *Evolution* in 1951. Like Burma's, this study explored the theoretical possibilities in applying statistics to more and more general evolutionary problems. Despite "severe restrictions" on many statistical problems due to insufficient data, the authors cited "many problems of evolution that cannot be treated adequately by classical approaches or by application of statistical procedures now commonly used." They also expressed their hope to identify "some unifying principle" to allow "more definite bases for determining how and why particular changes came about, than has been possible in the past" (Olson and Miller 1951, 325).

As Olson and Miller noted, most statistical approaches to fossil populations treated their data as static, whereas from an evolutionary perspective it is desirable to understand the extent to which characters are *changing* in relation to one another: in other words, to address "each changing part . . . in relation to its effects on other changing parts, whether these be closely related or partially or completely independent." They argued that the best approach to this kind of question was to use regression analysis, which shows the relationship between changes in variables, allowing the investigator to determine whether those variables are correlated. For example, given a table of values for variables x and y, regression would show that if x changes by some amount n, y changes by n^{-1}, and so on. Note that demonstrating correlation does not necessarily demonstrate *cause*; there is always the possibility that some additional factor explains the relationship better. But regression is extremely useful for answering questions about relative growth, since it ultimately allows one to predict the extent to which changes in a dependant variable would affect or constrain changes in the value of an independent variable (for example, how changes in the size of a skull would affect such variables as snout length, sinus capacity, orbital length, etc.). Of course, while running bivariate regressions is relatively straightforward, adding additional variables introduces substantial computational complexity, meaning that multivariate regression analysis is limited by the time and computing power available to the investigator.

Olson and Miller demonstrated their general thesis by running bivariate regressions on a series of characters in three samples of fossil captorinomorph reptiles, two of which had been assigned reliable generic and specific classifications, with the third being undetermined. After presenting the results of the analysis, Olson and Miller concluded that the regression model demonstrated significant correlation patterns that allowed "speculation concerning the factors that underlie the evolutionary changes" in the three species (Olson and Miller 1951, 334). This kind of conclusion could not, of course, be made independently of knowledge of other important biological and environmental factors, but Olson and Miller thought that properly executed statistical analysis could highlight evolutionary relationships likely to be missed using traditional qualitative methods.

This is the value Olson and Miller highlighted: while paleontologists did not have access to genetic data for their specimens, they could reason by analogy from morphological features to the kinds of genetic relationships seen in living populations, and the application of statistical analysis offered a powerful approach to "the study of evolution of extinct groups at low taxonomic levels" (Olson and Miller 1951, 336–37). In this way, statistical analysis could function "first as a guide in the development of hypotheses and second as a source of information to be tested in the light of paleozoological data independently gained." This is exactly the kind of "probabilistic approach to hypothesis testing" that

Hagen identifies as the hallmark of the statistical frame of mind among younger systematists in the 1950s and 1960s; Olson and Miller's paper demonstrates that this approach clearly had indigenous roots in paleontology as well (Hagen 2003, 365).

John Imbrie and Statistical Paleontology

An important aspect of the growth of quantification in paleobiology concerns the way these methods were taught and presented to students. Quantitative and statistical methods were not part of a normal graduate curriculum in paleontology at most universities until the mid-1970s, so for the most part students were not formally trained in these techniques. Resources for self-study, like Simpson and Roe's textbook, were available, but more sophisticated analysis required the proximity or encouragement of one of the few paleontologists actively pursuing this kind of work.

However, one figure features prominently in the recollections of many young paleontologists of that period, both as an indirect influence and as a teacher: the Columbia University and AMNH invertebrate paleontologist John Imbrie, who was hired at the AMNH in the early 1950s as part of Newell's program to expand invertebrate paleontology at the museum. Although he did not remain in the field for long (he moved to Brown University in 1966 and worked in the field of paleoclimatology for the rest of his career), during his short time in New York he made an extraordinary impact on students who passed through the geology program at Columbia, and his few published pieces had an even wider influence. Raup cites Imbrie as an early inspiration, noting that he even traveled to New York as a graduate student in 1954 or 1955 to learn Imbrie's "biometrics and general applied statistics" directly from the source. Niles Eldredge, who studied with Imbrie as both an undergraduate and a graduate student towards the end of Imbrie's tenure at Columbia, recalls him as an "enormously entertaining," intellectually "dynamic" teacher, whose approach to statistical analysis was innovative and inspiring (Eldredge interview). But even students who did not encounter Imbrie directly could make use of his "little treatise on biometrical methods" (as Raup calls it): a 40-page paper in the *Bulletin of the American Museum of Natural History* published in early 1956 (Imbrie 1956). Imbrie published very little during his career in leading journals like *JP* and *Evolution*, but his few publications dealing with statistics quickly became classics. Raup calls the *Bulletin* essay "magnificent," noting that its methods

had a "spark of freshness" lacking in other statistical work (Raup interview), and Roger Thomas recalls making "extensive use" of the essay as
a graduate student at Harvard in the late 1960s (Roger D. M. Thomas,
personal communication 23 October 2003).

Imbrie's treatise opened with a general defense of statistical analysis
of invertebrate fossil samples. Citing the recent growth of recent interest in statistical methods, he identified three major reasons for continuing to investigate such applications. In the first instance, he noted simply that quantitative analysis is more precise than traditional qualitative
study. Second, he explained that current interest in the populational approach to evolution (i.e., studies of variability within a population) made
it desirable to combine observations of many representative individuals
rather than to merely document a single holotype specimen, as was traditional in paleontology. As he put it, "Descriptive devices are needed that
will summarize compactly the essential characteristics of a series of observations." He added that "just such devices, called statistics, have been
designed." Third, statistics offered the additional advantage of clarifying
the relationship between samples under study and known populations,
and could help taxonomists determine the margin of error for associat-
ing a particular sample with a population (Imbrie 1956, 217–18). Imbrie believed these reasons to be self-explanatory, but wondered why, despite encouragement from a variety of workers (e.g., Burma, Olson and
Miller, and Simpson and Roe), "such techniques have not in fact been
more widely adopted in invertebrate paleontology" (Imbrie 1956, 218).
It is interesting to note here that Imbrie focused specifically on invertebrate fossils, which he felt had been neglected by statistical studies.

Imbrie labeled the kind of statistical analysis he was promoting "biometrics," which he defined as "the statistical treatment of qualitative morphological data" (Imbrie 1956, 219). The term "biometry" was
coined in the nineteenth century by William Whewell and employed in
the early twentieth century by biologists such as Francis Galton, J. B. S.
Haldane, and Julian Huxley. It also became the preferred label for the
numerical taxonomy advocated by Sokal and Rohlf over the next decade.[1] Imbrie felt that biometry was particularly useful for dealing with
populations because "group tendencies" were highlighted, and because
it provided "an opportunity to record with reasonable precision some

---

1. The *Oxford English Dictionary* cites an 1876 letter written by Whewell as the first instance of the term. For other early uses, see Galton 1901, 9; Haldane 1927, 72.

of the results of dynamic evolutionary processes." Here his arguments echoed some of those expressed by Olson and Miller, Newell, and others: biometrical analysis is not just a method of quantifying and summarizing mean morphological characteristics of groups of organisms; more specifically, it captures the *dynamics* of populations as fluid entities subject to growth and evolutionary change. Hence, the populational and statistical approaches to evolution coincided: if taxa were understood to be continuous rather than fixed entities, statistical analysis offered the best way of precisely specifying the factors that justify designating certain portions of the continuum as distinct from others. A major concern for paleontologists was that samples would be erroneously assigned to taxa based on sampling errors—in other words, that the few representative fossils a paleontologist was studying would not indicate the true range of variability of the population. Statistics, Imbrie argued, offered the best chance of catching such mistakes, as it could assign probabilities of error with reasonable accuracy (Imbrie 1956, 220–21).

Imbrie's paper discussed only univariate and bivariate statistics, since at that time multivariate techniques were considered "too laborious and abstract for wide application." Still, Imbrie argued that much could be learned from analysis of only one or two variates. Univariate analysis was most useful for "calculation of certain qualities that summarize important features of a sample," such as the mean, standard deviation, "population parameter," and standard error (Imbrie 1956, 223–26). If two samples were compared for a single variate, a null hypothesis that the means of the two samples are equal (i.e., that they belong to the same population) could be tested, and the probability of error—"the chance of saying that the populations are different when they are really the same"—could be calculated. Bivariate analysis, on the other hand, was presented as the dynamic analysis of relative growth. With this method an investigator could characterize a sample "in a way that has general taxonomic value" whether or not the investigator could identify the growth stages represented in the sample—i.e., whether or not the sample contained mostly adults, mostly juveniles, or a mixture of stages. Equally important, "By shedding light on morphogeny, bivariate analysis provides a better understanding of the underlying genetic mechanism" that produces observed patterns of relative growth (Imbrie 1956, 227). The importance of this last feature is obvious, since paleontologists otherwise had no insight into genetic mechanisms and relationships in fossil samples.

The methods Imbrie described are straightforward and mostly

adapted from previous studies, in particular Huxley's 1932 *Problems of Relative Growth* (Huxley 1932). The procedure Imbrie outlined involved plotting morphological features x and y along the simple allometric curve $y = bx^a$, which when converted to a logarithmic plot shows a steady increase of the rate of growth ($\log y = a \log x + \log b$). The parameter a indicates the slope of the growth line, and b is the "initial growth index," or the absolute value of y when $x = 1$. The task, then, was to determine how well the observed sample fits this line; in many cases, the cluster of observations could be visually confirmed (Imbrie 1956, 230). However, in some cases "an objective algebraic solution to the problem is desired" if the "line of best fit" could not be determined visually. Here Imbrie offered four potential solutions. The first, determination of the "major axis" (the sum of squares of perpendicular distances of each point from the desired line of best fit) he rejected as being "unsuitable for taxonomic problems" because of variability in the line's slope. The next two involved regression, where one variate was taken to be independent and the other dependent, and changes in the dependent variate were measured as the perpendicular distance from the axis of the line of the independent variate. This can be performed as regression of x on y or vice versa; what it shows is the extent to which changes in one variate affect subsequent changes in the other. While Burma had championed this method, Imbrie concluded that it was seriously flawed, since it assumed that the dispersion of points was caused by deviations in a single variate only. As Imbrie put it, "From the biological point of view this assumption is never warranted, for biological variability and observational errors are always involved in both variates" (Imbrie 1956, 230).

Imbrie preferred to determine the "reduced major axis," based on the calculation of the area of a triangle formed by lines extending from a point perpendicular to both x and y axes (fig. 3.1). As Imbrie concluded, this method was superior because "1) it makes no assumptions of independence; 2) it is invariant under changes of scale; 3) it is simple to compute; and 4) results obtained from its use are intuitively more reasonable than corresponding results obtained from regression analysis" (Imbrie 1956, 231). Imbrie then demonstrated the superiority of this method using a concrete example drawn from a hypothetical data set, and showed that analysis of dispersion around a reduced major axis could reveal both absolute and relative variation in a sample, or "the amount of shape variation as a proportion of the average shape attained by the sample" (Imbrie 1956, 241). In his conclusion, he argued that so long as reliable

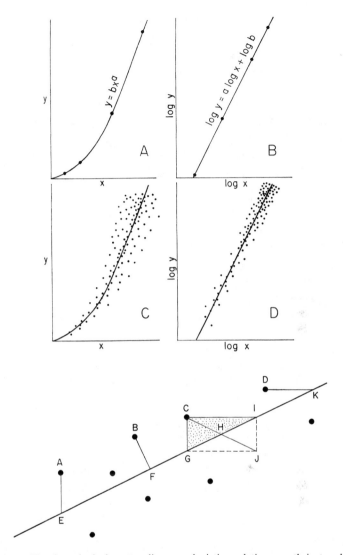

FIGURE 3.1. *Top*, hypothetical scatter diagrams depicting relative growth in two linear measurements, plotted both arithmetically and logarithmically. *Bottom*, Imbrie illustrates methods of line fitting for a scatter of points, including regression, major axis, and reduced major axis techniques. John Imbrie, "Biometrical Methods in the Study of Invertebrate Fossils," *Bulletin of the American Museum of Natural History* 108 (1956): 229, 231. Courtesy of the American Museum of Natural History.

measurements of a sample were taken, "the resulting statistical conclusions will be biologically valid" (Imbrie 1956, 242).

Imbrie did not invent this technique. The major axis was introduced by Karl Pearson in 1901, and its modified form, the reduced major axis, was first proposed by K. A. Kermack and J. B. S. Haldane in 1950 as a technique whereby calculations of the line would remain invariant under changes in scale (Pearson 1901, 559). But Raup recalls that "among other things [Imbrie] introduced to paleobiometrics the reduced major axis and that was highly important for when we were dealing with bivariate plots of two dimensions that had no clear independence. . . . And he made a good case for it and everybody used it for years in that situation. . . . It wasn't just somebody else's cookbook; here was something that John really introduced" (Raup interview). The only limitations on this method were that it could be applied only to two variates simultaneously, and that sample sizes would be limited to what could realistically be computed by hand. Imbrie estimated that, using a calculator, "a bivariate sample of 50 specimens can easily be characterized statistically in 30 minutes" (Imbrie 1956, 242). For many purposes this would indeed be fine. Imbrie did not mention, however, that if multiple pairs of variates were tested for that same sample the time involved would increase dramatically, nor that the technique would be helpless in the face of samples containing one thousand or ten thousand individuals.

Imbrie's only other major published contribution to quantitative paleontology was a chapter in a book he edited, along with Newell, in 1964 entitled *Approaches to Paleoecology*. As the next chapter will discuss, paleoecology was a major innovation in paleontological study of the interrelationships between fossil organisms and their environments, and as such it was a multidisciplinary field that drew from paleontology, ecology, population biology, and geology. In the 1950s and 1960s a major center for paleoecology was Yale University, where the ecologists G. Evelyn Hutchinson and A. Lee McAlester attracted many students who were interested in pursuing an approach to paleontology that was more explicitly biological. In a broader sense, though, paleoecology encompassed much of what I am describing here as "paleobiology," though its practitioners tended to focus more specifically on environmental structure and interactions and less on evolutionary dynamics.

One of the most important questions to develop out of 1960s paleoecology was the problem of how to define "natural" ecological units, since, as paleontologists and biologists had been aware for some time,

taxonomic categories shed little light on the actual ecological and evolutionary relationships between populations of organisms. Perhaps the
most significant response from paleoecology was the articulation of the
concept of the "community," which University of Chicago paleontologist
Ralph Gordon Johnson defined in 1964 variously as "the assemblage of
organisms inhabiting a specified space" or "an assemblage of organisms
which often occur together" (Johnson 1964, 109). Johnson urged workers
to "utilize the quantitative methods and the biological concepts that are
available" to promote this new community approach, and in the 1970s
and 1980s this proved to be a tremendously important line of research.
As will be discussed in later chapters, the problem of the succession of
communities over time was taken up by a number of the most prominent
paleobiologists of that period, including James Valentine, Richard Bambach, David Raup, Jack Sepkoski, and others.

From an analytical standpoint, the drawback of the community model
is that it is fearsomely complex due to its recognition of the interrelationships between multiple levels of the ecological system. As Imbrie noted
in his contribution to the paleoecology volume, "It is precisely *because*
of the complexity of the paleoecologic system and our meager level of
understanding of it that mathematical models are essential adjuncts to
other approaches" (Imbrie 1964, 407). Imbrie's chapter offered a statistical approach called "factor analysis" for dealing with extremely complex multivariate systems that helped to make sense of this complexity.
Essentially, what factor analysis does is to take a very complex system
with multiple "surface" variables that are of interest to the investigator
but whose correlation appears chaotic or random, and identifies meaningful patterns and correlations among them. It does this by assigning
to the system hypothetical "internal" attributes, which are unobserv-
able and unmeasureable but are thought to account for patterns among
the surface variables. These attributes are called factors. For example,
in psychometric research it might be desirable to know how a number
of measurable surface attributes like performance on various tests of in-
telligence are related for a large sample population; investigators might
posit factors like "innate" reading or mathematical ability to explain the
observed range of variation in tests, which are assumed to influence test
performance in a linear way (Tucker and MacCallum 1993, 3–4). In genetics, Sewall Wright pioneered the use of factor analysis in modeling
path coefficients for breeding populations (Imbrie 1964, 409; Provine
1986, 376–77).

Unlike some earlier paleontologists, however, Imbrie used statistics to provide a simplifying *model* for the explanation of a complex phenomenon, and not just as a method for calculating values of data. In other words, Imbrie believed that factor analysis assumes that "the raw data of any science are extraordinarily complex, and the prime task of science is to discover simple, general principles lying behind surface appearances," such as Newton's law of universal gravitation. This is where factor analysis differs most significantly from other statistical approaches: normally, statistics provides evidence of correlations between variables, which are then used to argue towards causes. However, in factor analysis the causes or factors are assumed a priori and the goal is to determine what the causal relationships are between factors and observed variables. The method, in principle, can tell us three things: (1) what minimum number of causes are required to account for a complex phenomenon, (2) what the causal influences are, and (3) the relative influence of each causal factor on each observed variable (Imbrie 1964, 411). Significantly, Imbrie was also very careful to acknowledge that the model's notion of causality "is defined strictly as a mathematical concept," and that it assumes causes to be linear.

The method Imbrie recommended is fairly straightforward: variables are taken as N observations or measurements of empirical data, and are arranged in tables of $n \times N$ rows and columns. M causal factors are assumed, and between each cause C and effect x a causal chain is constructed. Next these causal correlations are graphically represented as a "path coefficient model" in which the strength of each relationship between causes and effects can be calculated, and the resulting diagram shows both the number of cause-and-effect relationships and the relative strength of each one. In theory, the upper limit to the number of causes and variables is limited only by the time and computing capability of the investigator. The factor analysis itself involves "algebraically complex" procedures, but Imbrie noted that his calculation of a 13×13 matrix was performed "in less than three minutes" on an IBM 7090 transistorized computer (Imbrie 1964, 417). Indeed, the technique Imbrie proposed in this paper had been made practicable only by very recent developments in computing technology: the IBM 7090 had been released in 1959, when it was billed as "the most powerful data processing system now coming off production lines at International Business Machines Corporation."[2]

2. "IBM 7090 Data Processing System Press Technical Fact Sheet," accessed at http://www-03.ibm.com/ibm/history/exhibits/mainframe/mainframe_PP7090.html.

This was a fully transistorized version of the earlier vacuum-tube model 709, equipped with magnetic tape storage for programs entered through the card reader; according to the company, it was seven times more powerful than its predecessor. Columbia's machine, purchased and installed in 1963 to replace the IBM 650 the university had purchased in 1955, was part of a major initiative at the university to expand its computing facilities and operations.[3] The technology available to perform complex factor analysis was therefore quite new for Imbrie and other researchers.

Thus, Imbrie's short paper on factor analysis represented the state of the art in computer-assisted quantitative analysis available to paleontologists at the time. Over the next several years, however, his techniques would become more widely dispersed and available, eventually becoming standard in quantitative paleobiology. During the 1970s, pioneering studies of diversity by Raup, Sepkoski, and Imbrie himself would make extensive use of factor analysis as a part of what I will describe as the "generalizing" or statistical approach to reading the fossil record. This discussion will be taken up in later chapters.

Models of Growth and Form

D'Arcy Thompson and the Science of Form

The kind of statistical analysis of sample populations promoted by Simpson, Newell, Imbrie, and others was undoubtedly a major component of the quantification of paleobiology. Another mathematical direction, however, involved the geometric modeling of the form of organisms. This approach, which drew inspiration from D'Arcy Thompson's compendious *On Growth and Form*, originally published in 1917 and revised and expanded in 1942, produced a movement in "theoretical morphology" that sought to quantify and model morphology on the basis of fairly simple, geometric parameters that were assumed to convey basic laws of formal constraint in certain organisms. Here David Raup was the acknowledged pioneer, and the development of theoretical morphology became one of the first analytic approaches in paleobiology to make digital computers an essential research tool.

The fact that certain kinds of shells evidence a basic spiral geometry

3. Frank da Cruz, "Columbia University Computing History," accessed at http://www .columbia.edu/acis/history/index.html.

was known at least as far back as the seventeenth century, when Christopher Wren commented on the applicability of René Descartes' analysis of the equiangular spiral to coiled shells (Wolfram 2002, n. 1008). As early as 1838, the British mathematician Henry Moseley determined equations for describing the generation of several important geometrical characteristics of ideal coiled shells. In a paper published in the *Philosophical Transactions of the Royal Society of London*, Moseley commented: "There is a Mechanical uniformity observable in the description of shells of the same species, which at once suggests the probability that the generating figure of each increases, and that the spiral chamber of each expands itself, according to some simple geometrical law common to all." Moseley based his mathematical analysis on the assumption that "the surface of any turbinated or discoid shell may be imagined to be generated by the revolution about a fixed axis (the axis of the shell) of the perimeter of a geometrical figure, which, remaining always geometrically similar to itself, increases continually its dimensions" (Moseley 1838, 351). He suggested equations for a number of parameters of shell geometry, including volume and surface area, and determined, significantly, that "in every case" the shape described by the generating curve of the shell's growth "is a logarithmic spiral" (Moseley 1838, 352). Perhaps because of their complexity, Moseley's equations were not applied to actual specimens during his lifetime, but his paper provided much of the basis for Thompson's similar investigations of shell geometry nearly a hundred years later.

D'Arcy Thompson's *On Growth and Form* is a thousand-page treatise that purportedly explains how "the forms of living things, and the parts of living things, can be explained by physical considerations . . . in conformity with physical and mathematical laws" (Thompson 1992, 15). Thompson spent his long career as professor of natural history at Dundee and St. Andrews Universities in Scotland, and he was a firm believer in the Platonic or Pythagorean principle that "numerical precision is the very soul of science, and its attainment affords the best, perhaps the only criterion of the truth of theories and the correctness of experiments." His goal was to define the mathematical principles that govern the formal characteristics of all organisms in nature (Thompson 1992, 2). In its implicit assumption of the priority of physical constraints before function, *On Growth and Form* is also a substantial critique of the principle of functionalism in biology, and a forerunner to Gould's paleobiological critique of "adaptationism" in evolutionary biology (Gould and Lewontin 1979).

One of Thompson's central insights was that the form of an organism must be viewed not as a static property, but rather as a function of its growth over the organism's lifetime. As Archimedes knew, there is a necessary relationship between increase in the surface area, volume, and linear dimensions of a figure, and Thompson set out to apply this principle to various kinds of forms in plants and animals (Thompson 1992, ch. 2).[4] From the perspective of theoretical morphology in paleontology, the most important chapters come towards the end of *On Growth and Form*, where Thompson discussed various naturally occurring spiral forms in organisms. Chapter 11, "The Equiangular Spiral," is the classic discussion of the geometry of spiral shells in a wide variety of molluscan forms. Here Thompson's point was to show that a great diversity of form could be produced by manipulation of a few simple variables, since "the spiral curve of a shell is, in a sense, a vector diagram of its own growth; for it shews [*sic*] at each instant of time the direction, radial and tangential, of growth, and the unchanging ratio of velocities in these directions" (Thompson 1992, 768). Specifically, the form of many common radial shells follows the geometry of the *equiangular* or *logarithmic* spiral, which is a spiral produced when the generating point of the curve proceeds not with constant velocity (as in the simple Archimedian spiral), but rather with increasing velocity in proportion to the distance from the center of the spiral, so that successive whorls of the shell get larger the further they are from the center (fig. 3.2).[5]

However, in order to explain the great diversity of form actually seen in nature, Thompson also developed a broader theory that attempted to explain how basic, idealized forms (such as the curvature of a spiral shell, the dimensions of a vertebrate skull, the shape of a leaf, or the

4. For example, the surface area increases as the square, and the volume as the cube, of the linear dimensions of a figure. Growth in length will therefore produce exponential increase in volume, and since volume increases more rapidly than surface area, an organism will be constrained by certain physical restrictions on size.

5. The spiral is labeled "equiangular" because the generating point of the curve sweeps out equal angles in equal amounts of time, similar to Kepler's law describing the motion of a planet orbiting the sun in an ellipse. Another way of describing this is to say that "the vector angles about the pole are proportional to the logarithms of the successive radii," and that hence the spiral is "logarithmic" (Thompson 1992, 755).

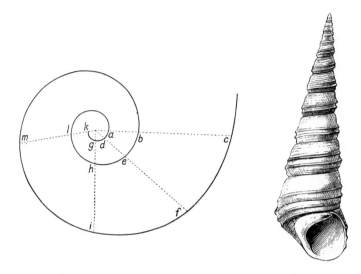

FIGURE 3.2. D'Arcy Thompson's illustration of the equiangular spiral. D'Arcy Wentworth Thompson, *On Growth and Form* (New York: Dover, 1992), 771.

body plan of a cephalopod) could be modified to produce the wide variety of basic shapes in nature. The method Thompson used was to project one such form onto a two-dimensional grid, and then to deform that grid by modifying parameters in order to produce a new but related form (Thompson 1992, 1033). By showing how simple modification of physical dimensions could produce variations in form between actual organisms, Thompson hoped to show how unmediated physical forces were sufficient to produce organismic variation, without the intervention of Darwinian mechanisms like adaptation and natural selection (Thompson 1992, 1033–34). Of course, Thompson recognized that mathematics could not perfectly account for nature in all of its diversity, and that simple mathematical rules do not *explain* every organic form. However, he viewed his approach as an important step towards mathematizing the biological sciences, a project he regarded as long overdue (Thompson 1992, 1029). Nonetheless, he argued that important conclusions could be drawn from the mathematical approach, including an understanding of the relationships between basic organismic types. Thompson believed that if the method were extended beyond structures empirically observed in organisms to "drawing hypothetical structures," it could be applied to "the particular case of representing intermediate stages between two forms which are actually known to exist, in other words, of reconstructing the

transitional stages through which the course of evolution must have successively traveled if it has brought about the change from some ancestral
type to its presumed descendant" (Thompson 1992, 1069–70).

Elegant as it may be, Thompson's theory is open to the critique that
his argument is a fundamentally flawed syllogism: geometrical laws govern the basic form of organisms, these laws reflect physical forces and
constraints on form, certain geometrical transformations are impossible,
hence certain lines of physical evolution cannot have taken place. Take
away the assumption that nature adheres to strict rules of geometrical
transformation and the argument falls to pieces. But despite its logical
flaws, Thompson's study was an important basis for the developing field
of theoretical morphology, and it even, ironically, provided indirect support for Darwinian evolution. As Gould points out,

> Thompson's error can be epitomized this way: He viewed physical forces as
> the efficient cause of form; they are, in fact, formal causes or blueprints of op-
> timum shapes that determine the direction which natural selection (the true
> efficient cause) must take to produce adaptation. . . . He was right to corre-
> late physical forces with organic forms and to claim that the correspondence
> was no mere analogy; but he was right for the wrong reason. D'Arcy Thomp-
> son thought he had a theory for the efficient cause of good design; he gave us
> instead the basis for a science of form—an analytic approach to adaptation
> (Gould 1971, 252–53).

Gould nonetheless admired Thompson, and he excused some of Thompson's conclusions for a very basic technical reason: according to Gould,
Thompson's mathematical analysis simply required the consideration of
too many variables to be practical when it was first proposed. As Gould
put it, Thompson's "approach to form was multivariate in conception;
hence it suffered the misfortune of much prophecy—it could not be used
in its own time"; however, "with the advent of electronic computers, the
situation has changed completely" (Gould 1971, 253–54).

David Raup and Theoretical Morphology

It is generally acknowledged that the modern field of theoretical morphology owes its existence largely to the work of David M. Raup during the mid- to late 1960s. While the term itself was first coined by
E. S. Russell in his 1916 book *On Form and Function*, it was Raup's pio-

neering analysis in a series of highly original papers, beginning in 1961 with "The Geometry of Coiling in Gastropods" and carrying through the end of that decade, that first exposed the power of rigorous analytical, computer-aided techniques for the study of idealized shell geometry. Because there are several related approaches to questions involving the quantitative study of morphology, it is useful to define and distinguish some basic terms. *Morphometrics* involves the precise measurement and comparison of features in individual organisms with an eye towards fitting mathematical generalizations to empirical data collected from actual specimens. *Functional morphology* is the traditional analysis of the functional or adaptive basis for certain organic forms. It may or may not involve quantification; pioneered in the 19th century by Cuvier, Richard Owen, and others, its modern, more analytic branch is concerned mostly with the biomechanical analysis of form. Finally, *theoretical morphology* is specifically that branch of morphological study "concerned with the simulation of the principal aspects of form with a minimum number of geometric parameters, or with the simulation of the morphogenetic process itself that produced the form under study, and is not concerned with the production of a precise mathematical characterization or picture of any given existent form" (McGhee 1999, 4). Theoretical morphology is distinguished from the other approaches in that it is often directed solely towards the production of "hypothetical" forms via computer simulation; the assumption is that by comparing simulated models with actual data an understanding of the constraints on actual form can be better understood. As Raup and Arnold Michelson put it in 1961, "One advantage of this approach is that it puts each morphological type in a conceptual framework which makes possible comparison with all other possible types" (Raup and Michelson, 1961, 1294).

Raup is a key figure in the development of paleobiology, and his contributions will be discussed in much detail in subsequent chapters. He was a central figure among the "Young Turks" who took over the field during the 1970s (along with Gould, Tom Schopf, Niles Eldredge, Steven Stanley, Jack Sepkoski, and others), but chronologically his career also overlaps significantly with the generation of Imbrie and Newell. He was one of the first paleontologists to seriously apply computers to paleontological problems, and he was a key innovator in two major areas of quantitative analysis: 1) understanding the geometry of morphological constraints by modeling mollusk shells using computer simulations, and 2) examining the fossil record using advanced statistical techniques. In the latter area, Raup collaborated on several pioneering projects during the 1970s and 1980s, including simulations of clade diversity with Gould, Schopf, and Daniel Simberloff, and the study of "periodicity" in mass extinctions, with Sepkoski. Raup's studies of morphology and biocrystallography earned him a substantial reputation while he was still a young paleontologist in the 1960s, but later in his career he established himself as one of the leading theorists of the effect of extinction, particularly mass extinction, on macroevolutionary dynamics. Additionally, he was the coauthor (with Stanley) of the highly influential textbook *Principles of Paleontology*, which was read by generations of paleontology students from the 1970s to the 1990s (Raup and Stanley 1971, 1978).

Raup grew up in a scientific family—his father, Hugh Raup, was a noted Harvard botanist—and he began his scientific education at Colby College and the University of Chicago. Having begun study of geology at Colby, Raup transferred to Chicago to complete the BA. The timing of this move was significant. At Chicago, Raup's primary mentor was Everett (Ole) Olson, an eminent vertebrate paleontologist with significant interests in quantitative and biological approaches to paleontology (Rainger 1997). Surprisingly, however, Raup reports being unenthusiastic about the novel approach being taken at the time at Chicago: "I developed this very strong interest in paleontology and wanted to continue, but was completely disgusted by the looseness of the program there and wanted something a little bit more classical. It was all about numbers and models, which I had no use for. I wanted very traditional paleo and biostratigraphy: collecting, describing, and classifying fossils." Unsure of what to do following his graduation in 1953, Raup opted for the easiest choice—"I ended up going to the hometown school" (Raup interview).

The hometown school, in Raup's case, was Harvard University, where Raup's father had recently met a young paleontologist named Bernhard Kummel, who had been Norman Newell's first doctoral student, and who became Raup's advisor. Up to this point Raup had shown little interest in mathematics or quantitative paleontology, which he saw as "shallow" and "sloppy," and "too theoretical for use." Otherwise, he recalls, "I had very good high school math and a little college math. . . . Beyond that I didn't have any formal training that I can remember, certainly no math courses. Not even a statistics course. I took a population genetics course at Harvard which gave me some applied statistics." At Harvard, Raup found an inspiring teacher in Kummel, who, despite being fairly traditional, was "a tremendous cheerleader, inspirer." While Kummel did not

share Newell's enthusiasm for paleobiology, Raup remembers that Kummel "was a disciple of [Columbia geologist] Marshall Kay and was pushing the models of tectonics and sedimentation and geosynclines and so
forth, which were exciting and very new. It was a way of synthesizing a
whole lot of geologic and paleontologic data and this was exciting. This
got me back to models" (Raup interview).

In addition to working with the fairly traditionally-minded Kummel,
Raup also began studying with Ernst Mayr, who introduced him to Newell and encouraged him to ask more general evolutionary and biological questions that could be applied to paleontology. Raup remembers
Mayr's influence as "much greater" than Kummel's, and also remembers
that during the mid-1950s he visited the AMNH, where he studied biometry with Imbrie. He recalls that "this was when John [Imbrie] was just
getting going. So in terms of biometrics and applied statistics, I really
got started with John." Raup's dissertation was a study of the geographic
speciation and phyletic change in fossil echinoderms, and he used his
new experience with biometry and applied statistics to analyze his data.
While still finishing his PhD at Harvard, Raup was hired by the Califor-
nia Institute of Technology in 1956, but after only a year he was lured
away to Johns Hopkins University. It was at this point that he was introduced to computers, when he plotted data from a quantitative study
of sand flea populations using FORTRAN on an IBM 1620 mainframe
(Raup interview).

This initial exposure led to Raup's first paper on theoretical morphology, "The Geometry of Coiling in Gastropods," published in *Proceedings of the National Academy of Sciences*. The paper was essentially a survey and update of Thompson's work on shell morphology, to
which, Raup commented, he was offering "a fresh approach." Despite
acknowledging the salutary influence of Thompson's work, Raup noted
that "there still remains a paucity of application of geometric characteristics to practical problems of gastropod systematics and evolution"
(Raup 1961, 603). One indication of the dearth of such study is evident
from the literature the paper cited: aside from Thompson's *On Growth
and Form*, only one other reference was from the 20th century (a paper by
G. C. Martin from 1904!), and the earliest citation was of a paper published by the French Royal Academy in 1709.

According to Raup, Thompson's most important contribution "was
to emphasize a few characteristics of shell form which may be built upon
in the formulation of a workable analytical scheme." He cited "four

generalizations" about coiled shells made by Thompson that are particularly valuable: (1) the constancy of the shape of the shell's generating curve during growth; (2) the constancy of the curve's rate of expansion; (3) the constancy of the overlap between successive whorls of the
shell; and (4) the constancy of the ratio between the size of the generating curve, the distance between the coiling axis, and a point on the
curve (Raup 1961, 604). Raup concluded, however, that Thompson's parameters were too imprecise for quantitative analysis, and he instead
proposed four new parameters, which would feature repeatedly in successive papers on the topic: the shape of the generating curve (s), the rate
of whorl expansion (w), the distance of whorls from the axis of coiling
(d), and the translation or speed of the generating curve (t) (Raup 1961,
606; modified slightly in Raup and Michelson 1965, 1294). These parameters, he proposed, defined the basic shape of all gastropods, and he
tested this using not a computer but an old-fashioned camera lucida to
reconstruct the basic geometry of several sample shells. Raup suggested
that this preliminary analysis might eventually have application to studies of ontogeny, ecology, and phylogeny and evolution, since "evolutionary series may thus be recognized which have been obscured by the conventional methods of description," and "a fundamental question here is
whether the four parameters that have been defined have genetic reality and would thus be expected to change independently and in regular
fashion" (Raup 1961, 608).

Raup found a broader audience for his next foray into theoretical morphology, "Computer as Aid in Describing Form in Gastropod Shells,"
which appeared in 1962 in *Science*. Publishing in such a widely read
venue was somewhat risky because, as Raup recalls, "throughout the
1960s, computers were anathema to most paleontologists," and he was
unsure of the response he might get from reviewers. As it happened, one
of the reviewers "turned out to be John Imbrie, who gave it high marks,
and the other was Ellis Yochelson who recommended rejection on the
grounds that the work was not science." Fortunately for Raup, "thanks
to John, the paper was accepted" (Raup interview). This paper revisited
Raup's modification of Thompson's parameters for reconstructing shell
morphology, except that instead of relying on pen-and-paper methods,
Raup now used a computer to assist in drawing the figures and to enhance the speed and accuracy of the process: the same IBM 7090 used
in Imbrie's 1964 study. This computer was hooked up to an x–y plotter,
which allowed Raup to test his parameters by graphically reconstruct-

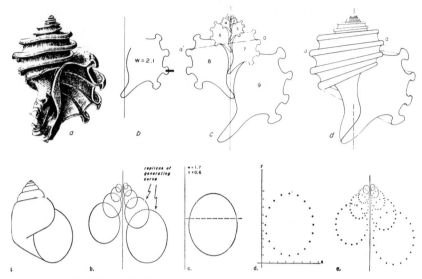

Fig. 1. Stages in the reconstruction of snail form from the four basic parameters.

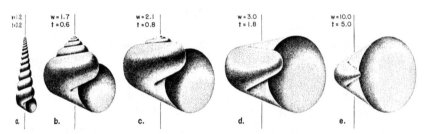

Fig. 2. Hypothetical snail forms drawn from cross sections made by the computer method.

FIGURE 3.3. Examples of Raup's early modeling of shell morphology. *Top*, Raup's reconstruction of the basic form of a gastropod from its coiling characteristics in his 1961 PNAS paper. *Center*, an example of Raup's generation of hypothetical snail forms using a computer, from his 1962 Science paper. *Bottom*, the snail forms were hand-drawn by an artist and not produced by the computer. David M. Raup, "The Geometry of Coiling in Gastropods," *Proceedings of the National Academy of Sciences of the United States of America* 47 (1961), 606. David M Raup, "Computer as Aid in Describing Form in Gastropod Shells," Science 138, no. 3537 (1962): 151; reprinted with permission from AAAS.

ing "five hypothetical snails drawn from computer produced cross sections" (fig. 3.3). The images reproduced in the paper were striking, and while this study is considered a classic in the field of theoretical morphology, there may have been some initial confusion about the extent of the computer's contribution. As Raup recalls, a number of years later Gould cited it in one of his popular *Natural History* columns: "He praises my 1962 *Science* paper for showing computer-produced perspective drawings of snails. Unfortunately, these were done by a hired artist who based the images on computer-produced cross-sections." Here Raup was simply pushing up against the limitations of available technology: "At the time, I was trying hard to get the computer to make nice-looking pictures but had not succeeded" (Raup, personal communication, 27 March 2007).

At the conclusion of the 1962 paper, Raup hinted that "a project is under way to 'map' the total spectrum of possible snail form" using a computer (Raup 1962, 151). He discussed this task in greater detail in his 1965 collaboration with Arnold Michelson (a colleague at Johns Hopkins in electrical engineering), "Theoretical Morphology of the Coiled Shell," also published in *Science*. This paper's major innovation was the introduction of the concept (but not yet the term) of "theoretical morphospace":

> The four variables [that regulate shell coiling] can be combined to define a "four dimensional" space which contains most of the theoretically possible shell forms. When the geometries of naturally occurring species are plotted in this space, it becomes evident that it is not evenly filled. Evolution has favored some regions while leaving others essentially empty. In the empty regions, we are presumably dealing with forms which are geometrically possible but biologically impossible or functionally inefficient (Raup and Michelson 1965, 1294).

The authors concluded that such theoretical mapping might give clues about the actual morphology of shells by pointing to "gaps" in the morphospace that were either evolutionarily disadvantageous or physical impossibilities. The paper also refined the computer techniques used to depict theoretical morphologies: whereas the earlier study with the IBM 7090 could only produce cross-sections of the hypothetical shells, Raup and Michelson found that by using a PACE TR-10 analog computer and an oscilloscope, they could produce much more striking images of shells (fig. 3.4).[6] In fact, these images were so innovative that one found its way onto the cover of the issue of *Science* in which the article was published,

---

6. Basically, the oscilloscope would generate a fast-moving circle that traces a pattern that looks like a shell—only backwards (from the large end to the small end).

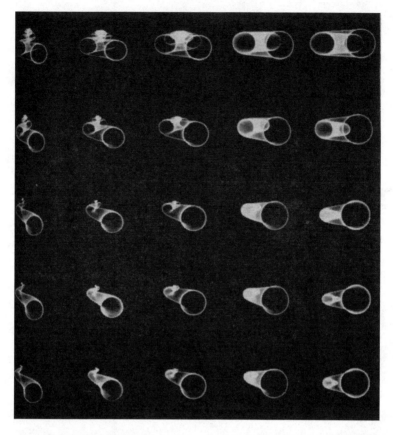

FIGURE 3.4a. Raup and Michelson's "coiling geometries," generated using an oscilloscope and an analog computer, from their 1965 paper. David M. Raup and Arnold Michelson, "Theoretical Morphology of the Coiled Shell," *Science* 147, no. 3663 (1965): 1295. Reprinted with permission from AAAS.

giving Raup considerable exposure for his new technique and young career.[7]

The most complete articulation of Raup's agenda, however, came in two papers published in the *Journal of Paleontology* in 1966 and 1967 respectively: "Geometric Analysis of Shell Coiling: General Problems," and "Geometric Analysis of Shell Coiling: Coiling in Ammonoids." In the first of these papers, Raup described the project explicitly using the term "theoretical morphology," a concept he had hinted at in the 1965

7. Raup's image is on the cover of the 12 March 1965 issue of *Science*, vol. 147, no. 3663.

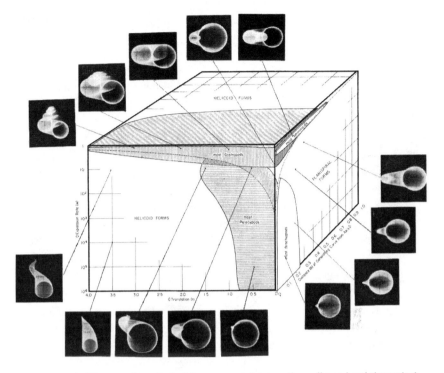

FIGURE 3.4b. The same hypothetical forms mapped onto a three-dimensional theoretical morphospace. Note that only a small portion of available morphospace (shaded area) is actually colonized. David M. Raup, "Geometric Analysis of Shell Coiling: General Problems," *Journal of Paleontology* 40 (1966): 1184; courtesy of the Society for Sedimentary Geology.

paper with Michelson but now named and defined in considerably more detail: "Using this approach, a conceptual or mathematical model is established for some aspect of morphology. The model in turn is used to formulate or describe the total spectrum of physically possible forms. This spectrum then serves as a framework through which actually occurring forms may be interpreted" (Raup 1966, 1178). Raup also explained that whereas previously "the analysis of shell coiling has been little more than an intellectual exercise," this paper was more explicitly attuned to relationships between hypothetical morphologies and the forms of actual organisms. Like the pioneering work of Moseley and Thompson, it began with the a priori assumption that coiled shells are variations on the basic logarithmic spiral, but in addition to refining the mathematical treatment of the parameters involved in coiling, it also addressed the relatively unexamined "basic relationships between shell geometry and shell function [that] have not been clearly presented" (Raup 1966, 1178).

Conceptually, the purpose of the 1966 paper was to much more thoroughly examine the "total spectrum of geometrically possible shell forms," and "to show how this hypothetical four-dimensional space is filled with actual species, living and fossil" (Raup 1966, 1181). Raup's analysis of this problem showed that only a small portion of available morphology has been utilized by nature, begging the important question: Why? He suspected that the area of the theoretical morphospace that is actually used is not arbitrary, and guessed that by studying the relationships between gaps and filled spaces, one could reach important conclusions about the effects of physical constraint and function:

> The bulk of the species in the four taxonomic groups are confined to non-overlapping regions which if taken together comprise a relatively small part of the block. Clearly, the distribution of actual species is not random. Do the relatively unused regions represent physiologically impossible shell forms or has the evolution of these taxa simply not had sufficient time in which to populate the entire block? Are some regions of the block suited only to one set of shell functions? Are swimming forms, for example, necessarily limited to certain discrete regions? The answers to these and similar questions are necessary if we are to understand fully the evolution of coiled organisms (Raup 1966, 1185).

Raup did not attempt to definitively answer any of these questions, but he posed a number of intriguing possibilities involving physical constraints, ontogenetic development, and functional requirements. Noting that "quite a high degree of plasticity in geometry during growth" was possible, he suggested that changes during growth offered an insight into ideal ratios between area and volume for shelled forms. In bivalves, for example, relationships between the size and thickness of a shell, on the one hand, and the strength of muscles required to hold the shell closed and available internal space for muscles and organs, on the other, were necessarily constrained by the basic physical limits Thompson had described. Univalve organisms (gastropods and cephalopods) face related, but different, problems: aperture size must balance the functional needs for both feeding and protection, while whorl dimensions have consequences for the strength and integrity of the animal's shell. In all of these cases, Raup hypothesized that answers to these physical and adaptive

needs helped account for the colonization of particular areas in the theoretical morphospace, and that this in turn could illuminate basic questions about the evolution of shelled organisms.

Having laid out these basic problems, Raup's follow-up paper in
1967 explored the particular case of planispiral ammonoids (cephalopods whose shells coiled in a single plane). His goals were (1) to map
the distribution of ammonoids within the theoretical morphospace described in the earlier paper, but also (2) to evaluate "the functional factors covering the coiling geometry" of ammonoid shells, and (3) to "explain why planispiral ammonoids are restricted geometrically to a small
part of the geometric spectrum" (Raup 1967, 43). Using a sample of 405
ammonoid genera drawn from the *Treatise on Invertebrate Paleontology*, Raup found that, as in the earlier study, the shells examined fell
into a fairly narrow range of the available morphospace. Again, Raup
asked why only a small portion of available geometry was used, and he
observed that both chance and function likely played important roles.
While "the point of view exists that animals have a given form simply
because of phylogenetic chance," he concluded that "the only reasonable
approach is to assume that the observed morphology has had, in evolution, a selective advantage over other possible morphologies" (Raup
1967, 55). In other words, there had to be both a selective basis and a genetic explanation for the observed morphological range of ammonoids,
which casts doubt on Thompson's assumption that physical constraint
alone explained morphology. Essentially, Raup's study advanced the
Thompsonian program of mathematizing the study of growth and form
(using the considerable advantage of computer technology never available to Thompson), but also recast it in more explicitly Darwinian evolutionary terms. If, as Raup contended, the study of theoretical morphology demonstrated "the physiological and genetic capability of exploring
a wider geometric range than is normally occupied," then something else
had to account for the limits on actual observed forms (Raup 1967, 55).
That "something" could only be the functional, adaptive needs of organisms, which could be explained primarily by natural selection.

Computers in Paleontology

As many of the examples discussed in this chapter demonstrate, computers were becoming a basic research tool for paleobiologists by the mid-

1960s. While there was (and occasionally still is) active resistance to computerization and quantification in some quarters of paleontology, by the late 1960s it was becoming more readily apparent that paleontologists could not afford to ignore the major advances in technology that had revolutionized other disciplines. In 1969 the broad question of applying computers to geology and paleontology was the subject of two meetings. The first was a symposium held at the University of Kansas in June of that year, entitled "Computer Applications in the Earth Sciences." The second was in a session called "Computers in Paleontology," at the First North American Paleontological Convention held at the Field Museum of Natural History in Chicago in September. The published proceedings of both meetings offer significant discussions of both general problems and specific techniques in computer application; in fact, many of the predictions made by speakers at both meetings quite presciently summarized some of the major directions that would be pursued over the next decade.

The most interesting paper from the Kansas meeting was presented by William C. Krumbein, a professor of geology at Northwestern University, who gave the concluding summary for the symposium, "The Computer in Geological Perspective." Krumbein's discussion is of special interest because it drew active attention to computers as a tool for producing models: "Along with the development of programs for conventional statistical analysis of data came the realization that the computer provides a powerful tool for experimentation in geology by way of simulation programs" (Krumbein 1969, 251–52). In particular, Krumbein discussed a type of model called a Markov chain. Markov chains are stochastic statistical functions in which at each "step" in a series of events, the probability that a certain outcome will occur is independent of the results of all the prior steps. For example, in a series of coin flips, the probability of tossing heads or tails remains 50% for each successive flip, regardless of the outcome of previous tosses. Markov processes will be discussed in greater detail in chapter 7, but Krumbein's discussion is notable because of his willingness to consider the importance of nondeterministic processes in the modeling of complex geological systems. According to Krumbein, "A random variable has as valid a base in scientific investigation as the conventional nonstochastic (systematic) variable that forms the basis of classical mathematical physics," and "it is partly because of the complexity of geological processes that some geologists have introduced the idea of random processes into geology"

(Krumbein 1969, 261–64). Overall, Krumbein concluded that the necessity and utility of pursuing more complex statistical analyses in geology would require training students with "a much firmer foundation of mathematics," since "the recent introduction of probabilistic models in geology suggests that mathematical training include probability theory as well as classical mathematics" (Krumbein 1969, 260–61).

These issues were discussed in considerably more detail that September at the 1969 Paleontological Convention in Chicago, where Raup coordinated a set of highly interesting papers discussing aspects of computer application to paleontology. James F. Mello, a paleontologist with the US Geological Survey, presented the first paper, "Paleontologic Data Storage and Retrieval." In addition to surveying the important topic of computerized fossil databases, this paper made a number of fairly prescient observations about future directions in the field. Mello proposed two ways in which computers could be applied to paleontological work: first, as a tool for performing "tedious or time-consuming tasks more efficiently," and second, "and much more important," as a means of conducting "a much freer analysis of individual variables, which are presently treated collectively." Mello added a prediction: "There can be little doubt that the computer will become a most potent tool in paleontology and that exploitation of this potential will give new impetus to paleontology for many years to come" (Mello 1970, 57). Mello explained that potential computer applications ranged from simple indexing of fossil data—much like the physical index-card system most paleontologists used to keep track of data sets—to processing of that data, which he called "a fundamental departure from the indexing function" (Mello 1970, 58). Among his other predictions, Mello speculated that computer storage capacity and processing speed would continue to grow, while the availability of computers would increase to the point where "the small laboratory computer and the desk side time-sharing console will become standard fixtures for the paleontologist" (Mello 1970, 70). He did caution that computerized analysis could not substitute for poor data, and he raised a concern, which would be shared by others, that basic research along traditional lines might be eclipsed by enthusiasm for new technology and techniques. Nonetheless, he confidently concluded:

> Within the next decade, the blending of these internal and external developments will place much more information in the hands of paleontologists. This must result in the more efficient conduct of presently established lines

of research, and will also open up many new areas of investigation based on comparison and evaluation of data which are presently too widely scattered, ill-defined, or voluminous to be properly dealt with now. The potential for experimentation with heretofore unmanageable data opens exciting vistas for progress and should revitalize paleontology for many years to come (Mello 1970, 70–71).

Following Mello's broad summary of basic applications, Raup and Williams College paleontologist William T. Fox each presented papers investigating particular methods for simulating and modeling complex morphological features and ecological dynamics. These papers moved beyond imagining the computer as a simple data storage and retrieval device, and examined the potential for computers to assist in the actual formation and testing of new hypotheses. Raup's paper covered the now familiar territory of theoretical morphology, and he put forward simulated morphologies as an attempt "to replicate the genetic instructions, or genetic 'program,' used by an organism" (Raup 1970, 72). Advancing beyond simulated morphology, Raup also extended his discussion to other approaches to simulation, including the "grazing track" model he had introduced that same year in a paper published in *Science*, co-authored with the German paleontologist Adolph Seilacher. This model analyzed the fossil burrowing tracks left by some invertebrates as evidence of foraging behavior, and simulated the track left by an animal in two-dimensional space (Raup and Seilacher 1969). What was notable about this model is that, unlike Raup's fairly deterministic simulations of shell morphology, which were closely controlled by basic geometrical constraints, the grazing track model "is partially stochastic, meaning that certain aspects are controlled by chance" (representing choices made by the organism as it foraged for food), and thus "successive simulation runs using the same input yield different, though comparable, output" (Raup 1970, 76–77).

As we will see, nondeterministic or stochastic simulations would provide an important impetus for new approaches to understanding evolutionary patterns in paleobiology, although Raup's early treatment represented only a glimmering of awareness of the breadth of their eventual application. This awareness is also present in Fox's contribution, which considered the considerably more complicated problem discussed in "Analysis and Simulation of Paleoecologic Communities through Time." As Fox noted, "In the past, the large mass of data available on fossil and sediment distribution made a quantitative paleoecologic analysis prohibitively expensive in both time and money." He reported, however, that "with the introduction of high speed digital computers, it is now possible to process large amounts of data and establish a paleoecologic simulation model." Fox emphasized that simulations could not be used to *prove* individual hypotheses, but by serving as null hypotheses they could eliminate hypotheses "which do not fit the observed fossil distribution" (Fox 1970, 117). Intriguingly, Fox described the example presented in his paper as "a mixed deterministic and probabilistic model" (Fox 1970, 117–18). As will be discussed shortly, the use of computer simulation for null hypothesis testing would have a significant role in 1970s paleobiology, and it points to the conclusion that by the late 1960s, computers were beginning to be seen as much more than simply powerful calculating machines.

Raup's session at the Paleontological Convention concluded with a summary of developments in the field by University of Chicago paleontologist Everett Olson, an early and longtime supporter of quantitative paleontology. At the start of his paper, Olson explained that his remarks were based not only on his own experience but also on a comprehensive review of recent literature as well as a survey "of written and verbal opinions from many of my colleagues," including James Beerbower, Niles Eldredge, William Fox, Stephen Jay Gould, Ralph Johnson, James A. Peters, David Raup, Bobb Schaffer, Francis Stehli, and Leigh Van Valen (Olson 1970, 135). Olson's paper was therefore informed by a cross-section of attitudes arrayed across a wide spectrum of generational and methodological perspectives, and it can be taken as a partial sampling of attitudes in the profession at the time.

From his survey, Olson concluded first that computerization had expanded over the previous decade to become an important phenomenon in paleontology. He noted, however, that research in which computers were central still represented a fairly marginal fraction of all paleontological work, and that he doubted whether this proportion would change much in the immediate future. Nonetheless, he pointed to the potentially vital role computers could play in paleontological contributions to evolutionary theory, since "biological causality can be read into the complex events that emerge from studies of fossils," and "one of the major aims of paleontology since the acceptance of evolutionary theory has been the investigation of the causes of the relationships revealed by the fossil record" (Olson 1970, 138). Olson cited "immense" possibility for pa leontologists to contribute to this understanding, since fossil data were "eminently suitable for study by electronic computers," but he cautioned against overenthusiasm about technology leading to "the tendency to credit beyond their worth the results that issue from such studies. . . . Somehow the quantification and the manipulations and the neatness of the readout seem to compensate for the inadequacies of the data that were initially thrown into the hopper" (Olson 1970, 139–40).

Olson's general assessment of the potential future significance of computers in paleontology turned his discussion towards historical and philosophical considerations. He concluded first that the use of computers was not "a faddish thing that will die out," and predicted that "the greatest advances will come from students who have obtained competence in several fields involved in the course of their formal education" (Olson 1970, 146). Here Olson recognized the importance of revising the traditional paleontological curriculum to include increased emphasis on quantitative techniques, as well as greater familiarity with biology and allied disciplines—a call first made by Newell and others in the late 1940s that would finally begin to become a reality during the 1970s. However, on the question of whether "the development of computers [could] be put into the same category as other technological advances which have been basic to the development of new ways of thought," Olson was more ambivalent. While he saw devices such as the telescope or microscope has having had the primary function of "extend[ing] our limited sensory spectrum to more detailed levels of observation," and thus opening new vistas of objects and events, he argued that computers had the more subsidiary role not "of supplying data but of aiding in its analysis" (Olson 1970, 146–47).

If computers were to genuinely effect "a major reorganization in concepts of paleontology," Olson concluded, it would be either "by assumption of a role of supplying new data, or of making up for inadequacies in treating the data that we now have available." One such possibility would be for computers to "unlock" the content of the vast repository of fossil information collected since the 19th century, which had been "obscured by the mass of data and the intricate intertwining of different systems of variables from diverse disciplines." Olson asked, "Can the computer so viewed act as an analytical tool that will serve as an analogue to those technical achievements that gave insight into regions beyond our immediate sensory world? . . . To date, it would seem to me, there are very faint glimmerings to suggest that this might be the case" (Olson

1970, 146–47). Here, Olson found that computers held the greatest potential to treat problems involving many variables in complex dynamic systems such as ecology and biogeography. These possibilities included giving "deeper insight into the patterns and causal aspects of evolutionary change"; clarifying evolutionary sequences "as a series of 'if-then' statements"; "establishing laws of change in complex situation" by simulating and evaluating "all reasonable, logical contingencies" in major hypotheses via "game playing"; and finally using computers to predict the probabilities of "the most likely outcome of several contingent variants of modifications of climate and organisms during a period of some millions of years" and comparing those results to the actual record (Olson 1970, 149–50).

It is striking that, as will be seen in later chapters, all of these possibilities would explored—and to some degree realized—by paleobiologists during the next decade or so. The concluding paragraph of Olson's essay is an appropriate way to close this chapter, since it asks precisely the questions that the younger generation of paleobiologists—including Raup, Gould, and Eldredge—were preparing themselves to answer:

> Would it be possible, taking some such direction as suggested in the preceding paragraph, to set up a program to be applied to some extremely well documented part of the record, to test the efficacy of the modern theory of the nature and causes of change? Could we establish several predictive, contingency models, using alternative causal schemes, from several theories of evolution for example, and, knowing the outcome in the record, test the effectiveness of each in predicting this outcome? Perhaps this is getting too close to the invisible "window" and is just pipe-dreaming, but it does seem to have some interesting possibilities which should tax the ingenuity of the paleontologist and programmer (Olson 1970, 150–51).

Indeed, over the next decade the emerging generation of younger paleobiologists would attempt to transform these goals from pipe dream to reality. However the ultimate success of that project is measured, the attempt—which centrally involved the related themes of biological integration, quantification, and pedagogical reform—would transform the discipline of paleontology.

CHAPTER FOUR

From Paleoecology to Paleobiology

By the late 1960s, paleontologists like George Gaylord Simpson and Norman Newell had laid much of the groundwork for the central themes that would emerge in paleobiology's "revolutionary" period in the 1970s. The new paleobiology that would be championed by paleontologists such as Stephen Jay Gould, Niles Eldredge, Steven M. Stanley, David Raup, Thomas J. M. Schopf, and others was heavily influenced both by the quantitative analysis of rates and trends in origination and extinction of taxa pioneered by paleontologists like Simpson and Newell and by the studies of the genetic and geographical mechanisms of selection and speciation promoted by synthetic biologists such as Ernst Mayr, Theodosius Dobzhansky, and Sewall Wright. But in the eyes of many paleontologists who were interested in evolutionary and paleobiological questions, a crucial ingredient was still missing: a more nuanced appreciation for the ecological and biogeographical context of the history of life.

As this chapter will argue, a crucial prerequisite to the modern paleobiological synthesis was the development and refinement of the field of paleoecology during the 1960s and early 1970s. Paleoecology itself was not a new field; by the mid-1960s paleontologists had been using the term for decades, and its subject matter—the relationship between fossil organisms and their environments—had been studied since at least the middle of the 19th century. But paleoecology went through its own revolution during the 1960s, in which the subject was redefined and reconceptualized by a number of mostly younger paleontologists, many of whom would also become leaders in the paleobiology movement. The new approach to paleoecology emphasized two important conceptual revisions. First, rather than focusing on the physical parameters of the environment as a geological problem, the new paleoecology placed much greater emphasis on the biological aspects of populations of fossil organisms, and it often explicitly compared fossil assemblages to living counterparts. Second, the new paleoecology drew heavily on important work in theoretical ecology and biogeography that was having its own revolutionary impact on the broader study of ecology. Ecologists such as G. Evelyn Hutchinson and Robert MacArthur were opening new conceptual frontiers in ecology through the use of mathematical models designed to offer generalizations about community structure, colonization and extinction, and evolution. The new paleoecology was, in large part, an outgrowth of this movement in theoretical ecology.

Viewed from this perspective, important similarities between the growth of paleoecology and paleobiology emerge. Both movements drew inspiration from disciplines outside of paleontology, both involved self-conscious reflection about the disciplinary identity of paleontology (being drawn towards biological rather than geological questions and methods), and both emphasized mathematical generalizations and theoretical models. But rather than treat paleoecology and paleobiology as separate movements, I will view paleoecology as an important part of the overall development of paleobiology. Many methods and objectives of paleoecologists in the 1960s became central to paleobiology in the 1970s, and many of the most active young advocates of the new paleoecology—such as James Valentine, Richard Bambach, and Steven Stanley—became outspoken proponents of paleobiology in the next decade.

This chapter, then, will explain the ecological foundations of modern paleobiology. It will also provide a vital insight into why paleobiology "exploded" when it did. A sustained paleobiology movement had been brewing since the early 1950s, as the first chapters of this book have shown. But the revolutionary phase of paleobiology cannot be explained simply as an outgrowth of that earlier movement, important though the work of Simpson, Newell, and others may have been. Population ecology and biogeography provided the crucial inspiration for paleontologists to begin thinking beyond the static physical record of the geological column or the linear and apparently progressive evolution of individual fossil lineages, towards viewing the fossil record as a dynamic history of change that could be modeled using many of the tools used to study living populations. The model-oriented, explicitly mathematical orientation of theoretical ecology would also directly inspire paleobiologists to more aggressively pursue their own mathematical models. By the late

1960s, paleontology was primed for a revolution, but without stimulus from ecologists and biogeographers, paleobiology might never have entered its revolutionary phase.

MacArthur, Wilson, and Island Biogeography

One of the central features of the growth of paleobiology was a tension between traditional empiricism and quantitatively oriented theory. During the 1960s, the discipline of ecology was also experiencing tension between traditional field-oriented and more abstract theoretical approaches. G. Evelyn Hutchinson provided much of the impetus for a new theoretical agenda in ecology, and the movement he inspired had an enormous impact on the development of paleoecology and paleobiology. Much of this activity was centered at Yale University, where Hutchinson influenced students in a variety of different disciplines, and where his own former student, Robert H. MacArthur, carried on his tradition of theoretical ecology. As Sharon Kingsland puts it, "Hutchinson and MacArthur helped to make ecology intellectually exciting" (Kingsland 1985, 176).

In the 1950s, many practitioners considered ecology very much a field science, and generally disdained or avoided overly mathematical treatment of ecological problems or models. However, Hutchinson influenced a new, more quantitative and theoretical branch of the discipline, which, Kingsland argues, helped introduce a "schism" into ecology, "between those interested in ultimate causes and those interested in proximate causes." Hutchinson's approach to the study of populations promised to provide a "unifying ground for ecology" (Kingsland 1985, 175), but it needed activists to promote it, and "it was into this milieu that Robert MacArthur stepped in 1953" (180). As William Dritschilo explains, by the 1960s two ecological traditions were in collision: "One was the study of ecological communities based on their material and energy flows; the other was through mathematical description of interactions between species. One, then associated with Eugene P. Odum, came from a Midwestern tradition of classification, experimentation, and field manipulation; the other, associated with Robert H. MacArthur, took the observations of naturalists . . . and added mathematical and Ivy League refinement to it" (Dritschilo 2008, 357).

The excitement in MacArthur's approach to ecology lay in his willingness to advance elegantly simple mathematical models that served as heuristic tools to describe the behavior of complex ecological systems. One example was his so-called "broken stick" model of relative species abundance, in which "environment was compared to a stick broken simultaneously at randomly chosen intervals, with each segment representing the abundance of one of the species in the community" (Kingsland 1985, 185). MacArthur proposed that if there were s species in a community, the environment could be compared to a stick randomly broken at $s - 1$ places, and the lengths of the resulting fragments would represent relative abundances of each species in the community (MacArthur 1957).

While MacArthur eventually abandoned this particular model, it became a paradigm example of a new approach to ecology in which complex ecological systems could be described by fairly simple models despite limited or imperfect empirical data. In this view, as Kingsland puts it, "the point of ecology was to discern the repeating patterns of nature and to interpret these in terms of general principles" (Kingsland 1985, 190). MacArthur's methodological revision carried a philosophical or even aesthetic undertone as well. Kingsland argues that Hutchinson imparted to his students the message that "the search for larger patterns in nature . . . is an artistic achievement" (Kingsland 1985, 180). This moral was absorbed by MacArthur. Hutchinson and E. O. Wilson recalled that "in conversation, MacArthur would say that the best science comes, to a great extent, from the creation of *de novo* and heuristic classification of natural phenomena. 'Art,' he enjoyed quoting Picasso, 'is the lie that helps us see the truth'" (Wilson and Hutchinson 1989, 321).

The ecological model that had the greatest influence on the further development of analytical paleobiology was MacArthur and Wilson's insular theory of biogeography, which grew out of Hutchinson's investigations of species abundance. In a classic 1959 paper, "Homage to Santa Rosalia," Hutchinson asked what the relationship was between ecological niche space and species abundance. MacArthur's broken stick model attempted to provide an abstract solution by representing relative species abundances in an equilibrium community. But what ecological factors control the maintenance of populations at equilibrium? In 1960 MacArthur addressed this question in his essay "On the Relative Abundance of Species," in which he observed that "the relative abundances of equilibrium species are of considerable ecological interest and frequently can be deduced from the assumption that increase in one species population results in a roughly equal decrease in the populations of other species" (MacArthur 1960, 33). In other words, given the finite resources in a given community, one population cannot grow without corresponding decline in another. The same relationship would be true for immigration, as Hutchinson had already noted, For a new species to successfully colonize an existing equilibrium community, an existing niche would have to be either unoccupied or available for partitioning; otherwise some species would be squeezed out (Hutchinson 1959, 142). In relatively unbounded geographical areas, however, where populations migrate fairly freely and species' ranges overlap two or more ecological communities, it is difficult to accurately measure or model the effects of immigration because there are simply too many variables in play. What was wanted—especially by an ecological theorist enamored of clean, idealized mathematical scenarios—was a simplified picture in which to test general rules of ecological population dynamics.

This is where islands came in. As naturalists had long observed, islands are in many ways ideal natural laboratories for investigating ecological and evolutionary relationships. Darwin's observations about the geographic distribution of birds and tortoises among the Galapagos Islands became a major part of his argument in *Origin of Species*. During the period of the modern synthesis, Ernst Mayr had focused on islands in a number of publications where he investigated the role of peripheral isolation in the emergence of new species (Mayr 1942, 154–73; Mayr 1947). And, in the 1950s and 1960s, ecologists like E. O. Wilson had begun to study the dynamics of island species in earnest, to try to answer questions about the relationship between geography, ecology, and evolution. In particular, Wilson's studies of the Melanesian ants of the subfamily *Ponerinae* had led to his proposal of the "taxon cycle," an evolutionary-ecological model that explains the evolutionary consequences of successive cycles of invasion in marginal habitats like islands (Wilson 1961). During the taxon cycle, a generalist species invades a marginal habitat, where it quickly acquires adaptations that enable it to dominate and displace the species that already live there. Over time, subsequent invaders from the same ancestral lineage or from different lineages will eventually displace the initial invader, and so on. Thus the taxon cycle describes a dynamic equilibrium over successive waves of invasion, adaptation, and displacement.

Shortly after Wilson began publishing the results of his Melanesian studies, he and MacArthur began collaborating on a more general theory of island equilibrium dynamics that would combine Wilson's empirical insights with MacArthur's gift for mathematical models. The first articulation of what would be known as their theory of island biogeography was their essay "An Equilibrium Theory of Insular Zoogeography," published in 1963 in *Evolution*. The paper began by describing the species-area curve, which states that as the geographical area of sampling increases, so does the number of species found in that area, a rule that applies to islands as well as mainland environments (MacArthur and Wilson 1963, 373). MacArthur and Wilson next introduced the distance-area effect, which they described as a quantitative demonstration of a qualitative observation made by Mayr, that as distance increases from the mainland, "island faunas become progressively 'impoverished'" (MacArthur and Wilson 1963, 373). This phenomenon, the authors noted, had usually been explained as a function of the length of time needed for colonists from a mainland to colonize remote islands. Here they raised an objection: the distance-area effect would imply that over a sufficiently long period of time, remote islands would accumulate the same number of species as closer ones, but this does not happen.

Instead, MacArthur and Wilson proposed their famous equilibrium model, which holds that "the number of new species entering an island may be balanced by the number of species becoming extinct on that island" (MacArthur and Wilson 1963, 374). As the number of species already inhabiting an island increases, the number of new colonizers decreases, and the rate of extinction also increases. This pattern can be represented graphically as two intersecting curves; where the immigration and extinction curves intersect at s, the number of species present is stabilized at equilibrium (fig. 4.1). The entire system functions as a dynamic equilibrium: if the rate of immigration rises, the rate of extinction also increases to compensate, and vice versa. MacArthur and Wilson expressed this equilibrium model as an equation in which Δs (number of species) = M (number of successful immigrants) + G (number of species added by speciation) – D (number of extinctions), and explained that at equilibrium, $M + G = D$ (MacArthur and Wilson 1963, 378). In conclusion, the authors argued that "the establishment of the equilibrium condition allows the development of a more precise zoogeographic theory than hitherto possible" (MacArthur and Wilson 1963, 386).

In 1967, MacArthur and Wilson expanded their theory in a monograph, *The Theory of Island Biogeography*, which is now considered a foundational text in the fields of both ecology and biogeography. In

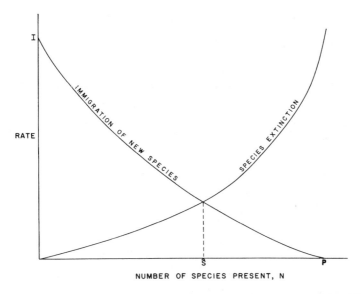

FIGURE 4.1. Intersecting immigration and extinction curves from the MacArthur-Wilson equilibrial model. Robert H. MacArthur and Edward O. Wilson, "An Equilibrium Theory of Insular Zoogeography," *Evolution* 17 (1963): 376. © Society for the Study of Evolution.

addition to presenting many of the elements first found in the 1963 paper in considerably more detail, they also took the opportunity to explore more general principles. To begin with, they emphasized the relationship between biogeography, ecology, and evolutionary theory, arguing that "the study of insular biogeography has contributed a major part of evolutionary theory and much of its clearest documentation," since "islands provide the necessary replications in natural 'experiments' by which evolutionary hypotheses can be tested" (MacArthur and Wilson 2001 [1967], 3). They noted, however, that the study of biogeography had long lacked a quantitative and theoretical basis, and they presented the book as being aimed at promoting a greater interest in theoretical and mathematical generalizations. Here MacArthur's heuristic approach was clearly evident: in acknowledging that they did "not seriously believe that the particular formulations advanced in the chapters to follow will fit for very long the exacting results of future empirical investigation," MacArthur and Wilson stressed that they "hope instead that they will contribute to the stimulation of new forms of theoretical and empirical studies, which will lead in turn to a stronger general theory" (MacArthur and Wilson 2001 [1967], xi).

At the heart of the book was the equilibrium model, which was not substantively altered from its 1963 formulation. However, MacArthur and Wilson explored its theoretical consequences in considerably more detail. In particular, they discussed equilibrium dynamics explicitly in terms of survivorship (the longevity of particular groups or species over time), and introduced a version of the "logistic equation" invented in the 1920s by Raymond Pearl and Lowell Reed to describe population growth (Kingsland 1982). This equation states that the growth of populations over time is a function of birth rates, death rates, and environmental constraints on intrinsic growth. A standard form is

$$\frac{\Delta N}{\Delta t} = rN\left(\frac{K - N}{K}\right),$$

where ΔN is change in number of individuals (either increase or decrease), Δt is change in time, r is the intrinsic rate of increase (also known as the Malthusian parameter), and K is the population limit (Wilson and Bossert 1971, 16–19, 92–106). The equation yields a distinctive, sigmoidal (S-shaped) logistic growth curve in which, after an initial burst in growth, the population levels off as it approaches its limit, the "carrying capacity" of the environment (fig. 4.2). MacArthur and Wilson explicitly connected this logistic equation to their equilibrium model by expressing the "success" of an individual colonizer in terms of its survivorship (the probability that any particular species would still be present at a given time x) as the likelihood of a colonizing species reaching its population limit K. In relating island population dynamics to a well-established demographic principle, MacArthur and Wilson gave their model additional theoretical support and generality. By expanding the parameters of the logistic equation to include entire species, and not just individuals, it also opened the door for an expanded use of the equation that would be seized on by paleobiologists for describing radiation and survivorship patterns, as we will see in succeeding chapters.

In the penultimate chapter of the book, MacArthur and Wilson discussed the implications of island biogeography for competition, natural selection, and evolution. Here they emphasized the usefulness of islands as "natural laboratories" for the study of evolution, since islands provided "a simpler microcosm of the seemingly infinite complexity of continental and oceanic biogeography" (MacArthur and Wilson 2001 [1967], 3). In a number of important papers, Mayr had shown that peripherally isolated communities were more likely to undergo periods of

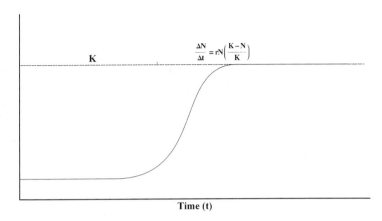

FIGURE 4.2. The logistic growth curve.

rapid speciation; he had also shown that island communities had particular dynamics that made them especially favorable to accelerated evolution (Mayr 1942; Mayr 1947; Mayr 1954). MacArthur and Wilson followed Mayr in asserting that "a new insular population is set in one of the most favorable circumstances imaginable for rapid evolution," and they used their theory of dynamic equilibrium to explore the mechanism by which this occurred (MacArthur and Wilson 2001 [1967], 152). Of the many features that make islands different from mainland environments, one of the most significant is that they are less favorable to predators, allowing greater evolutionary emphasis on traits that do not favor escape. Drawing on the variables r and K in the logistic equation, MacArthur and Wilson identified two distinct kinds of selection: r-selection, which favors rapid exploitation of vacant niches through rapid reproduction, and K-selection, which favors competition in crowded niches and invests in fewer offspring.

They then extended this to propose three "phases" of evolution on islands. During the first phase, a newly-arrived population will experience rapid expansion by placing higher selective value on genotypes with high r-fitness. This produces what Mayr had called the "founder effect," in which random mutations in an initially small but rapidly expanding population are magnified, thus leading to greater initial genetic diversity (Mayr 1954). The second phase begins after the newly arrived species has established a foothold, during which time "the population begins a long-range adaptation to the peculiarities of the local environment"

(MacArthur and Wilson 2001 [1967], 179). In this stage, r-values will be diminished, and a number of changes often occur that enhance adaptation to the local environment and promote competition with other species. Finally, during the third phase, speciation and adaptive radiation take place. MacArthur and Wilson argued that "evolution on islands and archipelagos can eventually lead to the formation of new, autochthonous species," provided that the island is large and stable enough to support the population for a sufficient amount of time. Since islands are often geographically located "near the outer limit of the dispersal range of a given taxon," speciation within a given archipelago can outstrip the rate of immigration from outside the islands, and lead to an "accumulation of species on single islands." Despite sharing a common ancestry with a mainland population, "such species tend to be adaptively quite different from each other," and as in the case of Darwin's finches on the Galapagos, they will disperse to new islands within the archipelago or even across new stepping stones. At this point, MacArthur and Wilson explained, "adaptive radiation in the strict sense" has taken place (MacArthur and Wilson 2001 [1967], 180).

In the book's conclusion, MacArthur and Wilson summarized the importance of their contribution and outlined further steps to be taken. They maintained that new study and experiment must be conducted on individual species to move beyond some of the generalizations they had presented, but nonetheless proclaimed that "biogeography appears to us to have developed to the extent that it can be reformulated in terms of the first principles of population ecology and genetics." The authors also argued that for such a synthesis to take place, "population and community ecology" should be cultivated "in a way that contains much more evolutionary interpretation than has been traditional" (MacArthur and Wilson 2001 [1967], 183). To any young paleobiologist who might have picked up this book in the late 1960s or early 1970s, the effect must have been powerful. Here were two biologists who dared to tackle broad conceptual problems using mathematical generalizations. MacArthur and Wilson were also crossing disciplinary boundaries by conflating biogeography and ecology, and by connecting the problems of population ecology with the study of evolution. Perhaps most important, they were outlining questions that a paleontologist could study. As Lawrence Heaney explains, the MacArthur-Wilson model discarded the traditional perspective in ecology in which communities "were viewed as being either static . . . or changing only slowly and unpredictably due to the forces of geology and perhaps climate," in favor of a model "in which species richness was the dynamic outcome of ongoing colonization and extinction" (Heaney 2000, 60). Over the next two decades, paleobiologists would increasingly focus on problems such as diversity, species richness, survivorship and extinction, and adaptive radiation explicitly as problems of historical ecology and biogeography. And in a number of prominent cases, as we will see in later chapters, the MacArthur-Wilson model would be an important touchstone for paleobiological theory.

Global Diversity and Evolutionary Paleoecology

In the mid- to late 1960s, an influential group of young paleoecologists emerged from the Geology Department at Yale. This so-called Yale school, which included Steven Stanley, Geerat Vermeij, Jeremy Jackson, Richard Bambach, Jeffrey Levinton, and others, had a tremendous impact on the way the fossil record was studied from a biological point of view. As Raup recalls, "Yale became the center of it all because of that wonderful flock of students there all about the same time, all working very closely with G. Evelyn Hutchinson" (Raup, as quoted in Princehouse 2003, 198–99). Hutchinson may have been the intellectual godfather, but the administrative and organizational mastermind was the paleontologist A. Lee McAlester, himself a Yale graduate, who according to Bambach "catalyzed the success of the program" between 1960 and 1973 (Bambach 2009, 401).

In the late 1960s there were few programs that could match Yale's for commitment to integrating biology and ecology into paleontology, or that could boast a comparable critical mass of faculty and students with paleobiological and paleoecological interests. Columbia's program continued to thrive through the 1960s under the leadership of Newell, but by the early 1970s it began to fall off as key faculty members retired or left for other appointments. Harvard had a solid reputation for paleontology, but despite producing graduates like Raup, it was fairly conservative towards the major currents in paleobiology until the early 1970s. At the University of Chicago, Everett Olson and Ralph Johnson had strong paleobiological interests, but it was a decade or more before the "Chicago school" of paleobiology would rise to become dominant in the field. And while there were certainly a number of geology departments at other

/

universities with strengths in paleontology—Kansas, Michigan, and Wisconsin, to name a few—most of those programs were still quite traditional in orientation.

Valentine on the Analysis of Diversity Trends

In 1973, James W. Valentine published a book entitled *Evolutionary Ecology of the Marine Biosphere*, which in many ways epitomized the synthesis of paleoecology and paleobiology in the early 1970s. Valentone himself was not a product of the elite paleontology programs of the 1950s or 60s, and his paleoecology was mostly self-taught. After receiving a PhD in geology from UCLA in 1955, he began reading ecological literature by Hutchinson and MacArthur, and was particularly impressed by MacArthur's use of theoretical models: "If you looked at his graphs, they wouldn't even have numbers on them—they were just things, they were concepts, but . . . they were very powerful" (Valentine interview). At the time, many paleontologists, like Simpson and Newell, believed the key to advancing evolutionary paleobiology was to focus on populations, following the lead of geneticists and population biologists like Dobzhansky and Mayr. From his reading of ecological literature, however, Valentine became convinced that the fossil record could only provide limited insight into population dynamics, but was ideal for examining ecological relationships such as community associations and distribution patterns. Over the next several years he increasingly came to see the fossil record as a broad, interconnected ecological hierarchy, arranged in descending orders of resolution from the global biosphere down to the individual community.

A particular insight Valentine developed through his work in the 1960s and early 1970s was the relationship between evolution, ecology, and biogeography. As he put it, evolution occurs "in response to the ecological network of interactions that determines the hierarchy of eco-geographic units that is the biosphere." Biogeography, "which is just an extension of ecology," gave insight into how ecological relationships are distributed across space, and Valentine now considers MacArthur and Wilson's work on island biogeography to have had "a profound effect" on "the movement of theory in paleobiology." Before MacArthur and Wilson, population ecologists had been accustomed to ignoring the historical dimension of ecology, but Valentine realized that "if you do biogeography, it gives you the fourth dimension." As he remarks, "The movement of theory from biology into paleontology meant that the ecologists were looking at populations and things today. And if you want to take the ecological theory and run it to the standpoint of looking at diversity of the biosphere, you have to give it another dimension" (Valentine interview). In other words, *time* is the distinctive viewpoint paleoecological analysis gives to the study of ecological relationships.

One of the major ecological questions Valentine focused on was how patterns of diversity changed over the history of life. Hutchinson had earlier discussed the problem of why there are so many kinds of animals in his 1959 essay "Homage to Santa Rosalia," in which he concluded that diversity was a response to the colonization of ecological niches, and that global diversity should increase over time as organisms radiated and established community structures (Hutchinson 1959). This perspective stimulated interest among both ecologists and paleontologists, including MacArthur's 1960 paper "On the Relative Abundance of Species" and Simpson's 1964 "Species Density of North American Recent Mammals" (MacArthur 1960; Simpson 1964). Simpson's paper, which attempted to explain longitudinal species gradients for mammals in the Recent period (in other words, it attempted to ask why species are distributed with greater density of diversity as one proceeds from the polar regions to the equatorial tropics) was noteworthy for using Hutchinsonian-style niche analysis as a central theme. Simpson cited not only Hutchinson's 1959 essay, but also three additional papers authored or coauthored by Mac-Arthur (Simpson 1964, 63). Another important paleontological contribution was Alfred G. Fischer's 1960 "Latitudinal Variations in Organic Diversity," which argued that diversity is controlled mostly by environmental factors like temperature. Fischer's conclusion that "biotic diversity is a product of evolution, and is therefore dependent upon the length of time through which a given biota has developed in an uninterrupted fashion" marked this shift in paleobiology towards biogeographical questions of ecological diversity that would persist for several decades (Fischer 1960, 80).

The development of Valentine's interests in evolutionary paleoecology, then, coincided exactly with a broader growth of interest in the historical analysis of diversity patterns among some paleontologists and ecologists. One of Valentine's first major theoretical contributions to this discussion was his 1968 paper "The Evolution of Ecological Units Above the Population Level," published in the *Journal of Paleontology* (Valentine 1968). As Valentine put it, his purpose was "to outline the principal

ways in which ecologic units may evolve; to discuss the operational definitions of fossil representatives of these units; and to examine the special ways each of these units is represented in the fossil record. Stress is
placed upon the interrelations of the units. Such an approach involves
integrating modern evolutionary concepts with current ecological concepts" (Valentine 1968, 254). Here Valentine built a double-stranded
hierarchy of ecological and biological classification that integrated the
units of population biology, as defined by biologists like Mayr and Dobzhansky, with the units of ecological organization described by Hutchinson. For example, at the lowest level of the hierarchy is the niche, which
Valentine described as "the functional aspect of a population." Next are
ecosystems, which are occupied by communities of organisms, or "collections of populations located in space and time." After this comes the
biome, which is the ecological level occupied by provinces, or "collections of communities associated in space and time." Finally, the highest level of ecological hierarchy is the biosphere, which contains all of
the lower levels of classification, and whose population units are collectively known as the biota (Valentine 1968, 255–58). Underlying this double classification was Hutchinson's synthesis of population ecology with
genetics: Valentine described the stability of an environmental regime
in terms of the equilibrium of the gene pool of the population that occupied it, and argued that the evolution of units above the population
level (e.g., communities) was a product of (1) shifts in the occupancy of
the niches that constitute a particular ecosystem, and (2) concomitant
changes in the gene pools of the populations that made up the community, leading to "novel descendant populations with novel niches" (Valentine 1968, 256). Valentine's point was, essentially, that evolution above
the basic level of the population could not be understood without attention to ecological relationships, and he drew a direct analogy between
the "change in the state of or relationships among realized niches" and
changes in gene frequencies understood by population genetics (Valentine 1968, 257). While his argument was a general one, Valentine's specific message for paleontologists was that "the fossil record furnishes the
most direct type of evidence" for ecological and evolutionary transformations of population units at the scale of the community, the province,
and the biota (Valentine 1968, 253).

A year later, Valentine applied these concepts to the empirical record of changes in diversity of marine communities across the Phanerozoic. The result was "Patterns of Taxonomic and Ecological Structure

of the Shelf Benthos During Phanerozoic Time," which challenged the conventional wisdom about the evolution of marine diversity and began a controversy in paleobiology that would last more than a decade. We can recall that the first major analysis of Phanerozoic diversity was John Phillips's 1860 study of British strata that produced his famous diversity curve, showing an accelerating growth in diversity from the Paleozoic through the Cenozoic. Although Phillips's estimate was based on admittedly incomplete fossil evidence, the general trend he described was largely accepted by later workers, and similar analysis by Newell in the 1960s appeared to confirm this pattern of increasing diversity (Phillips 1860; Newell 1967). One of the central problems in estimating diversity based on fossil evidence was that of overcoming the problem of bias: it was not known whether organisms had been preserved with equal frequency in different periods of geologic history, meaning that estimates of diversity might be skewed by biases in preservation. Phillips realized that individual stratigraphic intervals examined were not of equal thickness, which meant that diversity values had to be adjusted to take into account the discrepancies between thinner and thicker intervals. He therefore adjusted his estimate by recalibrating diversity based on the number of species he found per unit thickness, rather than per stratigraphic interval. This correction seemed to solve the problem of preservational bias (Miller 2000, 57).

By the middle of the 20th century, knowledge of the fossil record had improved greatly, as had methods of analyzing fossil data, thanks to the accumulation of new fossil evidence and the advent of computers. Valentine was able to make use of the compendious *Treatise of Invertebrate Paleontology* compiled through the 1950s, which contained many more occurrences of taxa than had been known in the 19th century, and which enabled him to produce a much more refined tabulation than Phillips had achieved. Valentine was able to distinguish between patterns of diversity at different levels of taxonomic hierarchy, and he found that these patterns were not identical over geologic history. For example, taxonomic diversity at the levels of phylum, class, and order was greatest in the early Phanerozoic, peaking through the Ordovician, and then declining as extinct higher taxa were not replaced. However, later in the Phanerozoic—especially following the great end-Permian extinction—there was an increase in diversity among families and genera. In other words, the taxonomic diversity became more "bottom heavy" as lower taxonomic representatives from a smaller sample of higher taxa diversified more rapidly (Valentine 1969, 699). Overall, Valentine found in this paper (and in a 1970 follow-up) that standing diversity did genuinely increase over the Phanerozoic by about one order of magnitude, in general agreement with Phillips's broad estimation.

However, the most interesting aspect of Valentine's study was its attempt to explain diversity trends in terms of the ecological hierarchy he had presented in his earlier essay. He began his 1969 paper by observing: "It is possible to visualize the ecospace of any genus as composed of the ecospaces displaced by all its component species, and the ecospace of a family as composed of all the generic ecospaces, and so on. Thus defined, the ecospace of a higher taxon displaces the actual regions of the lattice that have been occupied by the members of that taxon. Thus the taxonomic hierarchy possesses a precise structure at any time. This structure changes through time in well-defined patterns" (Valentine 1969, 688). In other words, changes in diversity at a higher taxonomic level have consequences for lower levels because of the duality of biological and ecological hierarchies: a given taxon occupies a portion of theoretical ecospace that can be subdivided into smaller units of both taxonomic and ecological organization. Early in the Phanerozoic, when a comparatively large number of higher taxa had not yet differentiated into many taxa of lower rank, ecospace was partitioned into fewer, broader units. Niche size was relatively large, since there were fewer specialized kinds of organisms competing for resources. However, since many of these early higher taxa had lower diversity, they had a statistically higher probability of extinction, since they contained fewer individual lineages (Valentine 1969, 707).[1] Over the course of the Phanerozoic, as some of these higher taxa became extinct, portions of ecospace became available. Instead of being replaced by new taxa of equivalent rank, the free ecospace was colonized by lower taxa of surviving classes and orders, which diversified and evolved to fill smaller and more specialized niches. The pattern of Phanerozoic diversity, then, has seen roughly proportionate diversification between family and generic levels during the Paleozoic and early Mesozoic, but disproportionately high generic diversification during the Cretaceous and Cenozoic (Valentine 1969, 697–98; fig. 4.3).

So while Valentine concluded that indeed "a major Phanerozoic trend among the invertebrate biota of the world's shelf and epicontinental seas

---

1. Here Valentine cites Simpson's principle that the longevity of a taxon is correlated to the diversity of its constituent lineages. See Simpson 1953.

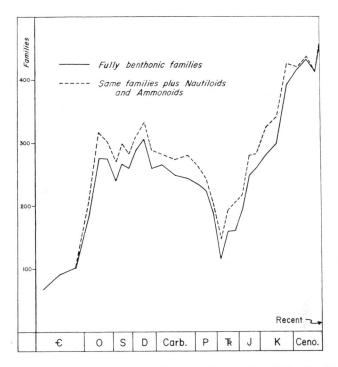

FIGURE 4.3. Valentine's graph of diversity of Phanerozoic marine benthic families. James W. Valentine, "Patterns of Taxonomic and Ecological Structure of the Shelf Benthos during Phanerozoic Time," *Palaeontology* 12 (1969): 692. Courtesy of the Palaeontological Association.

has been towards more and more numerous units at all levels of the ecological hierarchy," the ecological-evolutionary picture he painted was much more subtle and complex than what earlier studies had proposed. Valentine's analysis revealed that beneath the measure of total diversity are a series of independently fluctuating patterns at different levels of taxonomic resolution. Overall diversity throughout the Phanerozoic "has been achieved partly by the progressive partitioning of ecospace into smaller functional regions, and partly by the invasion of previously unoccupied biospace. At the same time, the expansion and contraction of available environments has controlled strong but secondary trends of diversity" (Valentine 1969, 706). Ultimately the tendency towards greater diversification at the lower taxonomic ranks is a product of what Valentine called the "canalization of ecospace": over time, the greater specialization necessitated by smaller and smaller divisions of ecospace makes it harder for departures from existing functions and morphologies to evolve. This trend favors small modifications of existing forms rather than the major evolutionary novelties that were possible earlier, when larger volumes of ecospace were available.

The most direct initial response to Valentine's proposal was Raup's 1972 paper "Taxonomic Diversity During the Phanerozoic," published in *Science*. This essay sparked a debate that played out through the 1970s, eventually drawing responses not only from Valentine but from Bambach and Jack Sepkoski as well. That debate and its ultimate resolution will be treated more thoroughly in chapter 8, but Raup's objection basically involved his suspicion that systematic bias in rock volume available for sampling from over the course of Phanerozoic history had led to the illusion that diversity had increased. Raup acknowledged that diversity had probably grown modestly, but importantly argued that an equilibrium pattern—similar to the sigmoidal logistic curve—probably best represented Phanerozoic diversity patterns (Raup 1972). Thus, in contrast to Valentine's "exponential" model of Phanerozoic diversification, Raup argued in favor of an "equilibrium" model.

While the ensuing debate between Raup and Valentine is outside the scope of this chapter, there is an important message to draw from Raup's analysis. Valentine's original diversity study was motivated by his concern that paleontological models should take ecological relationships into account, and he attempted to show that a widely accepted pattern of life's history could be explained using ecological theory. Valentine also helped to refine the way quantitative methods were used in paleobiology. While others, especially Newell and Imbrie, had explored the use of statistics to analyze large databases, Valentine was among the first to adopt sophisticated methodology for recognizing and overcoming biases. David Jablonski, himself a pioneer in quantitative paleobiology, describes Valentine's 1969 paper as "a joy to read even now," and says that it "showed that you could use statistics, or at least quantification, for things besides distinguishing two species morphometrically. . . . Unlike Newell, Valentine was much more open about where the data came from, what might be wrong with them, and what you could say despite the things that might be wrong with them" (Jablonski interview).

Raup's response went even further by casting the problem—in line with what was becoming the prototypical approach of evolutionary paleobiology—as a case study in which abstract mathematical principles, including a computer simulation, could provide a heuristic model for

testing the interpretation of real data. This was a method that Raup himself had already investigated, as he had done in his studies of simulated
morphology, but his critique of Valentine was his first major analysis of
fossil data to answer a significant question in evolutionary history. Over
the next few years, Raup and other colleagues expanded this program of
heuristic simulation into a thriving research agenda within paleobiology.
This use of abstract models as heuristic tools for interpreting real data
was quite explicitly an outgrowth of the Hutchinson-MacArthur school
of theoretical ecology. Over the next decade, paleobiologists interested
in the kind of analysis performed by Raup would increasingly turn directly to models from ecology to shed light on paleobiological questions,
particularly those that dealt with equilibrium problems relating to diversity, evolution, and extinction. As Daniel Simberloff put it in a 1972 essay about the application of theoretical ecology to paleontology, "The
apparent success of the equilibrium model ought to convince paleontologists of the utility of abstraction, then mathematical formalization, and
finally deduction of testable consequences as a method of generating and
answering illuminating questions" (Simberloff 1972, 189).

Evolutionary Paleoecology of the Marine Biosphere

When Valentine published *Evolutionary Paleoecology of the Marine
Biosphere* in 1973, many of the concepts he had introduced in his papers of the late 1960s had already infiltrated paleobiological thinking.
Looking back on the book more than 30 years later, he commented, "If
the book has a legacy, it will be simply that it represented a deliberate
and concerted attempt to bring a wide range of biological theory into
paleontology and to view the fossil record in a global framework or at
least context" (Valentine 2009, 389). The book itself was unrestrained
in its enthusiasm for the changes underway in the profession: "Advances
in population biology and in our understanding of the structure, function, and natural regulation of ecosystems now permit paleontologists to
study the qualities of fossil populations and their patterns of association
and of distribution from a new perspective." That perspective was paleoecology, "once a loose collection of techniques and environmental interpretations," but now "a focused discipline with its own problems, values, and goals" (Valentine 1973a, xiii–xiv). Having overcome a "lack of
common scientific goals among its practitioners," Valentine argued that
paleoecology had experienced "growing cohesion of the field [that] now

suggests that a common community of interest is developing which will lead to those traditions necessary for research continuity" (Valentine 1973a, 336). He also drew particular attention to the new importance of deductive models in paleobiology, noting the "widely demonstrated power" of "mathematical, or qualitative, or even actual physical working models" made possible by computers for analyzing paleontological data (Valentine 1973a, 11–12). These techniques were so powerful, in fact, that he reported that "it appears likely that we shall soon be able to reconstruct the sequence of past environmental regimes from first principles, independently of the fossil evidence," resulting in "a quantum advance in our knowledge of ecological history" (Valentine 1973a, xiv).

One of the central arguments of Valentine's book was that paleontology, ecology, and evolutionary biology are intertwined, and that evolution is fundamentally an ecological process. Beginning with Hutchinson's observation that ecospace occupation is regulated by the scarcity of environmental resources, Valentine noted that individual phenotypes can be seen as particular responses to different environmental conditions. Individual phenotypes are just realized portions of a larger "genotypic hypervolume," while the biospace occupied by actual organisms is just the portion of the n-dimensional ecospace hypervolume that is actualized. Thus biospace and phenotype represent the actual realized volumes of their respective hypervolumes (Valentine 1973a, 59–61). Ecological and genetic hierarchies, he argued, can be represented as overlapping theoretical volumes, providing a functional description of organism-environment relationships. Importantly, these hierarchies form a nested geometrical structure, meaning that changes at any given lower level affect all of the levels above. For example, "a change in gene frequencies may occur causing a niche to evolve, and therefore any community, province, or biosphere containing that niche changes also. The size of the unit being affected obviously increases greatly at each higher level, so that a given change affects a much larger proportion of the smaller units at lower levels than of larger units at higher levels" (Valentine 1973a, 81).

Valentine argued that "the geometric model of environment-organism relationships" is preferable to other conceptual ecological models—such as Simpson's "adaptive zone" model—for several reasons. First, he claimed, it "can be formalized and at least partially quantified so that the ecospace patterns may be analyzed numerically," although he did not devote much space to describing this kind of mathematical analysis. Second, it separates organismal and environmental properties, and demonstrates how they overlap and interrelate in a clear, pictorial manner. Finally, the model can be applied to the traditional taxonomic relationships studied by paleontologists: "It is a model in which the properties of taxa as well as of ecological units can be conceptualized," an advantage that is "especially significant for our purposes." It is in this regard that evolutionary relationships can be illuminated by the model, since "the ecospace of a phylum throughout its existence will somewhat resemble the family tree of that phylum, branching according to its evolutionary history." Therefore, "the fundamental significance of adaptive divergence and radiation is thus verified in ecospace—that is, the morphological divergences reflected in taxonomy represent functional divergences and radiations" (Valentine 1973a, 87).

In the book's final chapter, "An Approach to an Ecological History of the Marine Biosphere," Valentine elaborated on his vision of a synthesis between paleoecology and paleobiology. The central task of this synthesis, he argued, was to analyze "trends in ecological processes across the whole course of the Phanerozoic." Though much "paleontological and biological" work remained to construct "a solid theoretical framework in which to place the fossil evidence," Valentine maintained that "a beginning has been made and the discipline has advanced to the point where some of the highlights in the history of life may be examined from a paleoecological perspective" (Valentine 1973a, 409). Using the case of Phanerozoic diversification as an example, he argued that in addition to an increasing evolution of taxonomic specialization, evolutionary trends across the Phanerozoic also suggest an increase in "the average complexity of organization of organisms," which he attributed as a consequence of adaptive specialization. The understanding—and validation—of directional trends (such as increase in size, or Cope's law), had been a central feature of evolutionary paleobiology from the very outset. Here Valentine presented both a set of ecological explanations for these apparent trends, as well as (through his use of quantitative methods) a means for determining their veracity. For instance, he proposed that diversity has increased despite major extinction events because surviving "complex stocks" are able to adapt to low-diversity ecosystems in post-extinction regimes, quickly diversifying and expanding to fill newly vacated ecospace. "Therefore," he argued, "the extinctions do not reduce the average level of phyletic complexity, but merely pave the way for its increase or actually cause its increase by subjecting organisms to new adaptive challenges" (Valentine 1973a, 470). Ultimately, Valentine contended, "the evolution of both marine invertebrates and marine environments has proceeded hand-in-hand, and the history of their interrelation is recorded in the progressive and episodic changes in the ecological and taxonomic hierarchies, as represented by marine fossils." He thus painted a vision in which "a sort of moving picture of the biological world with its selective processes that favor increasing fitness and that lead to 'biological improvement' is projected upon an environmental background that itself fluctuates, with major changes in configuration occurring over long periods or at long intervals, and with what is probably a spectrum of shorter-term changes superimposed thereon. The resulting ecological images expand and contract, but, when measured at some standardized configuration, have a gradually rising average complexity and exhibit a gradually expanding ecospace" (Valentine 1973a, 471).

This kind of vision, with an overriding concern for painting a more complete picture of the history of life that was sensitive to the interrelation of environment, adaptation, and evolution, was fundamentally a paleobiological one. As Valentine described it years later, "evolutionary paleoecology, then, would for a start use an ecological theory as a framework within which to examine and evaluate paleoecological processes, which famously form the theater of the evolutionary play, over time. The evolutionary events revealed in such studies are chiefly macroevolutionary, involving scales appropriate to the fossil record" (Valentine 2001, 10).[2] In articulating this broader framework, Valentine's book effectively translated the lessons of theoretical population ecology into a paleobiological idiom, and helped to fundamentally change the definition of what paleobiology was. Authors now routinely distinguish "evolutionary paleoecology" (understood in Valentine's terms) from simple "paleoecology" (understood as environment reconstruction), and even from "community paleoecology," which focuses on the diversity, setting, and change in individual paleocommunities. In asserting the importance of macroevolutionary consequences of ecology at large scales, Valentine's approach also influenced the development of paleobiology in a macroevolutionary—and ultimately hierarchical—direction.

2. The title of Hutchinson's most important book was *The Ecological Theater and the Evolutionary Play*. (Hutchinson 1965).

Conclusion

In 1976, Stephen Jay Gould published an essay titled "Palaeontology plus Ecology as Palaeobiology." Gould began his essay by repeating the traditional lament of biologically-minded paleontologists that paleontology "has been the poor sister, and often the laughing stock, of the sciences," often depicted, "not unfairly, as the dullest variety of empirical catalouging practiced by the narrowest of specialists." Happily, however, "a remarkable transformation has taken place during the last decade," wherein "palaeontology has allied itself with evolutionary biology, and has experienced all the change and excitement of its most rapidly developing subdiscipline—theoretical ecology" (Gould 1976b, 218). In fact, Gould cited "the direct impact of theoretical population ecology" as the most important influence on the transformation of paleontology into "a conceptual science worthy of the name palaeobiology" (Gould 1976b, 220). In particular, he pointed to the way in which "modern ecological theory opens the whole field of adaptive strategies, and its criteria of quantitative population dynamics" to paleontological investigation, citing as examples the stability-diversity theory, the testing of survivorship curves, and MacArthur's equilibrium approach to understanding species diversity (Gould 1976b, 221). "Ultimately, however," he concluded, ecology's greatest influence had been in "the reorientation of basic attitudes towards evolutionary questions":

> Ecological time is rooted in this Darwinian present; ecological themes encourage palaeontologists to study adaptation for its immediate significance. Traditional palaeontology rarely worked at this level; it focused instead on the meaning of adaptation as a contribution to long-term evolutionary trends. And when it considered the immediate significance of adaptation at all, it did not venture beyond morphology in a "physicalist" perspective—i.e. to what aspect of the physical environment is this structure fitted? The ecological themes of population dynamics, life history strategies and species interactions were simply not categories for consideration (Gould 1976b, 233).

In its traditional mode, paleontology could describe the time series of adaptations of physical form in individual lineages, but it rarely thought of the subject matter—fossils—as actual organisms living in dynamic interaction with one another, nor did it deeply pursue the *reasons* for the adaptations themselves in the dynamic interplay of life habits, survival strategies, and population pressures that had once shaped them. Ecology, "rooted in this Darwinian present," gave that perspective. Therefore, paleoecology inspired a new perspective on "rereading" the fossil record as a history of ecological interactions between once-living organisms that could be described using the heuristic theoretical vocabulary of Hutchinsonian and MacArthur-Wilson theoretical ecology and biogeography.

Of course, as Gould conceded in closing his essay, the unique perspective of paleontology is its historical view, which opens analysis and interpretation of broad evolutionary patterns and trends. For paleobiology to explain the trends and not just document them—in other words, for paleobiology to achieve its ambition of contributing to evolutionary theory— "ecological time will remain the fundamental level of palaeontological analysis. We give our thanks to modern ecology for teaching us the rudiments of perception in Darwin's own sphere of operation" (Gould 1976b, 236). These comments, which were published during the active phase of the "paleobiological revolution" in the 1970s, certainly reflect Gould's proselytizing zeal for paleobiology. What they demonstrate, however, is the degree to which that paleobiological agenda was explicitly connected to theoretical ecology and biogeography. Theoretical ecology of the MacArthur-Wilson variety provided a crucial inspiration for paleobiologists to explore generalized theoretical models of their own. Some paleobiologists, like Tom Schopf and Jack Sepkoski, would directly apply ecological models to paleontological data. For others—such as Gould and Raup, who had only limited ecological interests—studies like those of MacArthur and Wilson were more of an inspirational model. In either case, though, as the remaining chapters in this book will show, the importance of theoretical ecology as a source of inspiration and example to paleobiologists was profound.

CHAPTER FIVE

Punctuated Equilibria and the Rise of the New Paleobiology

As the 1970s began, paleobiology was transformed from a loose movement with a fairly uncoordinated set of goals to a bona fide subdiscipline with an active institutional and intellectual agenda. The dynamics of this shift were complex, and they involved theoretical, institutional,
and pedagogical factors. The next several chapters will discuss the movement to establish paleobiology as a genuine subdiscipline within paleontology—paleobiology's activist or revolutionary phase—which saw the
promotion of a clearly articulated theoretical agenda, the establishment
of a new professional journal, and the centralization of paleobiological
activity in a few institutional settings. A unifying feature of these diverse
activities is that they were consciously planned and directed by a small
group of young, like-minded paleontologists who shared a vision for a
new subdiscipline within paleontology that would make paleontology an
independent voice within evolutionary biology, and would also acknowledge the broader theoretical landscape that had developed across biological disciplines.

One of the striking features of the history of paleobiology is the extent to which the movement came to be so closely associated with a few
signature theories. As we have seen, over the first half of the 20th century, "paleobiology" emerged as a fairly casual label to describe a broad
approach to an array of questions surrounding the application of fossil data to problems in evolutionary biology. While there was a suite
of practices that increasingly came to be associated with this approach
(such as quantitative analysis, modeling, and ecological/biogeographical
theory), there was no unifying theoretical basis for paleobiology, nor did

the term "paleobiology" connote a distinct disciplinary identity within paleontology. During the 1970s, however, paleobiology became closely associated with a prominent theory—Niles Eldredge and Stephen Jay Gould's theory of punctuated equilibria—and the study of macroevolution as a dominant conceptual framework. The arrival of this new theoretical perspective did not produce disciplinary coherence; rather, conceptual innovation and institutional framework were simultaneously and self-consciously engineered by a small group of paleontologists committed to disciplinary activism on behalf of paleobiology.

This engineering came during an initial burst between 1970 and 1975, and included both the establishment of a set of theoretical initiatives and a supporting framework of institutional support structures. These objectives went hand in hand; the theoretical agenda of 1970s paleobiology cannot be separated from its institutional one. The unifying factor was the group of younger, ambitious invertebrate paleontologists who shared a commitment to advancing the status of quantitative, theoretical, and evolutionary paleobiology to a central place within paleontology. Members of this group—including Gould, Eldredge, Steven Stanley, and David Raup—would produce some of the most significant theoretical innovations in paleobiology over the next decade, and many achieved fame and notoriety for this work. But from a programmatic standpoint, the driving force behind the paleobiological 'revolution' of the 1970s was not Gould or Eldredge or Raup, but rather a figure who was less well known (at least outside of paleontology) and who worked tirelessly behind the scenes to engineer this transformation. That person was Thomas J. M. Schopf.

Tom Schopf and *Models in Paleobiology*

At first blush, Schopf seems an unlikely revolutionary. Although he came from a prominent paleontological family—his father was the eminent coal geologist and paleontologist James Morton Schopf (winner of the Paleontological Society Medal in 1978), and his younger brother, the micropaleontologist J. William Schopf, has made some of the fundamental discoveries in Precambrian paleontology—Tom, as all his friends called him, is not remembered for any significant discoveries or major theoretical contributions. Unlike some of the more celebrated members of his generation, including his brother, Schopf did not attend an Ivy League graduate program, nor was he trained by one of the innovators in the previous generation of paleobiologists. Rather, his early career was marked by solid if somewhat unremarkable progress, from a PhD in the very traditional program in geology at Ohio State University in 1964 to a first appointment as assistant professor at Lehigh University in 1967, to an eventual professorship at the University of Chicago, where he remained from 1969 until his death in 1984. In many ways, however, Schopf seems to have been unusually driven, throughout his career, to demonstrate his professional worth, and very early on his professional ambitions were channeled into discipline building. This was fortuitous both for paleobiology and for Schopf personally: by the late 1960s the field had plenty of innovators, but what it lacked was an individual who, by force of effort and personality, was willing to devote himself single-mindedly to organizing and promoting the agenda of paleobiology. This would become Schopf's primary role.

In any event, by the time Schopf took his first teaching position at Lehigh University, many of his central personal and professional motivations were firmly in place. Certainly, his sense of drive and self-discipline—combined, equally, with painful moments of self-doubt—were deep-seated, and may have had their origin at the Schopf family dinner table. In a letter to Gould written many years later, Schopf revealed the persistence of his conflicting feelings of inadequacy and personal drive:

> I've "always" thought that the measure of a man involved four things. The questions he asked, the answers he got, whether he lived up to his abilities, and, no less importantly, whether he had opportunities for work. . . . And so I feel it is incumbent upon those in favored positions to do more. But because of this odd perspective on life (the Protestant ethic), I will probably end up trying to do more than any one ever had any right to expect. Indeed my formal academic record is a testament as to why I have no right to be at the University of Chicago. But working from small beginnings with no misconceptions of overpowering ability, leads to one strength. I have no fear of failure (Schopf to Gould, 27 January 1973: Schopf pap. 5, 14).

In the letter, Schopf confessed that he was "preconditioned to believe that most of the time if my science is going to be any good, it won't be much fun." Nonetheless, he admitted to taking great personal satisfaction from his accomplishments, although he acknowledged that "my science is mostly a question of self-discipline." In this letter, Schopf seemed to accept—writing in 1976, after a number of accomplishments that will be detailed in the next two chapters had already taken place—that he and Gould were destined for different roles in the development of paleobiology: "I perceive our abilities as being quite different, our strategies for doing science as different, and our potential contributions as likely being different" (Schopf to Gould, 27 January 1973: Schopf pap. 5, 14). Schopf was certainly correct in making this distinction, but I will argue that without his role the history of paleobiology would look very different.

Schopf's first major contribution to the agenda of paleobiology began shortly after he arrived at the University of Chicago's Department of Geophysical Sciences in 1969. There is very little information about the circumstances that brought Schopf to Chicago; presumably he was hired to replace the department's resident invertebrate specialist, J. Marvin Weller, who retired at about the same time. Schopf's arrival corresponded with the relocation of the department to a new building—the Henry Hinds Laboratory for the Geophysical Sciences—and the beginning of a period of sustained growth in paleontology at Chicago. When Schopf arrived, he was one of only three paleontologists in a department of 23 regular faculty members (the others were Ralph Johnson and Alfred Ziegler); by the early 1980s the number of paleontologists had more than doubled, and the so-called Chicago School of paleobiology had become a dominant force in international paleontology.

Schopf appears to have arrived at Chicago bent on making an immediate splash. At the 1969 meeting of the Geological Society of América (GSA), he approached James Valentine with an idea about co-organizing a GSA symposium (and eventually a book) entitled Models in Paleobiology. Valentine was enthusiastic about the idea, and he encouraged Schopf to take over leadership on his own. As the agenda for this symposium took shape in letters between Schopf and Valentine in late 1969, Valentine suggested "emphasiz[ing] the great power of models to explain or at least rationalize the fossil record, which as you mention doesn't seem to be generally realized" (Valentine to Schopf, 15 December 1969: Schopf pap. 5, 23). Schopf responded by elaborating his view of the importance of models, along with the topics that might be covered. "In paleontology," he wrote, "such models would be useful wherever the situation was sufficiently complex or obscure so that a generalization is hard to achieve." Appropriate occasions for modeling "occur in problems of growth, morphology, ecosystems, community structure, speciation, and biogeography, to name areas where work of this type comes immediately to mind." Schopf's list of potential speakers included Gould (on growth), Raup (morphology), Anthony Hallam (populations), Ralph Johnson (communities), Valentine (ecosystems), Frank Stehli (paleoecology), and E. O. Wilson (speciation) (Schopf to Valentine, 24 December 1969: Schopf pap. 5, 23). Oddly, Schopf did not at this point assign himself a topic; apparently at this stage he conceived of his role primarily as organizational.

Planning for the project went forward quickly. In February of 1970 Schopf approached the chair of the Technical Program Committee for the GSA with a proposal for the session, enlisting the support of Paleontological Society President William Easton to ensure success (Schopf to Lewis M. Cline, 16 February 1970; Schopf to William H. Easton, 18 February 1970: Schopf pap. 5, 23). In the end it was determined that the symposium would take place at the 1971 GSA meeting. By early March, Schopf began contacting potential participants and possible publishers for the eventual proceedings. In his solicitation letter to participants, Schopf laid out the intent of the symposium very clearly:

What I would like the papers in this Symposium (and published volume) to accomplish is to identify and evaluate the theoretical models which are guiding (by accident or design) the development of various parts of our science. We now have both an extensive and a modern documentation of life in the past in the *Treatise [on Invertebrate Paleontology]*, in discussions about specific groups in the *Journal [of Paleontology]*, and elsewhere. The theoretical framework, however, dictates where one looks and how one goes about the descriptive process. . . . It is also my conviction that the volume will strike a responsive cord [*sic*] in many of our colleagues who, having mastered the taxonomy of a group, now find it of interest to see additional ways in which their data can be significantly interpreted. In this way we can encourage the analytical 'problem oriented" approach to paleontology (Schopf to Raup, 7 March 1970: Schopf pap. 3, 30).

First and foremost, then, Schopf viewed this project as a campaign against "merely descriptive" paleontology. As he described the agenda to one prospective publisher, the book was "centered on the radical theme of making invertebrate paleontology into science" (Schopf to Robert J. Tilley, 2 March 1971: Schopf pap. 10, 18). To another he wrote, "The tone of the volume . . . is meant to be fresh and exciting, with the analyses more cartesian [*sic*], or hypothetico-deductive in emphasis than inductive" (Schopf to Robert H. MacArthur, 10 March 1970: Schopf pap. 10,

18).[1] The essential point here is that Schopf entered the project with a clear vision not only for the symposium, but for transforming the field of paleontology. These themes—that traditional descriptive methods must give way to theory—would resurface time and again over the next several years, both in published literature and private correspondence between Schopf and other members of his inner circle, such as Gould and Raup. Gould, in particular, would become a great champion of this program, as we will see shortly. However, no active campaign to establish theoretical paleontology existed before Schopf entered the field. As much as the modern paleobiology movement may have depended on the work of other, more prominent figures, it is fair to argue that the movement itself was Schopf's invention.

From early 1970, when Schopf began to enlist participants, to the preparation and publication of *Models in Paleobiology* in October 1972, Schopf set a demanding pace for himself and his contributors. Authors were initially given a year to produce drafts, which were required by July 1971, in advance of the GSA meeting, so that the book could appear as soon after the symposium as possible. In the meantime, Schopf set about trying to find a publisher. Between 1970 and 1971 he contacted a number of presses, including the university presses at Princeton, Columbia, Chicago, and Harvard, as well as textbook publishers John Wiley and Sons and Prentice-Hall. Schopf optimistically reported in the spring of 1971 that "we will have no trouble placing this book with a reputable firm," but explained that he wanted to "sit tight" until manuscripts had been completed and reviewed before pursuing a contract (Schopf to Raup, 2 April 1971: Schopf pap. 3, 30). While the process did not turn out to be as smooth as Schopf had hoped (the book was rejected, sometimes scathingly, by several presses), eventually the project found a home with the small academic publisher Freeman, Cooper, and Company.

Punctuated Equilibria and the Paleobiological Agenda

Niles Eldredge and the Origin of the Theory

The most famous contribution to *Models in Paleobiology* was unquestionably Eldredge and Gould's essay "Punctuated Equilibria: An Alter-

---

1. Schopf had contacted MacArthur in the latter's capacity as editor of the Princeton Biological Monographs series for Princeton University Press.

native to Phyletic Gradualism."[2] The theory this paper introduced had a transformative influence on the development of paleobiology, but in hindsight its history has often been distorted. Almost everything about punctuated equilibria has become mythologized to some degree: the circumstances of its invention, the alleged scientific and political radicalism of its agenda, the subsequent advancements and retractions of its major claims, controversies surrounding its relative originality, and debates about its empirical claims. The authors of the theory themselves have had a major hand in the retelling of this story, and it is fair to say that without punctuated equilibria the careers of Eldredge and Gould would have turned out quite differently. One fact that is not in question, however, is that punctuated equilibria is one of the more important contributions to recent evolutionary theory, from both a conceptual and a sociological standpoint. Even Ernst Mayr, one of the most vocal critics of punctuated equilibria, acknowledged this: as he wrote in 1989, "Whether one accepts this theory, rejects it, or greatly modifies it, there can be no doubt it had a major impact on paleontology and evolutionary biology" (Mayr 1989, 139).

Given the stature its authors achieved over the decades after the theory was first presented, it is easy to forget how young they were, and how tenuous was their grasp on the paleontological establishment, at the time of the first paper's publication in 1972. Gould, the senior of the two authors, had taken a position at Harvard in 1967 before his thesis had even been completed, but he owed this striking early success not to an established body of accomplishment, but rather to the glimmerings of promise evident in his few publications and the workings of the old-boy network. One of Harvard's invertebrate specialists, Harry Whittington (of eventual Burgess Shale fame) had recently departed, and Bernhard Kummel (Norman Newell's first PhD student at the University of Wisconsin and now one of the senior paleontologists in Harvard's geology department) simply contacted Newell and asked him to recommend a suitable candidate for replacement. As Gould later recalled, Kummel "wanted to get a young guy who did the opposite of what he did. . . . he wanted someone who could do quantitative evolutionary work." Gould had demonstrated

---

2. There is some debate about whether the theory is properly termed "equilibria" or "equilibrium." In his later career Gould favored "equilibrium," which undoubtedly accounts for its popularity. However, Eldredge has always maintained that "equilibria" is proper, since the theory describes multiple periods of stasis punctuated by rapid evolution. I follow Eldredge in using the plural form.

promise in this area with a 1966 review of theories of allometry in *Biological Reviews* which, combined with an unfinished dissertation on the evolution of Bermudan land snails, apparently constituted sufficient credentials for his appointment. Gould somewhat sheepishly acknowledged the irregularity of this arrangement in an interview years later: "I came by the job somewhat dishonorably by modern standards, or at least by my own. [But] I kept it totally honorably" (Gould, quoted in Princehouse 250–51).

While Gould had status as one of paleontology's up-and-comers, Eldredge was a virtual unknown. Like Gould a native New Yorker, Eldredge fell into geology as an undergraduate at Columbia University thanks to an early interest in evolution spurred by an introductory paleontology course with John Imbrie. In a classic case of being in the right place at the right time, he was quickly adopted by an exciting group of newly arrived graduate students as a kind of unofficial junior member. As Eldredge recalls, "Gould and about eight other graduate students showed up when I was a junior, I think. They were entry-level graduate students and right away they formed a seminar in evolution. Probably Steve did it. And I was allowed to participate" (Eldredge interview). Eldredge began supplementing his standard courses in the geology curriculum with long discussions about macroevolution with Gould and other graduate students, and with an extracurricular reading list that included Mayr's *Systematics and the Origin of Species* and Simpson's *Major Features of Evolution*. Eldredge stayed on at Columbia as a graduate student, where he turned to the study of the evolution of Devonian trilobites of upper New York State (Eldredge 2008, 109). From the very beginning, he set out to apply paleontology to the study of long-term patterns in evolution following the model of his major intellectual influences, Mayr and Simpson, and his mentors, Imbrie and Newell (the latter of whom would become his thesis supervisor). Eldredge recalls being impressed as a graduate student by the fact that he was living through "a renaissance of looking at evolution" because of the larger sample sizes available to invertebrate paleontologists and the effectiveness of new statistical techniques for their analysis pioneered by authors like Imbrie and Newell (Eldredge interview). At the time he was also very aware, thanks to the influence of Gould, of the potential for young paleontologists to make an immediate splash in the field. "Steve showed me that it was important to start publishing scientific papers right away. He thought it was absurd to think that theoretical matters should be in the hands of older, more mature scientists when really, if anything, it should be the province of the young, coming to their subjects with fresh minds and new insights. Why wait until you are 60?, he used to ask. And of course, he was right" (Eldredge 2008, 108–9).

These considerations led Eldredge to a dissertation on the Devonian trilobite *Phacops rana*, a species conveniently abundant in upstate New York which is so named because of its bulging frog-like eyes (*rana* is the Latin word for frog). Eldredge hoped that these trilobites would provide an evolutionary "experiment" in geographic variation, since the range of *P. rana* in the Middle Devonian extends over much of the ancient inland sea that covered the Great Lakes region 380 million years ago. Eldredge recalls being struck by the fact that although during the 1930s and 1940s Dobzhansky and Mayr had rekindled interest in the role of geographic isolation as a factor in speciation, "few modern studies had been done— especially with the idea in mind that geographic variation within species is an essential component of the evolutionary process" (Eldredge 2008, 110). Eldredge resolved to test the effects of geographic isolation on the evolution of his group of trilobites, and set to work collecting samples during the summers of 1966 and 1967. When he began examining his specimens, however, he noticed a problem: "My trilobites didn't seem to be evolving—at least [not] in the slow, steady sort of way that I was taught to expect" (Eldredge 2008, 108). The trilobites did not seem to be changing much in any measurable way—at least in standard morpho- logical measures like size and shape—and Eldredge became seriously concerned that he would not have any results to report in his disserta- tion. Fortunately, he was inspired to look more closely at the large, bul- bous eyes of *P. rana*, and he discovered that in fact the trilobites *did* vary in the number of individual lenses that made up their compound eyes. When correlated with geographic and temporal distribution, it appeared that *P. rana* had evolved through a series of forms containing succes- sively fewer lenses in their compound eyes.

Again, however, Eldredge ran into a problem: although *P. rana* had indeed evolved eyes with fewer columns of lenses, earlier forms of the animal (with eyes containing more lenses) had not given way gradually to their successors over long periods of time. Instead, Eldredge found that the earliest form of *P. rana* (which had 18 columns of lenses) per- sisted, unchanged, for millions of years, while a new species, which had 17 columns, appeared abruptly at the edge of the ancestral range. The old and new forms then existed simultaneously for a lengthy period of time, until the ancestral species was eventually replaced by its descendant and driven to extinction. Furthermore, this process was eventually repeated, with the 17-column species giving way to a species with 15 columns of lenses (Eldredge 2008, 110–12; Eldredge 1971). This pattern was not what the standard gradualistic picture of evolution that still held sway for most paleontologists had taught him to expect, but it did fit well with the so-called allopatric model of speciation proposed by Mayr in publications throughout the 1940s and 1950s.[3] Additionally, Eldredge was impressed by what he was *not* seeing: "The stability these species showed as they lasted long periods of time was just as fascinating as the patterns suggesting that allopatric speciation had been at work when evolutionary change did occur. . . . Once the eyes had changed, they remained stable—as did every part of the anatomy of these trilobites that was preserved" (Eldredge 2008, 111–12).

Armed with this insight, Eldredge completed his dissertation very quickly, although he did not propose a mechanism to explain what he had observed in the evolution of his trilobites: "It was my job as a fledgling paleontologist simply to point out that it was so" (Eldredge 2008, 112). Unimpressed with the availability of university jobs (he recalls interest from the universities of Pittsburgh and Kansas, but no formal offers), Eldredge happily accepted Newell's offer of a position at the AMNH as an assistant curator of invertebrate paleontology. In Eldredge's recollection, this appointment came as something of an anticlimax: while completing his dissertation at Columbia, Eldredge had been assisting Newell with renovations to the trilobite component of the Hall of Earth History at the AMNH. One day, while working in the hall, Newell simply asked him whether he would like to continue permanently as a member of the museum's staff. Eldredge isn't certain why Newell made this offer—"I don't think he ever really thoroughly read my dissertation"—but for Eldredge, who had spent years as an undergraduate and a graduate student at the museum, the decision was easy: "It was sheer luck as far as I was concerned" (Eldredge interview).

"Models in Speciation"

When, on 9 March 1970, Tom Schopf mailed a letter to Gould inviting him to participate in the Models in Paleobiology symposium, he could

---

3. Allopatric speciation takes place when a group of organisms is geographically isolated from an original population and undergoes rapid genetic change.

not have predicted that the series of events he had just set in motion would transform the history of paleobiology. Schopf did not know Gould well, but Gould certainly fit Schopf's targeted profile of a younger pale-ontologist positioned to challenge the prevailing orthodoxies in the pro-fession. Schopf would have been aware of Gould's 1965 essay, published while its author was still a graduate student at Columbia, provocatively titled "Is Uniformitarianism Necessary?" The essay argued that the con-cept of uniformitarianism, or the belief that natural processes remain constant and consistent throughout Earth's history, was both anachro-nistic and false. Nonetheless, Schopf appears to have had few preconcep-tions about what Gould might present on the proffered subject of "mod-els of speciation" (Schopf to Gould, 9 March 1970: Schopf pap. 5, 14). Gould's response to the invitation was enthusiastic: "A damned good idea, your symposium. I'm flattered by your invitation and will gladly accept." However, Gould also sounded a note of ambivalence: "My only hesitation is that you have given me a topic that ranks only third on your list [of possible symposium topics] in terms of my competence," he said, noting that he could "handle either morphology or phylogeny with much greater ease and would request, if either Dave Raup or Mike Ghiselin can't participate, that you allow me to cover one of their top-ics" (Gould to Schopf, 13 March 1970: Schopf pap. 5, 14).[4] In his reply to Gould, Schopf acknowledged that "speciation might be only number three on your list [of all possible topics], but number three on your list is still number one on my list as I wanted to include both Ghiselin [phylog-eny] and Raup [morphology]" (Schopf to Gould, 14 April 1970: Schopf pap. 5, 14). It can be argued, then, that of one of the most important and controversial theories in modern evolutionary biology owes its existence to a simple editorial compromise.

The problem for Gould was that he didn't have much to say on the topic of models in speciation. He initially suggested to Schopf that if he could be assigned one of his preferred subjects, then the topic of specia-tion should go to Eldredge, whom he recommended as "our best new thinker" and someone who "worries constantly about speciation in pale-ontology." Gould asked further that, failing that arrangement, he be al-

4. Schopf's invitation letter included a list of proposed topics for the symposium, with tentative authors attached to each topic. The topics were: models of morphology, biogeo-chemical paleontology, populations, speciation, biogeography, communities, ecosystems, paleoecology, paleooceanography, and phylogeny (Schopf to Gould, 9 March 1970: Schopf pap. 5, 14).

lowed "to expand it [the topic] somewhat to speciation and the origin of new taxa in general." However, at some point between spring and fall of 1970, Gould made the fateful decision to enlist Eldredge not as a replacement but as a coauthor on the piece. There is no correspondence that fixes the date of this arrangement: in a letter from Schopf to Gould in August, expressing "gratitude to you for agreeing to discuss in the symposium a topic that wasn't your first choice," no mention of Eldredge's participation was made; Schopf's next communication, dated 16 November, was addressed to both Gould and Eldredge (Schopf to Gould, 6 August 1970; Schopf to Gould and Eldredge, 16 November 1970: Schopf pap. 5, 14). Gould's motivation for partnering with Eldredge was fairly straightforward: as he had explained to Schopf, the topic of speciation was not his forte, and Eldredge just happened to possess the kind of innovative approach to the subject required for the symposium.

There has never been any question about to whom the inspiration for punctuated equilibria should be attributed: In both public and private recollections, Gould consistently credited Eldredge with the initial insight about stasis and allopatric speciation (Gould 1989a, 117–18; Gould 2002, 979ff). The real question, perhaps, is not why Gould needed Eldredge, but rather why Eldredge needed Gould. The first and obvious answer is that Gould was the one who had been invited to participate in the Models symposium, and it is clear from his correspondence with Schopf that he was not about to pass entirely on the opportunity to participate. While both Gould and Eldredge were relatively young and untested, Gould, with a position at Harvard and a growing list of publications, was by far the more established of the pair. In contrast, Eldredge had published only three mostly descriptive papers, and would have been unlikely to have made Schopf's list of "exciting" younger paleontologists on his own.

However, Gould was almost uniquely positioned to appreciate the potential importance of the insights Eldredge had made into patterns of speciation in his dissertation research, and Eldredge trusted and admired Gould from their days together at Columbia, where Gould had been a kind of "big brother" and informal advisor to Eldredge. At the time of the Models invitation, Eldredge was in the process of preparing a paper that encapsulated the theoretical findings in his dissertation for submission to *Evolution*, and he had shown the manuscript to Gould for criticism and advice. This paper, "The Allopatric Model and Phylogeny in Paleozoic Trilobites," was eventually published in 1971, but it received little attention from paleontologists or biologists. Nonetheless, Gould recognized that this paper proposed exactly the kind of novel approach to "modeling" speciation required for the symposium.

The Origin of Punctuated Equilibrium in Autobiographical Recollection

In Gould's somewhat idealized recollection nearly twenty years later, punctuated equilibria was the product of the meeting of kindred spirits:

> Niles and I went to study with Newell because we were primarily interested in evolution—a direction that was, at the time, still a rarity in palaeontology. We began our professional lives with a commitment to engage in the empirical study of evolution as illustrated by the fossil record. We were both interested in small-scale, quantitative research on species and lineages (a concern fostered by our other advisor, John Imbrie, who taught us multivariate statistical analysis). Now imagine the frustration of two hyperenthusiastic, idealistic, non-cynical, ambitious young men captivated with evolution, committed to its study in the detailed fossil record of lineages, and faced with the following situation: the traditional wisdom of the profession held (quite correctly) that the fossil record of most species showed stability (often for millions of years) following a geologically unresolvable origin. 'Evolution,' however, had long been restrictively defined as 'insensibly graded sequences'—and such hardly existed. Niles and I had one advantage in combating this frustration. We had been well trained in the details of modern evolutionary theory. . . . We had long discussions about whether insights from evolutionary theory might break the impasse that traditional explanations for the fossil record had placed before our practical hopes—for why would one enter a field where intrinsic limitations upon evidence had wiped out nearly all traces of the phenomenon one wished to study.

In Gould's recollection, he and Eldredge shared a frustration with the descriptive, untheoretical orientation of paleontology at the time, and with mentors like Imbrie, Newell, and (less directly) Simpson, they were well positioned to challenge basic assumptions about the fossil record. The essential elements of Gould's retelling of the history of his partnership with Eldredge were, then, that two young paleontologists, raised on the same diet of quantitative invertebrate paleontology and evolutionary theory, nurtured a mutual iconoclastic desire to find a new model for presenting the evolutionary significance of the fossil record. But in the connection Gould makes to the actual formulation of punctuated equilibria, his account becomes vague:

> Eventually we (primarily Niles) recognized that the standard theory of speciation—Mayr's allopatric or peripatric scheme (1954, 1963)—would not, in fact, yield insensibly graded fossil sequences when extrapolated into geologic time, but would produce just what we see: geologically unresolvable appearance followed by stasis (Gould 1989a, 118).

In this final sentence the parenthetical modification "(primarily Niles)" keeps Gould's account from being actually dishonest, but the word "eventually" glosses over most of the actual circumstances of the "discovery," and leaves any historical details to the reader's imagination. Eldredge's own account is, in contrast, much more direct:

> I had given the manuscript of the *Evolution* paper to Steve Gould to read and criticize before I dared submit it. A few months later, when the paper was in press, Steve asked me to join him in contributing a paper on speciation to a book project *Models in Paleobiology*, organized by Thomas J. M. Schopf. Steve had wanted some of the other topics—but they had already been assigned to other authors. Steve said that he could not think of anything else to say about speciation that I had not already said in my paper (Eldredge 2008, 113).

Are these two accounts substantially different? Does one get closer to the historical "truth" of the matter? I would argue that what these two recollections, composed long after the events they describe, reveal the concerns each respective author had with respect to constructing a historical memory of an event that had tremendous impact on the subsequent development of each man's career. To begin with, the rhetoric in these and other autobiographical accounts points to an unequal power dynamic between the two authors that existed from the very beginning of the collaboration. By beginning his account of the collaboration well before the actual circumstances of the paper itself, Gould establishes his role as essential and formative, though vaguely specified. He also emphasizes the connection that the particular details of the theory of punctuated equilibria—an application of allopatric speciation to the fossil record—had to the larger project of revitalizing paleontological study of evolution. This allows Gould to frame the 1972 paper in the context of his own career arc—which by 1989, when the account quoted was written, had broadened to encompass a wide array of macroevolutionary questions—rather than to focus on the paper itself as the origin or genesis of all or much of his later professional success. In other words, Gould's account makes the 1972 paper seem like the inevitable outcome of his own (and Eldredge's) long-established plan, not some fortuitous and chance event. In framing the story this way, Gould almost sounds magnanimous in crediting Eldredge as much as he does. While Eldredge has had an undeniably successful career, in any measurable way there is little question which of the two authors went on to greater fame, notoriety, and professional glory. From the outset Gould adopted the persona of senior author in this collaboration, and by reminding his readers later in his career of the importance of Eldredge's "contribution," he was simply doing what a generous senior author ought to do.

Eldredge's account (presented in a 2008 essay, "The Early 'Evolution' of 'Punctuated Equilibira') is quite different. In the first place, Eldredge accepts, in a blunt though nonconfrontational way, primary credit for the original idea. Though he has generally referred (as Gould did also) to punctuated equilibria as "our" theory, Eldredge's retelling brings Gould into the story at a much later point: Gould's participation effectively begins with Schopf's invitation. Nonetheless, Eldredge makes subtle gestures of acquiescence to Gould's implicit status as senior author: it is Gould's wise council Eldredge seeks before "daring" to submit the *Evolution* paper, and he credits Gould with a formative role in encouraging him to pursue theoretical, evolutionary topics. However, despite these gestures and a consistent avoidance of even the hint of bitterness regarding later apportioning of credit, Eldredge provides equally subtle indications about his perspective on the relative contributions of the two authors. In his 2008 essay he credits Gould primarily with providing the famous name for the theory: "Steve had a knack for names . . . and he thought a good name for my idea of stasis-plus-rapid evolution in geographic isolation would be 'punctuated equilibria'" (Eldredge 2008, 113). A careful reader will not fail to notice that while the result may have been "our paper," its basis was "my idea."

The Allopatric Model and Speciation

Gould's invitation to Eldredge came after Eldredge had already sent his manuscript on allopatric speciation to *Evolution* (it was received on

10 June 1970 and published the following year), so Eldredge's "Allopatric Model" paper is sometimes unofficially credited as being the first public presentation of punctuated equilibria. Indeed, Eldredge's 1971 paper presented much of the central conceptual vocabulary of punctuated equilibria (though not the name itself). Broadly speaking, the paper is a response to what I have called Darwin's dilemma: the notion that paleontology is prevented from making meaningful claims about evolution because of the imperfection of the fossil record. The opening lines of the paper asserted:

> Time is the one element of paleontological data which mitigates, to a degree, the disadvantages inherent in the fossilization process. Addition of this fourth dimension to evolutionary biology has greatly sharpened ou[r] perspective of both the rates and modes of evolutionary processes. Paleontologists have understandably emphasized the importance of time in the elaboration of evolutionary models, and have made particularly important contributions in the general area of the origin of higher taxa (Eldredge 1971, 156).

This opening passage made several key points: First, paleontological data has a unique empirical dimension, which is the depiction of the record of life over *time* (a distinction no doubt inspired by Eldredge's reading of Simpson). Second, this temporal dimension offers clarification of an important aspect of the evolutionary process, namely the rates and modes by which evolution takes place. Third, paleontology illuminates a distinctive element of evolutionary theory, the emergence of higher taxa. By implication, these are contributions not achieved by any other branch of evolutionary biology. Thus, Eldredge has invoked, from the start, the central arguments of the emerging paleobiological movement.

But the paper was much more than simply a reiteration of an agenda for paleontology. Eldredge quickly revealed that he was attacking a target from within paleontology: the "dominantly *phyletic* model of transformation," which "has underlain most paleontological discussions of the origin of new taxa, including species," and assumes that all evolution proceeds within lineages through "the aggregation of large numbers of small steps of morphological change" (Eldredge 1971, 156). In other words, despite being accepted for more than a century as a central doctrine of evolutionary paleontology, the "gradualistic view" was "clearly at odds with currently accepted views of speciation derived from studies of the recent biota." Such a radical claim from a young paleontologist to might strike readers as arrogant, but Eldredge had a clever strategy: over the first page of the paper he played two competing insecurities of paleontologists against one another to neutralize potential objections. Paleontologists (as this book has shown) had long chafed under the assumption, widely held in the community of evolutionary biologists, that their work was mere "stamp collecting"; Eldredge's initial flourishes about the unique role paleontology had in evolutionary study would have struck a sympathetic chord in many paleontologists. At the same time, paleontologists had internalized this negative assessment to the degree that an acculturated inferiority complex existed in which the techniques, models, and conclusions offered by biologists were often assumed by paleontologists to be innately superior to their own. Eldredge addressed both concerns by attacking the traditional paleontological assumptions of phyletic gradualism on the ground that it was incompatible with current *biological* understanding of speciation. His implicit argument was that for paleontology to achieve its aim of being taken seriously as an evolutionary discipline, it would have to adapt to accommodate up-to-date biological theory.

One suspects, then, that Eldredge chose his words carefully and counted on a certain degree of embarrassment among paleontologists when he informed his readers that "of the various models of speciation proposed and discussed over the past forty years, the allopatric model has gained nearly total acceptance among current evolutionary biologists" (Eldredge 1971, 156). He was not proposing to replace the hallowed assumption of gradual phyletic evolution with just *any* random model, in other words, but with the dominant biological theory. This was not only a rhetorical strategy, however. Eldredge was well aware that the allopatric model offered the best explanation for what he had observed in his Devonian trilobites and—following Gould's injunction to seize theoretical opportunities while still young—he saw the chance to use his humble dissertation to make a claim that would shake up all of paleontology: "I would suggest that the allopatric model (geographic speciation) be substituted in the minds of paleontologists for phyletic transformism as the dominant mechanism of the origin of new species in the fossil record, and that the allopatric model, rather than gradual morphological divergence, is the more correct view of the processes underlying cases of splitting already documented by numerous workers" (Eldredge 1971, 156–57).

The rest of the paper was fairly straightforward: in just under ten additional pages, Eldredge summarized the results of his *P. rana* study, which he used to test four plausible models "which may be in-voked to explain sequential occurrences of non-intergrading ancestral-descendant species" (Eldredge 1971, 158). The first model—saltation (A in fig. 5.1)—proposes that discontinuity between closely related species in adjacent stratigraphic horizons is produced by sudden speciation events via instantaneous "saltations." Eldredge rejected this model on the basis of its universal disfavor among biologists and paleontologists, but noted that it had an advantage in that it assumed "that little or no morphologi-cal change occurs within a species within its stratigraphic interval of oc-currence" (Eldredge 1971, 159). If the saltational model represented the most unorthodox view of speciation, the second model Eldredge con-sidered—"the standard phyletic model" (B in fig. 5.1)—was the most or-thodox paleontological assumption, that "linear selection pressures . . . produce a gradual (often statistical) change in a character or suite of characters" (Eldredge 1971, 160). The problem with this standard model, however, is that when lineages are present in the stratigraphic record, they are expected to display gradual, linear change in character traits—which Eldredge argued was not evident in his trilobites. Nor, he claimed, did *most* documented lineages in the fossil record display this trend: "The usual case, at least in Paleozoic epeiric [intracontinental] sedi-ments, is for the observer to document no change throughout the strati-graphic range of species." The central problem with the phyletic model, then, was not that it could not explain breaks in the stratigraphic record, but that it did not account for the "morphological stability"—what would be labeled "stasis" in the first paper on punctuated equilibria—that was the rule, not the exception, in paleontological data.

The second major problem with the phyletic model Eldredge identi-fied was that the persistence of an ancestral species alongside its descen-dent in the same strata was inadequately explained. Whether suddenly or gradually, both the saltational and phyletic models assume that spe-ciation involves the transformation and disappearance of the ancestral species as the descendent emerges. The only way to account for ances-tor-descendant coexistence is to consider speciation as the splitting of a lineage. This is accomplished when a small population of the ancestral species evolves distinctive morphological features in geographical isola-tion from the parent, and then at some later point migrates back to the range of the ancestral species. From the perspective of the fossil record, it would appear that a new species had suddenly appeared. The final two models Eldredge examined—the "gradual morphological divergence

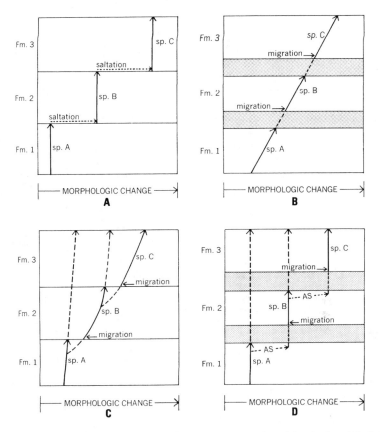

FIGURE 5.1. Eldredge's illustration of four models of speciation: (A) saltation, (B) phyletic, (C) gradual morphological divergence, (D) allopatric. Niles Eldredge, "The Allopatric Model and Phylogeny in Paleozoic Invertebrates," *Evolution* 25 (1971): 159. © Society for the Study of Evolution.

model" and the "allopatric model" (C and D in fig. 5.1)—are both models of splitting, and both assume that some degree of geographic isolation is a necessary component of the process. But Eldredge argued that what separates the two models is the inference about *"how* the splitting is actually effected" (Eldredge 1971, 161). Since the morphological divergence model assumes that splitting is accomplished by gradual morphological change, "lineage-splitting often becomes reduced to a special case of the phyletic model." Model C in fig. 5.1 is an adequate explanation of non-intergradation of ancestor and descendent species, but Eldredge felt its emphasis on the *gradual* nature of divergence "remains at odds with the general picture of morphological stability of Paleozoic invertebrates of the epeiric seas." As Eldredge concluded, the only remaining plausible account is the allopatric model, which explains stratigraphic breaks, nonintergradation, *and* the relative suddenness of the emergence of descendant species.

The final point to make about the 1971 paper is that in addition to proposing an allopatric model of species-lineage splitting as a general model for explaining the observed fossil record, it also made the equally important claim about the stability (or stasis) of lineages. There are really two features that distinguish Eldredge's models C and D, and both concern the relationship of horizontal (morphological) and vertical (time) axes. The first is what happens during periods of splitting: in model C horizontal and vertical axes change roughly equally during speciation (dotted lines), indicating that divergence is gradual. In model D, during splitting the horizontal axis moves considerably while the vertical moves hardly at all, indicating a period of very rapid divergence. The second feature is the depiction of the relationship of these axes between periods of splitting (bold lines). In model C, progression along the vertical axis is accompanied by some movement on the horizontal axis, indicating that a degree of gradual morphological change (phyletic evolution) takes place within lineages between speciation events. In model D, however, vertical movement is not accompanied by *any* horizontal change. The difference between the two models could not be more striking: in the first, change is always taking place, albeit slowly. In the second, change is very rare, and when it does happen it is rapid and significant. There is no reason why a model of splitting that invokes allopatric speciation (which, as proposed by Mayr, does indeed involve fairly rapid change) need also reject gradual phyletic divergence under normal circumstances. But this is where Eldredge's *Phacops* data played the crucial role: as he argued in the remainder of the paper, morphological stability is the general rule, likely because of the influence of stabilizing selection (Eldredge 1971, 166).

The Ontogeny of Punctuated Equilibria

From Eldredge to Eldredge and Gould

If Eldredge's 1971 paper contained virtually all of the important elements of what would come to be known as "punctuated equilibria," then why isn't it, rather than the 1972 Eldredge and Gould contribution to the Models proceedings, celebrated as the landmark publication? Why is the theory usually credited to "Eldredge and Gould" (though often to "Gould and Eldredge," and sometimes just to "Gould"), rather than to Eldredge alone? The simplest explanation is that, in Eldredge's own words, "that first paper sank without a trace" (Eldredge interview). The obvious question this begs, though, is *why* it failed to make an impact. Eldredge had published a paper containing all of the central elements of what would eventually come to be known as "punctuated equilibria," and in the major journal of evolutionary biology, *Evolution*, no less. Could it be, after all, that what made the greatest difference was simply Gould's "knack for names"?

There are a few additional factors to consider. First, there is authorship. Relatively few people in the wider community of evolutionary biology would have heard of Eldredge, then a young assistant curator with few publications; perhaps many readers of the journal had simply skipped that paper. Similarly, the title of the paper might have discouraged careful readership: if the allopatric theory had indeed been established for decades among biologists as the dominant explanation of speciation, a paper entitled "The Allopatric Model and Phylogeny" likely attracted little curiosity from biologists—especially when it promised to discuss Paleozoic invertebrates. The venue for the paper itself may have also worked against Eldredge. While a theoretical paper had an easier path to publication in *Evolution* than in a traditional paleontological outlet like *Journal of Paleontology*, many fewer paleontologists read *Evolution* than *JP*. Thus, those for whom the message of the paper was primarily intended—paleontologists—were largely bypassed. While no data is available on the demographics of *Evolution*'s subscribers in the early 1970s, a report commissioned by the Paleontological Society in 1974 as part of the planning for the PS's new journal *Paleobiology* estimated that only 13 to 14 papers on paleontology had been published in *Evolution* in the preceding year (Thomas J. M. Schopf, "Draft Report on the Feasibility of a New Paleontological Society Journal" [1974]. PS-*P* 14). If that figure is representative for the early 1970s, and if *Evolution*'s paleontological readership was proportionate to its contributors (not necessarily a given), then paleontologists may have comprised only a small percentage of *Evolution*'s readers.

Finally, there is one explanation that is almost blindingly obvious: Eldredge's paper appeared in the March, 1971 issue of *Evolution*. The

Models symposium took place at the GSA meeting in November of 1971, and the book *Models in Paleobiology*, in which the Eldredge and Gould paper appeared, was published in 1972. In other words, there was very little time for anyone to take notice of Eldredge's paper before becoming aware of its successor. We can never know what might have happened if Gould had not been invited to contribute to the symposium, or if he had been offered one of his preferred topics, or if he had decided not to enlist Eldredge as a coauthor. It is very likely, though, that Gould's career would have turned out somewhat differently. But what about Eldredge's? Eldredge himself has been very candid in acknowledging the importance of his collaboration on punctuated equilibria with Gould and the role that collaboration had in the subsequent success of his own career: Were it not for this partnership, "I think I would have lived a life of, probably, frustrated obscurity and it would have been very, very different. And I'm so happy it didn't go that way. Steve was a very important part of that—he helped make it happen" (Eldredge interview).

While counterfactual scenarios cannot be tested, we can directly examine the collaboration between Eldredge and Gould to more clearly reconstruct the process by which punctuated equilibria came to be. After agreeing to the coauthorship arrangement some time in the late summer or fall of 1970, Eldredge and Gould agreed on a division of labor for the production of the symposium paper and the published manuscript. Because Eldredge was anxious about giving public lectures, Gould was tasked with giving the talk at the GSA. Eldredge was assigned the task of writing the initial draft of the manuscript, which he sent to Gould for revision and expansion, and which served as the basis for Gould's presentation. From the outset it was agreed that Eldredge would be first author on the published paper. Because of this division of labor, and because the drafts and correspondence surrounding the original paper survive, the relative contributions of the two authors can, as I will show, be reliably determined.

In early December of 1970, Eldredge mailed his completed draft to Gould with the following instruction: "Change by deletion, expansion, or whatever. Change sentence structure, etc.; do your worst, I won't take it amiss. Hopefully though, the basic contents will prove acceptable as outlined and you won't have to rewrite the entire thing" (Eldredge to Gould, 3 December 1970: Eldredge papers). The manuscript itself was only 20 pages long and had the unassuming title "The Process of Speciation and Interpretation of the Fossil Record." It began by posing a question: "Are observed patterns of geographic and stratigraphic distribution, and apparent rates and directions of morphological change consonant with the retrodicted consequences (expected patterns) of a particular theory of speciation?" It argued that since "no mechanistic theory can be generated directly from paleontological data," paleontologists must turn to biology for theories of speciation (Niles Eldredge, "The Process of Speciation and Interpretation of the Fossil Record" [Punctuated Equilibria draft 1], 1. *EP*). Eldredge maintained that modern biology presents one overwhelmingly clear choice—the allopatric theory—which over the past 30 years "has gained ascendancy to the point where the vast majority of biologists view it as *the* theory of speciation" (Eldredge, draft 1, 2). The allopatric model—like the sympatric theory, which was also mentioned—assumes that speciation occurs by splitting lineages; "For most biologists," Eldredge explained, "speciation is synonymous with splitting."

The draft therefore began by presenting the interpretation of the fossil record as the central problem, and immediately proposed an apparently obvious and reasonable solution: adoption by paleontologists of the allopatric model of speciation. Nonetheless, Eldredge explained, most paleontologists did not favor this alternative, and instead supported "phyletic gradualism, where new species are thought to originate merely by the inexorable passage of time with a concomitantly constant, unidirectional change in morphology" (Eldredge, draft 1, 2). Thus far, Eldredge's introduction was, in content and even in wording, very similar to his 1971 *Evolution* paper. The next section of the draft presented a fairly detailed discussion of the consequences of adopting the allopatric theory as a model for interpreting the fossil record. This was a departure from the 1971 paper, which had immediately moved to a consideration of empirical evidence without dwelling in theoretical considerations. Eldredge's new draft contended, in the first place, that under the allopatric model new species wouldn't be expected to originate at the same times and places as their ancestral populations, and "it is therefore theoretically impossible to trace the gradual splitting of a lineage merely by following a certain species up through the local rock column" (Eldredge, draft 1, 2–3). Secondly, speciation would appear very suddenly in the fossil record—"close after, if not actually prior to, the onset of genetic isolation"—which reflects the allopatric theory's expectation that morphological differentiation takes place "early in the differentiation of that species, when the population is small and still adjusting more precisely to local edaphic conditions." Consequently, Eldredge explained, "we should not expect to find gradual morphological divergence between two species which are hypothesized to have an ancestral-descendant relationship" (Eldredge, draft 1, 3). Finally, he observed that the allopatric theory predicted that morphological stability would be the rule, since "most evolutionary events . . . are generally thought to occur in a short period of time," and "there will be little evolutionary change of these inter-specific differentia except at those times where there is good reason to believe that the two species became sympatric for the first time" (Eldredge, draft 1, 3).

The expected ("retrodicted") consequences of the allopatric theory, then, were (1) geographical/stratigraphic discontinuity between ancestor and descendant, (2) sudden appearance of new species, and (3) relative stability between speciation events. Or, as Eldredge puts it,

> Tracing the fossil species through any given local rock column, so long as no drastic changes in the physical environment are evident, should produce no pattern of constant change, but rather one of oscillation in mean values of morphological variables. New, closely related and perhaps descendant species observed to enter into the rock column should appear suddenly, and show no intergradation with the 'ancestral' species as far as the inter-specific differentia are concerned. There should be no gradual increase of morphological differentiation between the two species. . . . Quite the contrary—it is possible that the two species would show the greatest amount of differences at the time when the descendant species first appears. Finally, in exceptional circumstances, it may be possible to identify that general area of the known geographic distribution of the ancestral species where the new species most likely arose (Eldredge, draft 1, 4).

These consequences are quite different than those of phyletic gradualism, owing primarily to differences in the nature of the mechanisms proposed to produce speciation. By and large, Eldredge argued, the expectation under phyletic gradualism is a record of "gradual, uni-directional change." The mechanism for such change is "orthoselection," or "a constant adjustment to a uni-directional change in one or more features of the physical environment" (Eldredge, draft 1, 5).

According to Eldredge, both models offer predictions that can be tested against the fossil record. He noted, however, that there are deeper methodological and philosophical factors that have conditioned paleontology's preference for phyletic gradualism "from the earliest days of the acceptance of evolution as a serious scientific concept." Gradualism, Eldredge explained, first emerged as an alternative to "catastrophism," and later cemented its place in evolutionary thought with the defeat of the "mutationist" school of Hugo de Vries and others, which saw the establishment of natural selection as the guiding element in evolutionary theory. This had a particular impact on paleontology, since "one effect of the incorporation of the notion of gradualism into evolutionary theory was the eventual abandonment, by most paleontologists, of the notion of saltative evolution" (Eldredge, draft 1, 5–6). In effect, Eldredge argued that the entrenchment of phyletic gradualism in paleontology was both a consequence of the triumph of the selectionist perspective in evolutionary theory, and also a kind of atonement for paleontologists' lengthy flirtation with non-Darwinian theories mechanisms like saltation.

In Eldredge's view, the irony was that paleontologists' slavish devotion to phyletic gradualism had in fact kept them out of step with the modern biological understanding of speciation:

> Extrapolation of gradual morphological change under a selection regime to a complete model for the origin of new species fails to recognize the fact that speciation is primarily an ecological (including geographic) process. Natural selection in the allopatric theory is involved with adaptation to local edaphic conditions and to the elaboration of isolating mechanisms. Genetic isolation is at least as important as natural selection in allopatric speciation. Phyletic gradualism is in itself an insufficient model to explain the origin of diversity in the present, or any past, biota (Eldredge, draft 1, 6).

Eldredge's argument harkened back to Simpson: paleontological theory should reflect the understanding of modern biology, and in particular paleontologists should not use the limitations of the fossil record as an excuse for ignoring genetics. While the genetic history of a fossil is obviously unknowable, the assumptions of population genetics—particularly the role of isolation in speciation—can be brought to bear in interpretations of the fossil record. Modern genetics had advanced beyond a simple Darwinian model of orthoselection, and so too should paleontology.

In the remainder of the manuscript, Eldredge presented his case for the allopatric model based on the empirical evidence of the fossil record. He spent roughly three pages discussing classic "examples" of phyletic gradualism in paleontological literature, including a lengthy paragraph

on gradualistic interpretations of the hominid lineage. From this relatively brief survey he reached a tentative—and somewhat equivocal—
conclusion: If, on the one hand, phyletic gradualism is "poorly documented" and "largely based on a preconceived idea," on the other, "the
alternative picture of stasis punctuated by episodic events of allopatric speciation rests on a few general statements in the literature and a
wealth of largely informal data." And here we see the term "punctuated
equilibria" first used, though almost with a note of apology:

> The idea of punctuated equilibria is just as much a preconceived 'picture,' or
> 'model,' as that of phyletic gradualism. We readily admit our bias toward the
> former picture, and in the ensuing discussion of specific examples it must be
> remembered that our interpretations may be as equally colored by our own
> preconceptions as have been the interpretations of the champions of phyletic
> gradualism (Eldredge, draft 1, 10).

Why did Eldredge use such equivocal language in staking the paper's
major claim? In the first place, he and Gould were committed to making
the point that models are never neutral: *any* model rests implicitly on assumptions that cannot, ultimately, be acquired solely from the data. This
was an appropriate discussion for a contribution to a symposium entitled Models in Paleobiology, and it would receive much greater elaboration (as we will see) in Gould's reworking of the second draft. Secondly,
I believe this tactic was part of a rhetorical strategy to emphasize the
originality of the paper's thesis. One of the charges that has been consistently leveled against Eldredge and Gould is that punctuated equilibria
is merely a rehash of old ideas. Indeed, the paper itself explicitly states
that its central "theory" is simply the application of the dominant biological model of speciation to the fossil record, and Eldredge and Gould
have acknowledged many times over the years the debt they owed to
Mayr, Simpson, and others. But clearly the authors saw their paper as
more than just a remedial course in evolutionary biology for paleontologists. If punctuated equilibria was original and important at all—and I
argue that it was both—it was because it claimed not just that the modern biological model of speciation can be observed in the fossil record,
but also that the fossil record illuminates a genuinely novel feature of
evolution: the persistence of lineages in stasis for millions of years. It simultaneously struck at paleontologists' insecurities about their data, *and*
advanced a claim for paleontology's relevance to evolutionary theory.

To emphasize the novelty of their claim, Eldredge and Gould needed to do more than just reinterpret existing paleontological literature. They also wanted to advance an empirical claim—for allopatric speciation plus stasis—based on original research. Eldredge therefore summarized his own work on *Phacops rana*, as well as Gould's study of the cladogenesis of the snail *Poecilozonites bermudensis*, both of which examples, he argued, exhibited the characteristic pattern of rapid, discrete evolutionary bursts followed by lengthy periods of stasis expected under their new theory. Significantly, Eldredge described these patterns as "a more literal interpretation of the fossil record" (Eldredge, draft 1, 14). This added an important additional dimension to the presentation of punctuated equilibria: that as a more "literal interpretation" it takes paleontological data "more seriously" or "at face value." Again, this remark went to the heart of paleontologists' insecurities about the inadequacy of the fossil record.

Having established the data supporting the allopatric model, Eldredge brought the draft rapidly to a close. The paper concluded with a paragraph that is worth quoting in full:

> We thus propose that the allopatric model can be effective in explaining the results of the evolutionary process as seen in the fossil record. This suggestion is by known [sic] means original, yet it may still be fairly stated that the picture of phyletic gradualism, buttressed by the concept that phylogeny can be read directly from the rocks, is the predominant theme underlying most paleontologists' expectations upon initiating an investigation of a suite of closely related fossil species. Our alternative picture is merely the application, in gross outline, of the dominant theory of speciation in modern evolutionary thought to the fossil record. We firmly believe that the retrodicted consequences of this model are more nearly in accord with the fossil record of the vast majority of the fossil Metazoa (Eldredge, draft 1, 17).

Again, the language describing the originality of the paper's claims was modest, emphasizing that the conclusions were "by [no] means original" and "merely" an application of a well-known biological theory to paleontology. However, this modesty was deceptive; the more radical claim was the implied consequence that phyletic gradualism is *wrong*, and that stasis is common. But here the equivocal language has been stripped away: in contrast to the statement several pages earlier, Eldredge and Gould now "firmly believe" that the model is "more nearly in accord with the fossil record" of the Metazoa. The careful phrasing is understandable.

As Eldredge recalled many years later, "We were simply trying to avoid the not-unreasonable relegation to the lunatic fringe that some paleontologists in the past had suffered when they too saw something out of kilter between contemporary evolutionary theory, on the one hand, and patterns of change in the fossil record on the other" (Eldredge 1985a, 93).

Perhaps for this reason, Eldredge's initial draft included an appendix labeled "Some Comments on Trends." In Eldredge's cover letter to Gould he had written, "I hope you find some merit in my discussion of trends appended to the ms. Do with it as you see fit" (Eldredge to Gould, 3 December 1970: Eldredge pap.). It is not immediately obvious why he did not include this text as part of the body of the paper, although the most straightforward explanation is that he was not sure where to place the comments. However, in retrospect, the section on trends became one of the most important—and controversial—elements of the final paper, and it is possible that Eldredge was uneasy about sticking his neck out further than he already had. In either case, Gould later recalled being immediately struck by the potential importance of this section: "I distinctly remember my feeling that our section on trends (written by Niles) . . . was the most exciting and potentially significant in the paper—but I could not, for the life of me, quite figure out why" (Gould 1989a, 121).

The question Eldredge's appendix addressed, in short, was whether apparent trends in the fossil record are genuine. Eldredge introduced "the very real possibility that many, if not most, of the trends involving higher taxa may be based on a selective rendering of elements of the fossil record which are chosen because they seem to form a morphologically graded series coincident with a progressive biostratigraphic distribution" (Eldredge, draft 1, 18). In other words, he all but alleged that paleontologists had been manufacturing patterns in the fossil record based on cherry-picked samples that told the story that they expected to hear— e.g., that "orthoselection" or phyletic gradualism was the rule. However, Eldredge wasn't ready to reject trends entirely; he noted that there were simply too many apparent examples in the literature, and to dismiss them all as "unwarranted extrapolations based on a preconception" was "altogether too facile." What he did ask was whether, if some trends were genuine, they could be reconciled with punctuated equilibria.

This was potentially a major problem for the theory: Eldredge noted that trends had been defined by one contemporary author as "a direction which involves the majority of related lineages of a group" (Eldredge, draft 1, 18). This definition implied that some factor operates consistently on a very large number of organisms, over a potentially wide geographic area, and across a long period of time. This is why it was easy to explain trends as a consequence of phyletic gradualism: if the dominant mechanism for producing morphological innovations was orthoselection, then it was a reasonable inference that the same general forces would determine "directions" in speciation over the history of a lineage. But, as Eldredge put it, "We are left with a bit of a paradox, since to picture speciation as an allopatric phenomenon involving rapid differentiation within a general, long-term picture of stasis, is to deny the model of directed gradualism in operation" (Eldredge, draft 1, 19). To put it another way, a model of evolution in which speciation occurs as a response to local environmental conditions affecting small, peripherally isolated populations seemed to deny the very hope of generalizing about evolution at all. If evolution always, or nearly always, occurs in small, isolated groups, then—drawing an explicit analogy with Sewall Wright's interpretation of the role of mutations in a population—"adaptation to local edaphic conditions by peripheral isolates is stochastic with respect to long-term, net directional change (i.e. trends) within a higher taxon as a whole" (Eldredge, draft 1, 19).

Though it was expressed tentatively in this draft, the suggestion that "speciation is stochastic with respect to the origin of higher taxa" would feature prominently in the further development of paleobiological theory. At the time, Eldredge despaired that the problem was "incapable of solution" because of paleontologists' inability "to extrapolate mechanisms known to be operative in short-term processes and to apply them to events of very long duration in the fossil record." But in fact, this observation marked the beginnings of hierarchical macroevolutionary theory in paleobiology, or the attempt to explain how different levels of evolutionary hierarchy interact with one another. This line of inquiry would be boosted a few years later when Steven Stanley proposed a model of "species selection," and it flourished through the next decade in discussions of "exaptation" and "taxon sorting" in which Gould and Eldredge played prominent roles (Stanley 1975; Gould and Vrba 1982; Vrba and Eldredge 1984; Vrba and Gould 1986).

Eldredge's speculation that "we might envision multiple 'explorations,' or 'experimentations,' which are invasions, on a stochastic basis, of new environments by peripheral isolates" also seemed to offer a direct link to the thinking of theoretical ecologists like MacArthur, whose studies of the dynamics of island biogeography had begun to influence some younger paleontologists. However, Eldredge denies having been influenced by this ecological, biogeographical tradition when writing the 1972 paper:

> I became aware of the ecological work, of course, before I/we published. But I am pretty sure I am right that my/our thinking reflected the pretty much 100% dichotomy between evolution and ecology (still a problem, overt attempt to address it to the contrary notwithstanding). Paleoecology was of course all the rage—but nothing in either my *Phacops* or Steve's snail writings talk about other faunal components, communities, etc etc. In one fundamental sense, PE was extremely conventional—just talking about the components of a single lineage through time (and we added in space—following alloptaric speciation theory . . .) (Niles Eldredge, personal communication, 5 February 2008).

Nonetheless, others would very soon make these connections explicit, as we will see in the next few chapters. And finally, while Eldredge's intention was to preserve the validity of at least some trends, his discussion of trends did encourage some paleobiologists to question whether the very appearance of direction in evolution was an illusion. In particular, his speculation that while "the overall effect may be one of net, apparently directional change . . . the initial variations are stochastic with respect to this change" became, in the hands of a group of researchers that included Gould, Raup, and Schopf, the inspiration to analyze the broader role of stochastic events such as speciation and mass extinction in patterns in the history of life (Eldredge, draft 1, 20).

Gould's Reimagining of the Paper

The bottom line is that, despite its brevity, the first draft of the 1972 paper contained most of the central elements of the theory. It was also unquestionably authored by Eldredge alone—even the section on Gould's snails was written by Eldredge, who expressed to Gould the hope "that you will find my summary discussion of *Poecilozonites* acceptable" (Eldredge to Gould, 3 December 1970: Eldredge pap.). However, the final published version of the paper was, in total number of words, more than twice as long as Eldredge's original draft. What accounted for this difference?

The answer, of course, is Gould. At first glance, the second draft of the paper—reflecting Gould's extensive revision—is almost unrecognizably different from the first. The manuscript was now divided into five major sections, and each section labeled with a Roman numeral, a title, and in some cases an epigraph. Sections contain "a" and "b" subheadings, numbered lists abound, and explanatory footnotes are on almost every page. The bibliography was also greatly expanded: whereas draft 1 had listed only 15 references, draft 2 contained 74—a fivefold increase. In length, the manuscript went from 20 pages (exclusive of references) to 32; however, because the line spacing in draft 2 was much narrower than in draft 1, the new version was actually more than double the length of its predecessor. Finally, Gould gave the paper a new title: Eldredge's prosaic "The Process of Speciation and Interpretation of the Fossil Record" was replaced by the attention-grabbing "Speciation and Punctuated Equilibria: An Alternative to Phyletic Gradualism."

The deeper question, though, is whether Gould's modifications to the manuscript were more than superficial. That is to say, is draft 2 actually a different paper? Gould's revisions certainly established his authorial voice and stamp: anyone familiar with Gould's writing would instantly recognize the characteristic long winding sentences, the baroque rhetorical flourishes, the self-conscious displays of erudition, and even the untranslated foreign phrases, all of which made him famous (and infamous) in his later career. At the same time, however, the core of the paper was still the portion authored by Eldredge. A line-by-line comparison shows that much of draft 1 survives in the new version—in many cases verbatim—beginning in the middle of the manuscript and continuing nearly to the end. Eldredge's contribution, then, had been not so much rewritten as enfolded by Gould's prose.

This is not to say that Gould's contribution was unimportant. On the contrary, the paper was now in many ways clearer, more elegant, less equivocal, and more confident. Implications and arguments that had only been hinted at in the first draft were expanded, supported with references, and carried through to logical conclusion. Above all, the opposition between "phyletic gradualism" and "punctuated equilibria" now dominated the text: whereas the phrase "punctuated equilibria" had appeared only twice in the first draft—once on page 10, the second time in the appendix on trends—it was now found on nearly every page of the manuscript. If draft 1 had been a modest argument for consideration of the importance of allopatric speciation, draft 2 was the triumphant announcement of an important new theory.

The new manuscript began with a four-point "Statement" that summarized the major arguments of the paper. First, Gould argued, since "expectations of theory color perception" of data, "new pictures" are required "before facts can be seen in different perspective." Second, paleontology had been dominated by the picture of phyletic gradualism, which saw "unbroken fossil series" in "insensible gradation as the only complete mirror of Darwinian processes." Third, allopatric speciation—here acknowledged as "a picture developed elsewhere," though this phrase was omitted in the published version—"suggests a different interpretation of paleontological data" in which "the great expectation of insensibly graded fossil sequences is a chimera" and "many breaks in the fossil record are real." And fourth, "the history of life is more adequately represented by a picture of 'punctuated equilibria' than by the notion of phyletic gradualism" (Gould and Eldredge, draft 2, 1: Eldredge papers). Here Gould added the memorable line, "The history of evolution is not one of stately unfolding, but a story of homeostatic equilibria, disturbed only 'rarely' (i.e. rather often in the fullness of time) by rapid and episodic events of speciation" (Gould and Eldredge, draft 2, 2).

Gould's list effectively served as a table of contents for the manuscript. Section 2, "The Cloven Hoofprint of Theory," was an extended meditation on the role that theory or "preconception" plays in the interpretation of data. Beginning with a short epigraph from Peter Medawar ("Innocent, unbiased observation is a myth"), Gould attacked the "inductivist credo" that "facts" may be "discovered objectively." Here he drew on contemporary philosophy of science for support, citing papers by Paul Feyerabend, N. R. Hanson, and Thomas Kuhn (not to mention Hegel, Spencer, Newton, and Darwin). The central point was that scientists are too often unaware that theory comes prior to observation, and that "all our perceptions and descriptions are made in light of theory" (Gould and Eldredge, draft 2, 4). In paleontology, Gould argued, this "inductivist credo" translated to the expectation that fossils could be interpreted purely or objectively without the taint of bias or preconception. Its effect had been the long-standing influence of "an inadequate picture of speciation," because paleontologists had been unable to recognize that their preconceptions biased their interpretation of the fossil record (Gould and Eldredge, draft 2, 5).

Section 3, "Phyletic Gradualism: Our Old and Present Picture," elaborated on the traditional assumption of phyletic gradualism in paleontology (it also began with an untranslated German epigraph from Karl Alfred von Zittel). Here Gould directly invoked Darwin's dilemma, noting that Darwin's concern about the imperfection of the fossil record "set a task for the new science of evolutionary paleontology: to demonstrate evolution, search the fossil record and extract the rare exemplars of Darwinian processes—insensibly graded fossil series" (Gould and Eldredge, draft 2, 6). Gould also alleged that Darwin "muddled" the distinction between gradual phyletic evolution and true speciation ("splitting"), with the effect that all evolution was understood to happen via slow, "gradual transformation of entire populations." As a result, paleontology had inherited a doubly reinforced expectation of phyletic gradualism, which Gould presented as four "tenets": (1) that "new species arise by the transformation of an ancestral population"; (2) that "transformation is even and slow"; (3) that transformation involves most or all of the ancestral population; and (4) that transformation occurs over most or all of the ancestral population's geographical range. These tenets, he argued, had two important consequences for paleontology: first, the expectation that the fossil record should depict a continuous sequence of "insensibly graded intermediate forms linking ancestor and descendant"; and second, that breaks in this succession "are due to imperfections in the geological record" (Gould and Eldredge, draft 2, 7). This latter assumption, Gould argued, not only potentially distorted evolutionary interpretation of the fossil record, but "It renders the picture of phyletic gradualism virtually unfalsifiable" by prescribing a response to any anomalous data— 'the result of an imperfect record'—that reinforced its central assumption (Gould and Eldredge, draft 2, 8).

Gould also discussed some subtle but important (and potentially damaging) methodological consequences of the assumption of phyletic gradualism. In the first place, the language of paleontology was colored by the assumptions of that model: as Gould explained, "We are compelled to talk of 'morphological breaks' in order to be understood. But the term is not a neutral descriptor; it presupposes the truth of phyletic gradualism, for a 'break' is an interruption of something continuous." Secondly, the gradualistic model proscribed the objects of paleontological study, since apparently broken sequences would be viewed as "*ipso facto*, poor objects for evolutionary investigation" (Gould and Eldredge, draft 2, 9). Finally, Gould drew attention to the role of pedagogy in maintaining the ascendance of gradualism, and particularly the role of textbooks in promoting orthodoxy. Here he noted that the classic paleontology textbooks, such as von Zittel's *Text-book of Paleontology* and Henry

Woods' *Elementary Paleontology: Invertebrate*, all "present an orthodox version of phyletic gradualism." Gould found it more surprising that
newer texts, such as Moore, Lalicker, and Fischer's *Invertebrate Fossils*
(1952) and William Easton's *Invertebrate Paleontology* (1960) fared no
better. As Gould put it, "That these older texts hold so strongly to phyletic gradualism should surprise no one; harder to understand is the fact
that virtually all modern texts repeat the same arguments even though
their warrant had disappeared, as we shall now show, with the advent of
the allopatric theory of speciation" (Gould and Eldredge, draft 2, 10).

In all, the first three sections—comprising twelve manuscript pages—
were the bulk of Gould's original contribution to the draft. While some
of the content (on theoretical preconceptions) was hinted at in Eldredge's draft, Gould clearly expanded these arguments far beyond their
original scope. In the central section of the paper, however, most of Eldredge's draft survived in only slightly modified form, beginning with
section 4, "The Biospecies and Punctuated Equilibria: A New Picture of
Speciation."[5] This included Eldredge's original discussion "Implications
of Allopatric Speciation for the Fossil Record," his consideration of
"problems of phyletic gradualism," and his lengthy summary of Gould's
research on *Poecilozontes* snails and his own study of *P. rana*. Gould's
only changes were to lightly revise Eldredge's prose for style and clarity.
Equivocal phrases such as "there is good reason to believe," "can be expected," and "is possible" were also removed or altered. "Relatively few
classic examples" of gradualism became "very few"; it was now "likely,"
not just "possible" that the greatest morphological difference between
ancestor and descendant species existed on first appearance; and the
statement that interpretations "may be" colored by preconceptions was
changed to say that they "are."

Gould's only significant authorial input in this section was a summary,
at the end of the subsection entitled "Implications of Allopatric Speciation," where he inserted his agenda for reforming paleontology. After re-
visiting the central tenets of the allopatric model, as expressed in a form
virtually unaltered from that of Eldredge's original draft, Gould extended his remarks to a broader proposal for rereading the fossil record:

5. In one of Eldredge's few editorial modifications to Gould's draft, he proposed altering this title to read "A *Different* Picture of Speciation" (emphasis added). While a relatively minor alteration, I interpret this change as an attempt to avoid the perception that
Gould and Eldredge were taking credit for the theory of allopatric speciation. This change
was preserved in the final, published paper.

Many breaks in the fossil record are real; they express the way in which evolution occurs, not the fragments of an imperfect record. The sharp break in a local column accurately records what happened in that area through time. Acceptance of this point would release us from a self-imposed status of inferiority among the evolutionary sciences. Our collective gut-reaction leads us to view almost any anomaly as an artifact imposed by our collective millstone—an imperfect fossil record. But just as we now tend to view the rarity of Precambrian fossils as a true reflection of life's history rather than a testimony to the ravages of metamorphism or the lacunae of Lipalian intervals, so also might we reassess the smaller breaks that permeate our Phanerozoic record. We suspect that this record is much better (or at least much richer in optimal cases) than tradition dictates (Gould and Eldredge, draft 2, 16).

This paragraph is an excellent short summary of the overall importance of the theory of punctuated equilibria to the paleobiological movement. A reassessment of paleontological evidence—or, as Gould put it in a later insertion, "a *more literal* reading of the fossil record"—was the key to paleontology's emergence from the shadow of Darwin's dilemma. This is why the 1972 paper would prove to be so important, and such a lightning rod for controversy—and also why Gould's authorship was vital. On its own, Eldredge's insight was an interesting comment on models of speciation; with Gould's reworking it became a call for a paleobiological revolution.

After the central section composed primarily from Eldredge's initial draft, Gould reasserted his authorial presence in the paper's final section, "Some Extrapolations to Macroevolution." Here he framed Eldredge's discussion of trends by discussing the standard interpretation of macroevolution under the modern synthesis, an attitude Gould characterized as "extrapolation from species-level processes" (Gould and Eldredge, draft 2, 26). This allowed him to present punctuated equilibria as not just a corrective for paleontology, but as a major new theory of macroevolution in which patterns of evolution of the higher taxa were not explained simply as extrapolated microevolutionary processes. In this section, Gould marshaled "two phenomena of macroevolution" to illustrate the superiority of punctuated equilibria as a heuristic for explaining trends (Gould and Eldredge, draft 2, 27).

The first example he presented was the problem of the presence of "'Classes' of great number and low diversity." As Gould explained, by the Ordovician a large number of distinct classes of echnoderms had evolved, each with "a distinct Bauplan," although in most instances consisting of very few genera. Gould maintained that under phyletic gradualism, two aspects of this phenomenon were poorly explained. First, this pattern would entail "extrapolation to a common ancestor uncomfortably far back in the Precambrian if Ordovician diversity is the apex of a gradual unfolding" (Gould and Eldredge, draft 2, 28). In other words, there simply is not enough time for the level of disparity seen among echinoderm classes to evolve via phyletic gradualism. Second, under phyletic gradualism the expectation was that "successively higher ranks of the taxonomic hierarchy will contain more and more taxa." Gould argued, however, that this expectation was the result of imposing gradualistic assumptions on taxonomic organization: under phyletic gradualism, "a new class attains its rank *by virtue of* its diversity—an evenly progressing, evenly diverging set of branches cannot generate a class of very limited diversity, for a lineage 'graduates' from family to order to class only as it persists to a tolerable age and branches an acceptable number of times" (Gould and Eldredge, draft 2, 28). Happily, Gould reported, under punctuated equilibria both "problems" are resolved. In the first case, punctuated equilibria simply does not require the lengths of time required under phyletic gradualism. In the second, "classes of small membership are welcome" because rapid and episodic instances of allopatric speciation can easily produce higher taxa that are quite distinct yet contain relatively few constituent members of lower rank.

Eldredge's statement about trends now became the final argument of the paper, emphasizing the originality of punctuated equilibira as a potentially new model of evolutionary dynamics. Here Gould left the original appendix on trends verbatim in the last paragraphs before the conclusion. He also inserted a rough, apparently hand-drawn version of the now-famous three-dimensional sketch of punctuated equilibria (drafted by Eldredge), which was keyed to this final discussion (fig. 5.2). The conclusion itself, however, was entirely new. Gould's primary concern here was to argue that punctuated equilibria provided, in the modern genetic understanding of natural selection, a thoroughly adequate explanation for the observed pattern of fossils. Under normal conditions, he explained, species were "amazingly well-buffered to resist change and maintain stability in the face of disturbing influences." Leaning heavily on Mayr, Gould argued that in peripheral isolation rapid speciation can take place because gene flow is reduced: "Only here are selective pressures strong enough to produce the 'genetic revolution' that overcomes

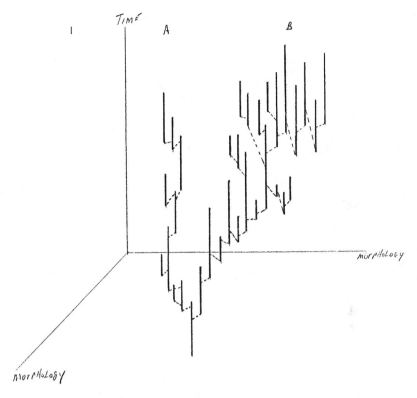

FIGURE 5.2. Eldredge's draft of the "three-dimensional" model of punctuated equilibria. Courtesy of Niles Eldredge.

homeostasis" (Gould and Eldredge, draft 2, 31). Therefore, the pattern one should expect to see in the fossil record is one in which stability of lineages is the rule and change is relatively rare, rapid, and dramatic.

The second draft prepared by Gould was nearly identical to the final, published version of the paper. By comparing the first two drafts, we can determine exactly what each author contributed: Eldredge's first draft accounts for approximately 36% of the roughly 11,000 words in the 1972 paper. Gould's contribution was indeed substantial and important—more than just supplying a "knack for names," he saw how to present the theory in a way that attracted attention and announced the appearance of an original and important new theory. At this point, drafts were circulated between all the *Models* contributors and were presented to outside reviewers as part of Schopf's efforts to secure a publisher. However,

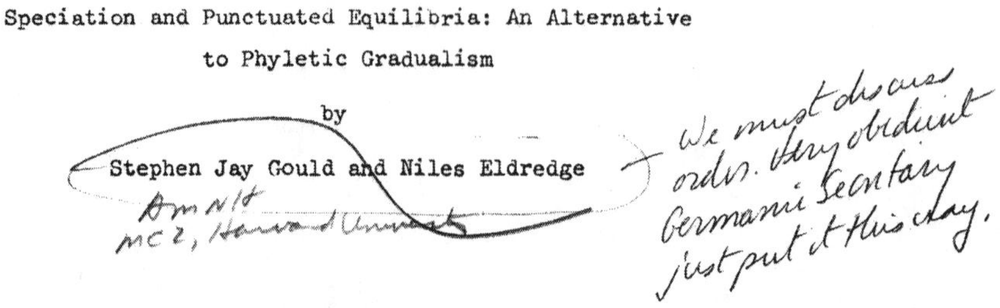

Speciation and Punctuated Equilibria: An Alternative
        to Phyletic Gradualism

                                by

    Stephen Jay Gould and Niles Eldredge

[handwritten annotations: "by" line; "Am N/t / MCZ, Harvard Univ" crossed out; right-side handwriting: "We must discuss order. Very obedient Germanic secretary just put it this way."]

FIGURE 5.3. The title page of the second draft of the 1972 punctuated equilibria paper, prepared by Gould. The handwriting on the right is Gould's, and it reads: "We must discuss order. Very obedient Germanic secretary just put it this way." Courtesy of Niles Eldredge.

a third draft of the paper does exist which contains few changes of any note—save one. A careful reader will have noticed that the second draft of the manuscript has been cited as "Gould and Eldredge" throughout this chapter. This has not yet been explained because one important dramatic element has been left out of this story so far—and indeed out of all published accounts of the theory's history.

Recall that the initial arrangement was that Eldredge would be first author on the paper. When Gould revised that first draft, however, he did more than just change the title: the draft that was returned to Eldredge now read "by Stephen Jay Gould and Niles Eldredge" (fig. 5.3). Accompanying this change was a note in Gould's hand that read, "We must discuss order. Very obedient German secretary just put it this way." While Gould did indeed have an officious German secretary (the legendary Agnes Pilot), this excuse sounds rather flimsy. If a prior agreement had been reached, one wonders what there was to discuss.

In fact, Eldredge remembers a more direct exchange that took place during a visit by Gould to the AMNH before the final manuscript was submitted to Schopf in early 1971. As Eldredge recalls, Gould walked into Eldredge's office and announced, "'I think I should be the senior author and you should be the junior author in this paper.' And I said to him, 'Okay.'" But later that night, after dinner with Gould and their spouses, Eldredge realized what he had done, and started "freaking out," telling his wife, "I'm done." Fortunately, Gould happened to return to the museum the next day. "He walked in," Eldredge remembers, "and I said 'We're not doing it. I'm senior author on the paper.' And he said, 'Fine'—he saw how intense I was" (Eldredge interview). This appears to have been the end of the story. The only hint of this episode in the "official" documentation is Eldredge's instruction to Schopf, in a letter from February 1971: "We have decided that I will be senior author on the paper for the book, but that Steve will be the senior author on the GSA abstract, and will deliver the paper in November" (Eldredge to Schopf, 11 February 1971: Schopf pap. 4, 12).

There is, of course, a larger context here. Eldredge acknowledges that some observers have, over the years, characterized Gould's appropriation of punctuated equilibria as unfair or self-serving. But he himself is philosophical about it. A lifelong jazz aficionado, Eldredge explained his relationship with Gould using the following analogy:

> Here's a model for you: People used to bitch about Joe Glazier, who used to be Louis Armstrong's manager. Louis played the trumpet and did all the creating. I'm not saying I did all the creating, but people said Joe took 50 percent of Louis's money. But Louis said, look at all the money I have. So, it's that kind of a feeling. It always basically worked out fine and I don't think we really had any personal problems about it. . . . Steve is the reason why I have a reasonably decent salary (Eldredge interview).

I am prepared to take Eldredge's assertion at face value. The fact that Eldredge and Gould were able to collaborate constructively on a major follow-up paper in 1977, and on various retrospective accounts in later years, suggests that whatever damage was caused to their personal relationship by this episode was not permanent.

The Immediate Reception

By and large, the internal and external reviews of the Eldredge-Gould manuscript were quite positive. Reviewers tended to single the paper out as among the best and most important contributions to the volume, and in several cases predicted that the argument would have a lasting influence on the field. However, several reviewers also expressed concern—and even mild offense—that Eldredge and Gould were claiming as a "new" theory an idea that had long been recognized in biology and even paleontology. For example, in an otherwise positive review, Stig Bergström noted surprise "to see that the idea of allopatric speciation at least partly is presented as something new to paleontology" (Bergström to Schopf, 16 May 1971: Schopf pap. 13, 10). Similarly, Richard Benson commented, "I cannot believe the section on the influences of Allopatric Speciation is original," adding, "I suppose there are more of us in the undeclared *punctuated equilibria* school than I had imagined" (Richard Benson, "Review of 'Speciation and Punctuated Equilibria'. . . .": Schopf pap. 13, 10). John Imbrie rather dismissively waived aside the central opposition between gradualism and punctuation, writing, "The pure concept of gradualism used by the authors went out near the turn of the century, yet they claim contemporary paleontology primarily adheres to this and no other philosophy—hogwash. A survey of my colleagues in paleontology shows a full spectrum of concepts of speciation—from pure gradualism to isolation and punctuated equilibria, with most believing both operate, gradualism being a slower and perhaps less important process" (anonymous [John Imbrie], "Specific Remarks of Individual Contributions," 6: Schopf pap. 13, 10).

Nonetheless, despite these criticisms, assessments of the quality of the essay were uniformly high. Bergström cited it as "a valuable contribution that may stimulate paleontologists to think along lines of some modern biologists" (Bergström to Schopf: Schopf pap. 13, 10); Benson lauded it for "a quality of introspection that makes even old subjects have renewed value" (Benson, "Review": Schopf pap. 13, 10); Leigh Van Valen called it "the most important paper in the book" (Leigh Van Valen to Michael Aronson, 8 October 1971: Schopf pap. 13, 10); and even Imbrie praised it as "a high-class contribution and ideal for the symposium" (Imbrie, "Specific Remarks," 5: Schopf pap. 13, 10). One of the most interesting remarks made by reviewers was Imbrie's comment that Eldredge and Gould did not make use of all of the available support for the model they were proposing: "Think of all the nice little island models they could use." This remark reveals that even the authors themselves had not fully absorbed the potential linkages between punctuated equilibria and other developing areas of paleobiological theory.

However, Schopf proved to be more critical of the manuscript than the other reviewers. In his initial response to Eldredge, who was handling editorial responsibilities for the paper while Gould was on sabbatical at Oxford, he noted that while reviews had found the essay "worthwhile," several "remain unconvinced that you have indeed discovered something new in punctuated equilibria." Schopf himself was among the unconvinced; he went on to recommend a section in Felix Bernard's 1895 textbook *The Principles of Paleontology* which "is certainly punctuated equilibria!" Schopf's entire letter was negative, and he criticized the authors' use of the term "biospecies," their assumptions about genetic iso-lating mechanisms, their discussion of homeostasis, and their presenta-tion of gradualism as a "straw man" (Schopf to Eldredge, 24 May 1971: Eldredge pap.). Years later, Eldredge summed up Schopf's response suc-cinctly: "You know, he really hated the punctuated equilibria paper. He really hated it" (Eldredge interview).

A week after receiving Schopf's letter, Eldredge wrote to Gould for advice about how to proceed. To begin with, he noted that despite Schopf's tone, "on the whole, reactions (negative type) were fairly mild." But he also expressed disappointment "to see that the positive reac-tions didn't really embody the essence of euphoria." Of Schopf's spe-cific criticisms, Eldredge called the allegation of unoriginality "the un-kindest cut of all"; indeed, Eldredge's copy of Schopf's letter includes Eldredge's handwritten comment in the margin: "We don't intend to im-ply it is anything *new*." Nonetheless, Eldredge informed Gould that he had reread the entire manuscript, searching unsuccessfully "for such a claim, or strong hint, or other such expression." The accusation of unoriginality—and even more, of falsely taking credit for a preexisting idea—has followed Eldredge and Gould to this day; from the very start, both authors have insisted that such criticisms misinterpret the intent of the essay. In the letter to Gould, Eldredge summarized his position on the matter:

> The problem here is that everyone who reads the paper says a) right on, and b) I knew that already, and c) ergo everyone knows that! But I maintain that c) is incorrect, and since the book is directed, according to Schopf, at stu-dents, our paper could well be the only place that 90% of the next crop of pa-leontologists first hear anything sane about how animals evolve. So I am un-moved by the criticism, but regret that people don't seem to see it that way (Eldredge to Gould, 31 May 1971: Eldredge pap.).

In his reply, Gould endorsed Eldredge's reasoning unequivocally:

> First of all, regarding the main and recurrent criticism (which I knew would come). I agree with you entirely; we need make no plea in our own defense beyond directing people to a more careful reading of the paper. Of course we do not claim that we alone knew about allopatric speciation; of course we know that every decent paleontologist has learned it and believes it; of course we know that they will state it properly in a lecture on neontology. We merely

claim that although they know it, a deeply entrenched picture (based on phyletic gradualism), all the more tenacious because it is scarcely perceived, pre-
vents them from applying it properly to the fossil record. . . . So let's not worry
about it (Gould to Eldredge, 6 June 1971: Eldredge pap.).

Gould's letter went on to make the important point that while the authors recognized that (a) biologists had long known about allopatric speciation, and (b) paleontologists had long argued that breaks in the fossil record may be real, it was the *combination* of these two concepts that
was original and worthwhile.

Acting on Gould's instruction "to write a tough letter" (Eldredge interview) in reply to Schopf, Eldredge sent Schopf a lightly revised manuscript along with a reply that flatly stated, "On the whole there are no
radical changes in content or organization of the paper." Eldredge repeated the defense that he and Gould had rehearsed, and reiterated his
contention that "intellectual appreciation of modern neontological theory is one thing—but its application is quite another, and I believe that
our paper pinpoints why this is so" (Eldredge to Schopf, 21 June 1971:
Schopf pap. 4, 12). He did concede a few instances where the manuscript's language might be softened to accommodate Schopf's concerns,
but concluded, "We don't feel any changes in the ms. are called for as far
as our 'discovering' anything is concerned." This reply did not sit well
with Schopf. Despite reporting that the paper had been accepted, he tes-
tily retorted, "Although you don't claim originality, you don't go out of
the way to give credit either!" Schopf went on to take Eldredge to task
for failing to mention those earlier paleontologists who "got it right,"
adding, "You do not, I feel, present a balanced view of the topic." Schopf
seemed particularly annoyed by the tone of the essay, and chided Eldredge: "To raise an argument for no good reason is, to my young mind,
immaturity at its most unrecapitulated [unreconstructed?] level. . . . I
only warn you that it is counter-productive and will achieve the exact
opposite of your proposed desire" (Schopf to Eldredge, 30 June 1971:
Schopf pap. 4, 12). Schopf then referred to the several reviewers who had
expressed similar criticisms, warning Eldredge, "You cannot be oblivi-
ous to the potential response." He concluded, "I do not think you want
that. However, the manuscript stays just as you have given it" (Schopf to
Eldredge, 30 June 1971: Schopf pap. 4, 12).

Eldredge's response to Schopf's letter—"written with a cup of coffee
and a cigar near at hand, in a relaxed atmosphere"—attempted to soften

the tone somewhat. He began by commiserating with Schopf about the "bitch of a job on your hands getting the book together, and that dealing with a couple of recalcitrant contributors who have resisted some of your suggestions which were deeply felt and well documented from your point of view, must be frustrating to say the least" (Eldredge to Schopf, 6 July 1971: Schopf pap. 4, 12). He also added his hope that "there will be no difficulties arising out of our 'official' relationship," and urged Schopf not to "think too ill of either of us." Eldredge concluded by stating, "We have arrived at a final version which we still feel, in all honesty, will do the best job in getting the message across. I am deeply sorry you don't agree, and that an impasse has developed, but I still think ours will be a valuable chapter in a book which I sincerely believe stands a good chance in reorienting much of your educational procedures in paleontology" (Eldredge to Schopf, 6 July 1971: Schopf pap. 4, 12). Needless to say, this would not be the last time Schopf would ruffle feathers as an editor, nor the only instance where Eldredge would be moved to propose having "a few drinks and maybe shoot some pool" as an attempt to smooth over differences (as we will see in the next chapter). However, as far as Eldredge and Gould were concerned, this was the end of the argument.

However, if Schopf could not force his authors to capitulate to his editorial authority, he still had one final recourse. In addition to writing a general introduction for the *Models* volume, Schopf also wrote short explanatory introductions to each chapter. These introductions were not shown to the contributors until the end of the copyediting stage, effectively giving Schopf the final word—and in the case of Eldredge and Gould's chapter, he took that opportunity literally. For every other chapter in the book, Schopf's introduction was no longer than half a page, and in some cases no more than a short paragraph. In the case of Eldredge and Gould's essay, it was nearly a page and-a-half. Here Schopf inserted the "corrective" he felt Eldredge and Gould ought to have included themselves. Noting that "some [paleontologists] have expressed a third interpretation, which views gaps as the logical and expected result of the allopatric model of speciation," he presented the lengthy quotation from Felix Bernard's 1895 *Eléments de Paléontologie* that he had asked Eldredge and Gould to consider. As Eldredge put it bluntly in a letter to Gould, "It ain't anything like what I would have penned myself, but I resisted complaining—it is his prerogative to write pretty much what he damn well pleases there, and he is merely using the intro to get

across some of the things that he couldn't get us to change in the ms."
(Eldredge to Gould, 19 January 1972: Eldredge pap.).

The Broader Legacy of Punctuated Equilibria

Overall, the *Models* volume was well received and had a respectable success throughout the 1970s. Over the next several years a number of reviews appeared in major journals like *Science, Nature,* and *Systematic
Zoology,* as well as in more specialized outlets like *Journal of Biogeography* and *Limnology and Oceanography.* J. A. E. B. Hubbard's review
in *Nature* was among the most enthusiastic, beginning, "At last we have
been provided with a thought-provoking text on invertebrate palaeontology," and concluding, "No geology department library can afford to
be without this volume" (Hubbard 1973, 208). A. J. Rowell reviewed the
book for *Systematic Zoology,* where he called it "a remarkable book. . . .
bound to have a substantial impact on paleontological thought for several years." Schopf must have been delighted with Rowell's conclusion
that "to those who have regarded paleontology as something of a nineteenth century relict . . . this book will be as refreshing as the dawn—
which perhaps it is" (Rowell 1973, 94–95). C. Barry Cox was similarly
effusive in *Journal of Biogeography,* where he singled out Eldredge and
Gould's "important paper" in a "book [that] should expand the horizons
of most palaeontologists" (Cox 1975, 229–30).

In a couple of instances, however, reviewers were somewhat less impressed. E. S. Deevey, in *Limnology and Oceanography,* wondered
whether the book's authors weren't guilty of trying "to embrace both
evolutionary and ecological biology without at all times carefully distinguishing between the two," and while he called Eldredge and Gould's
chapter "the most cogent contribution," he also dryly remarked that its
emphasis on allopatry "will sound novel only to earth scientists who
have not read anything in 20 years" (Deevey 1974, 375–56). The only review that really got under Schopf's skin, though, was Jonathan Roughgarden's piece in *Science,* which although not entirely negative did question the general applicability of the book's models to actual data. While
Roughgarden praised several of the book's chapters (including Eldredge
and Gould's), he also warned, "The emerging theoretical paleobiology
will soon be critically judged on whether models have been developed
which are a priori cogent and rigorous and on whether these models

have in fact led to new and successfully tested predictions" (Roughgarden 1973, 1225). Schopf took umbrage with this review, and dashed off a letter to Roughgarden in which he questioned the author's qualifications, clarified the intent of the volume, and sniped that "if you did not do your homework but rather reviewed the book as though it were directed at your field, then I feel you should rethink what a review is all about" (Schopf to Roughgarden, 7 April 1973: Schopf pap. 5, 11). Roughgarden's restrained response reassured Schopf: "I am extremely sympathetic to what you are trying to bring to paleobiology." However, he also warned Schopf not to get carried away: "As you know, many fields have been infused with theory and modeling (including ecology, geography, etc.), often introduced with much fanfare. The trumpeter alone does not win the battle" (Roughgarden to Schopf, 12 April 1973: Schopf pap. 4, 11).

In any event, it is fair to say that *Models in Paleobiology* had a lasting effect on paleobiology, although it may not have been the unambiguous "smash hit" Schopf had originally hoped for. A 1981 letter to Schopf from the book's publisher, William H. Freeman, summarized the book's fortunes: "In nine years, the book has sold about 2,100 copies. Economically, that is not good. But professionally, I am satisfied." Freeman even reported, "So many requests come to quote something from *Models* that I decided to ask any future enquirer to 'give a little'" (William H. Freemen to Schopf, 13 March 1981: Schopf pap. 8, 30). The date of the letter, 1981, may give a clue about the reason for Freeman's change in policy. It is now virtually impossible to find a copy of the first printing of the book, which indicates that its economic fortunes may have also been on the rise at the time of Freeman's letter to Schopf. Jim Valentine, an original organizer of the symposium, offers this anecdote: "I think it got most attention because of that one paper ["Punctuated Equilibria"]. I knew [Freeman] fairly well, casually, of course. And he had a garage full of those things, full of those books. They weren't selling very well and then Steve started making his points very publicly and getting into fights with people and stuff and it just flew out of his garage" (Valentine interview).

Valentine's comment alludes to the fact that, though it was first presented in 1972, the theory of punctuated equilibria did not gain its full notoriety until almost a decade later. One explanation for this is simply the fact that new scientific ideas take time to filter into the zeitgeist. Another is that the notoriety of the theory grew dramatically around the time that Gould achieved wider public fame, following the success of his first collections of popular essays in the late 1970s. The third expalanation—and the most important, in my view—is that punctuated equilibria became a hot topic only when paleobiology as a discipline had achieved sufficient attention to warrant its noteriety in the wider community of evolutionary biology. As this book will go on to describe, this was not accomplished until the early 1980s. More than anything, punctuated equilibria symbolized the paleobiological movement, even if it did not encompass it.

There is a final question to ask that bears on the historiography of punctuated equilibria: to what extent was the theory genuinely original? As Gould and Eldredge protested frequently, punctuated equilibria owes a substantial amount to earlier biological theories. The most frequently cited touchstones are Ernst Mayr's work on peripheral isolation and allopatric speciation, and G. G. Simpson's briefly held theory of "quantum evolution" (Mayr 1982a, 188–210; Simpson 1944, 206–7). Indeed, Mayr had written in 1954 that "many puzzling phenomena, particularly those that concern paleontologists, are elucidated by a consideration" of peripherally isolated populations, including "unequal (and particularly very rapid) evolutionary rates, breaks in evolutionary sequences and apparent saltations, and finally the origin of new 'types'" (Mayr 1982a, 206). In fairness, though, Mayr's discussion understates the genuine significance of punctuated equilibria: Mayr had understood allopatric speciation, and Simpson had acknowledged that evolutionary rates could vary greatly, but nobody had combined these insights to offer an explanation of the actual state of the fossil record—a solution to Darwin's dilemma. Furthermore, neither Mayr nor Simpson emphasized stasis, which Eldredge and Gould have consistently pointed to as the most important and "radical" element in the theory.

Generally, however, most critics of the theory have found reasons to implicate Gould in a more radical agenda: for example, Mayr himself commented that since "the Eldredge and Gould theory was expressly based on my 1954 paper one might think that I would be completely behind the Gould theory," but added, "This, however, is only partially the case. I strongly object to the Goldschmidtian interpretation of rapid speciation in founder populations and I likewise do not agree with the complete stasis of other species" (Ernst Mayr to Schopf, 9 February 1982: Schopf pap. 8, 32). What Mayr was suggesting was that Gould and Eldredge had attempted to smuggle in the Richard Goldschmidt's much-derided theory of genetic saltations under the cover of a legitimate

discussion of allopatric speciation. Mayr went on to argue in a 1982 essay that there are two possible readings of punctuated equilibria: first,
a "moderate" or "Mayr version" that involves only a "slight translation" of his 1954 theory "into vertical terms," and second, a "drastic" or
"Goldschmidtian version" that originated in Gould and Eldredge's 1977
follow-up to the original paper, which suggests that speciation is based
on major mutations (Gould and Eldredge 1977; Mayr 1982b).

Michael Ruse has argued that three distinct historical phases characterize Gould's thinking about punctuated equilibria, and I suspect that
Ruse drew heavily on Mayr's interpretation. The first phase of the theory was represented by the 1972 Eldredge-Gould paper, and according
to Ruse it offers "a fairly straightforward extension of orthodox Darwinism" (Ruse 1989, 120). Ruse contends, however, that between this first
phase and the later second phase (which he dates to Gould's 1980 essay
"Is a New and General Theory of Evolution Emerging?") something important happened: Gould was now "downplaying the role of natural selection," and accordingly "the father figure had changed from Charles
Darwin to Richard Goldschmidt" (Ruse 1989, 122). Ruse freely acknowledged that by 1989 (when Ruse's essay was published), Gould had
come to "categorically" deny "that he himself was ever a saltationist in
Golschmidt's or anyone else's sense." Nonetheless, Ruse concluded that
"one can fairly say . . . Gould (especially) was starting to think of evolution's processes through a lens or filter of discontinuity. . . . In his own
mind, he was starting to highlight the essential abruptness of evolution,
as opposed to its continuity" (Ruse 1989, 122). Finally, a third phase represents Gould's final position, which dates from about 1982. This, Ruse
comments, offers "a pull-back from extremism," though not a "retreat"
from hierarchical evolution. The final version of the theory offers a defense of "punctuationism," but at the same time disavows Goldschmidtian saltationism.

If one looks at the early history of punctuated equilibria, as this
chapter has done, it appears that Ruse's model holds water. Immediate reactions—both initial reviews and comments in the scientific literature—suggest that the theory was seen as mildly controversial but hardly
upsetting by most paleontologists and biologists who first encountered
it. Intense reaction would only begin in the early 1980s, shortly after
Gould's infamous essay in *Paleobiology* entitled "Is A New and General Theory of Evolution Emerging?" Where I differ with Ruse, however, is in his apparent assumption that these phases reflect actual, sub-

stantive modifications to Gould and Eldredge's conception of the theory. This is an assumption I believe Ruse has inherited from Mayr—he has made Mayr's "moderate" and "drastic" readings of punctuated equilibria into actual stages in the theory's development. In other discussions of the topic, Ruse maintains that while Gould's later stance was amicable towards natural selection, "one sees that other factors, including brute chance, come increasingly into play" (Ruse 1999a, 138). I argue, however, that brute chance was *always* a central component of the theory. If it was not mentioned as explicitly in the 1972 paper as it was in later publications, it was highly visible in many of Gould's significant publications throughout the early and mid-1970s.

The next several chapters will explore this more thoroughly, as we examine the further expansion of paleobiology in the 1970s. I believe that punctuated equilibria had two important roles in this era. First, its invocation of Wrightian "stochasticity" provided inspiration for a number of studies that more deeply probed the influence of random or nondirectional elements in evolution as seen in the fossil record. But secondly, and perhaps more importantly, it acted as a model of the kind of paleontology that could break the grip of Darwin's dilemma and could offer a route to bringing paleontology into the mainstream of evolutionary biology.

CHAPTER SIX

The Founding of a Research Journal

As I have maintained throughout this book, the paleobiological revolution of the 1970s built upon the foundation the generation of G. G. Simpson, Norman Newell, John Imbrie, and others laid down from the 1940s through the 1960s. During that earlier phase, however, paleontology never achieved genuine parity with biology and genetics within the community of evolutionary biology, and was ultimately denied the place within the neo-Darwinian framework Simpson envisioned when the modern synthesis was being established. This had partially to do with fairly deep-seated prejudices among biologists and even paleontologists about the adequacy of paleontological data and interpretations. Equally important, though, were a set of institutional constraints that prevented paleontologists from establishing and promoting the paleobiological agenda. These constraints included departmental organization (i.e., subordination of invertebrate paleontology to geology in universities and museums), pedagogical practices (the lack of biological training for aspiring paleontologists and inability to recruit the "best" young minds), and, as I will discuss in this chapter, the absence of an outlet for paleobiological research. The first two constraints have been and continue to be somewhat intractable; the last was solved, quite swiftly, in 1975 with the establishment of the journal *Paleobiology*.

The Birth of *Paleobiology*

As the last chapter discussed, a great deal of intellectual momentum was generated by Tom Schopf's organization and publication of the *Models in Paleobiology* symposium, which unveiled Eldredge and Gould's theory of punctuated equilibria and drew attention to theoretical paleontology. Schopf built on this momentum by organizing a series of informal meetings at the Marine Biological Laboratory in Woods Hole, Massachusetts, to discuss further opportunities for promoting paleobiology. This collaboration, with Gould, David Raup, Daniel Simberloff, and eventually Jack Sepkoski, would produce a very successful series of papers on the stochastic modeling of evolution, which will be discussed in detail in chapter 7. Despite these successes, however, by late 1973 Schopf was feeling gloomy. In particular, he worried that loose, informal collaborations like the "MBL group" wouldn't have the transformative effect on the field he hoped for. He recognized that without some kind of formal, institutional support, paleobiology would always be marginal. For that reason, in the fall of 1973, he hit on the idea of starting a new journal.

Conceiving the Journal

The documentary evidence surrounding the founding of *Paleobiology* does not pinpoint the exact moment the idea was conceived. Paleontological Society reports show that its council was engaged in an active re-evaluation of the society's goals and commitments between 1972 and 1973 (corresponding with the presidency of James Valentine), which included efforts to consolidate interest and membership ("Report to Paleontological Society: Activities of the Committee on Paleontology and the Paleontological Society during Last 6 Months." PS 24). One stated goal of the council was "the obligation to promote the cause of paleobiology," which acknowledged interest among its membership in "cross-fertilization and exposure time between biologists and paleontologists." Accordingly, the council recommended that society membership be polled regarding interest in establishing a new journal which would both promote work different from the "mostly descriptive articles with very restricted interest" currently published in *Journal of Paleontology*, and also attract readership and submission from workers in the biological sciences. At some point (likely because of his collaboration with Valentine on the Models symposium), Schopf became involved in this initiative. Interestingly, an earlier PS committee had considered the very same proposal in the late 1960s, but it never got off the ground. Raup recalls: "Some time along there I chaired a committee in the paleontological society to look into long-range plans . . . and one of the things we considered pretty thoroughly was whether to start a journal like *Paleobi-*

ology, and we unanimously recommended [against] . . . because we were afraid that to make it distinct from the *Journal of Paleontology* would be to lead to more strife . . . but Tom didn't accept that—he didn't accept anything that he didn't agree with" (Raup interview [Princehouse]).

In late 1973, Schopf circulated a draft report on the feasibility of a new journal among the membership of the Paleontological Society. Valentine enthusiastically supported the proposal, but struck a note of caution: "This represents a major undertaking for the PS, of course, and we need to evaluate the whole proposition accurately before acting. A goof could break the society" (Valentine to Schopf, 20 November 1973: PS-*P* 14). Valentine was especially concerned that the PS's general membership was more conservative than its council, and emphasized that the rank and file would have to be sold on the necessity for a new journal. He outlined several points he felt were important in selling the idea. One prominent issue was the fact that the PS did not itself own a journal: *Journal of Paleontology* was owned not by the PS but by the more conservative Society of Economic Paleontologists and Mineralogists (which is now called Society for Sedimentary Geology, although it has retained the acronym SEPM). Another was economic viability; Valentine estimated that a subscription list of 1,500 would cover expenses, particularly if a large number of subscribers were institutions paying the higher rate.

The major point of contention was that PS members might view the new journal as competition or replacement for *JP*, and Valentine urged Schopf to address these concerns: "The JP should indeed continue; the new journal would supplement, not compete. Indeed, the new journal would provide an ideal outlet for summary statements or salient biological conclusions reached in taxonomic and biostratigraphic studies, which might be in JP or in a monographic series." To address these issues Valentine prepared a referendum for insertion in an issue of *JP*, including a short description of the journal and a ballot. Valentine noted "I should expect your main problems of rating feasibility will be (1) how much it will cost; (2) can we get the subscribers; and (3) can we get the manuscripts. The last question is of course linked to problems of format and content. I imagine that a first issue with papers by such people as George Simpson, Th. Dobzhansky, Ernst Mayr and the like—in fact a first year with a heavy scattering of eminent biologists—would help solve both problems." Valentine closed by resolving that the matter would be taken up at the PS council meeting in San Antonio the next year (Valentine to Schopf, 20 November 1973: PS-*P* 14).

Schopf responded positively to Valentine's suggestions, noting only that the new journal should be made available at a reasonable rate to subscribers who were not PS members, since "this will add biologists" (Schopf to Valentine, 4 February 1974: PS-*P* 14). In a letter to his Chicago colleague Ralph Johnson (who would join Schopf as founding co-editor of the journal), he discussed possibilities for the journal's editorial board, noting, "I think the editorial board should be chiefly younger workers who represent the changing interests in our profession" (Schopf to Johnson, 17 December 1973: PS-*P* 14). Johnson replied with several concerns. In particular, he wondered whether *JP* would suffer alongside a new journal and be forced to "limit itself to papers relating to geological studies of the fossil record[.] Is systematics and phylogeny geo or bio?" He also worried that "the new journal will begin well and then in several years starve for good papers," noting that in the past, *JP* had also "attempted to attract ms. of broader interest but without success" (Johnson's handwritten reply in Schopf to Johnson, 17 December 1973: PS-*P* 14).

Schopf's initial queries to prospective members of the new journal's editorial board elicited mixed responses. One prominent paleontologist voiced concern that the new journal "may encourage speculation that is not firmly based on geological and paleontological documentation" (Richard E. Grant to Schopf, 4 January 1974: PS-*P* 13). This, he noted, could have unfortunate effects: "One will be to foster the notion of a first-class journal that deals in great unifying principles, and a second-class journal of hackwork [i.e., *JP*]." He did, however, note that "megathinking is 'in' now," and ultimately voiced his support. Steven Stanley questioned the validity of Schopf's comparison between the new journal and the established journal *Genetics*, noting that "the field of genetics has been booming since the turn of the century," while "the current renaissance of paleontology is opening up entirely new avenues of research and, in my view, much garbage is floating about" (Stanley to Schopf, 21 February 1974: PS-*P* 14). Nonetheless, he too accepted appointment on the editorial board. Another member wrote to express support for the journal while at the same time cautioning that a new journal might widen the gap between geologists and biostratigraphers on the one hand, and paleobiologists on the other. He closed his letter with a suggestion: "Perhaps instead of founding a new journal, we need to do an education job. Don't *reduce* the paleobiological coverage in *JP*. Instead, *increase* it. Show the neontologists that *JP* is worthy of their consideration (and, hence, subscription). And, show the paleobiologiphobes that they, too, need the very paleobiology they despise—that they really can't do without it" (R. A. Davis to Schopf, 11 March 1974: PS-*P* 14).

Among the responses Schopf received, a surprisingly ambivalent reply came from Niles Eldredge. While stating that he of course wanted "to see our science become more enmeshed with theoretical organismic biology," he noted that contributions of paleontologists to biology had been largely ignored by biologists, and wondered whether a new journal would actually reverse the trend. Specifically, he cautioned that "any increase in the literature is a glut," since "our existing journals are already publishing a large quantity of crap, mixed in with worthwhile things" (Eldredge to Schopf, 1 February 1974: PS-*P* 14). Eldredge expressed doubts that there would be a sufficient quantity of quality manuscripts to fill the new journal, and feared that the new journal would inevitably suffer from too few papers with "potentially theoretically interesting implications" and be forced to rely on "conventional descriptive" papers in traditional systematics. But Eldredge deferred the question to the general membership of the PS: "If they are for it, I am."

In response, in early February 1974, Schopf put together the "Draft Report on the Format and Feasibility of a New Paleontological Society Journal" which was explicitly intended "for distribution to members of the Council and proposed members of the Editorial Board" of the new journal (Schopf, "Draft Report on the Format and Feasibility of a New Paleontological Society Journal": PS-*P* 14). This document provided a detailed summary and analysis of the factors to be considered by the council, and explained the rationale for the new journal. It included sections surveying competing publications, detailing journal specifications, estimating costs and subscription requirements, formulating editorial policy and board structure, and offering general recommendations. On the question of the availability of appropriate material, the report noted that "2–3 times the number of suitable articles required to support *Paleobiology* [as the new journal was now being called] were published" during the preceding year, spread out over more than 23 journals. Schopf also made the optimistic prediction that this number would grow, due to the increase in graduate training in paleobiology and the increasing numbers of biologists willing to consider paleontological topics (Schopf, "Draft Report," 2). The top five outlets for paleobiological work had been the journals *Lethaia*, *Journal of Foraminiferal Research*, *Evolution*, *Palaeontology*, and *Journal of Paleontology*. Interestingly, among these journals the calculations for "approximate percentage of volume taken up by material suitable for *Paleobiology*" showed *Lethaia* as the clear leader with 56%, while *JP* lagged at only 18% (compared with 21% for *Evolution*). Schopf presented this as ammunition against the claim that *JP* would suffer greatly from the establishment of a new journal. He also argued that consolidating most paleobiological work in a single journal would attract additional readership and submissions from biologists who did not normally read the other publications.

Outside of financial considerations, the other major topic in the report was proposed editorial policy and structure. Schopf defined the editorial policy as follows:

> We seek in *Paleobiology* a broad spectrum of paleobiological thought in both paleobotany and paleozoology, including studies of invertebrates and vertebrates, from both neontologists and paleontologists. We envision that most volumes will deal with biological or paleobiological aspects of morphology, biogeochemistry (organic and inorganic), populations, faunal provinces, communities and ecosystems. Emphasis should be on biological or paleobiological processes. This includes speciation, extinction, development of individuals or of colonies, natural selection, evolution, or patterns of variation, abundance and distribution, in space and time. Historical analysis of paleobiological themes is also welcome (Schopf, "Draft Report," 10).

The primary qualifications for members of the editorial board would be "sympathy" with the philosophical orientation of the journal, but also having their "best research years ahead of them instead of past them," highlighting Schopf's desire that *Paleobiology* be an outlet for young, creative iconoclasts like himself. His suggestions for the board's initial membership reflected this agenda: while the more senior Ralph Johnson was tapped as editor, other members included James Schopf (Tom's brother), Simberloff, James Hopson, Richard Bambach, Karl Flessa, Gould, Raup, Steven Stanley, and of course Schopf himself. The report closed with a general recommendation to the council that highlighted the issues that would most likely appeal to PS membership. The journal would fill a gap in literature and was financially feasible, it would belong solely to the PS and would represent independence from SEPM, and it would broaden PS membership by appealing to biologists. Overall, "*Paleobiology* would establish a new image for the paleontological society in a way that could never be done by simply enlarging the *Jour-*

nal of Paleontology. Paleobiology should be the focal point for ecologists, evolutionary biologists, and other neontologists who are interested in the history of life" (Schopf, "Draft Report," 19).

Organizing Support

In March 1974, PS members were mailed ballots inviting them to vote on the society's intention to start a new journal. The accompanying letter made its case for support by citing a number of familiar factors: (1) it reiterated the need for the PS to have its own journal that would be autonomous from SEPM, (2) it dismissed the concern that *Paleobiology* would weaken the *Journal of Paleontology*, (3) it cited the frequency with which paleobiological work was currently published in other journals, and (4) it emphasized the financial viability of the venture (Schopf and Valentine, "Draft Memo": PS 24). This memorandum also noted that while there was increased interest on the part of paleontologists to communicate their research with biologists, such rapprochement was hampered by the lack of a biologically oriented paleontological journal that both paleontologists and biologists would want to read. Therefore, it said, "the proposed journal *Paleobiology* is designed to compliment the *Journal of Paleontology* by meeting the expanding biological interests of our profession." The statement went on to explain that "this expansion is partly due to young paleontologists who are increasing[ly] well-trained biologically," and it promised that "contributions from paleontologists on matters of biological interest will be joined by contributions from bioscientists on matters of paleontological interest." The letter concluded with the "hope that *Paleobiology* would penetrate into biology libraries and into the hands of bioscientists, leading to an expansion of the dialogue that is now established between paleontology and relevant biological fields."

By the 1 April Council meeting in San Antonio, more than half of all members had responded, and the overwhelming majority supported the venture (81%, or 861, to 17%, or 142) ("News and Notes of the Paleontological Society," 6 June 1974: PS 24). Additionally, an encouraging 544 members indicated on their ballots that they would subscribe to the new journal. Many members had included comments.[1] One member stated,

---

1. Photocopies of ballots are preserved in the *Paleobiology* archives, Box 14. I have suppressed names of most individual respondents.

"The paleobiology bandwagon is for the most part embarrassingly far from where modern biology is at," and claimed, "The best way to communicate with biologists is to publish in biological journals." Another ballot labeled the proposal "a dangerous dilution of effort," still another called the letter itself "offensive," and others complained about the proposed annual subscription fee of eight dollars. Several members voted against the proposal but nonetheless grudgingly acknowledged that they would subscribe if the measure passed. On the positive side, one member (Gould) enthusiastically remarked, "As Jerry Rubin said—Do It!" Another spoke for "young people" who otherwise would "publish innovative work elsewhere." Yet another wondered whether *Paleobiology* would become the "journal for *real* Paleontology," leaving *JP* to become "a Journal of Paleontological Trivia." Finally, one female member who supported the proposal responded, "I think it will be good for the paleontologists. I don't know what one can *do* with biologists! They seem to have exactly the same opinion of paleontologists that the Victorians did of women—i.e., nothing written by such a low form of life could possibly be taken seriously."

Acting on this information, on 1 April 1974 the PS Council voted 6–1 in favor of establishing the journal *Paleobiology* with a first issue planned for the following year. At this meeting it was also decided that Schopf would join Ralph Johnson as founding coeditor, and that both men would serve for an initial three-year term (PS Council meeting notes, 1 April 1974: PS 24). Despite the apparently clear sentiments of the society's membership, two local branches of the PS—the Pacific and North Central sections—objected to the vote and requested that formal ballots be mailed to membership asking not for *support* for the journal, but for *approval*. While he noted the danger in delaying momentum for the new journal, Valentine acceded, and ballots were duly re-mailed. When the returns came in membership was still in favor—this time by a 78% majority (491 of 632 received). Following this, at the 17 November meeting of the PS Council in Miami Beach, the initiative to publish *Paleobiology* was finally ratified (PS Council meeting notes, 17 November 1974: PS 24).

Meanwhile, Schopf and Johnson jointly wrote to Valentine to formally accept coeditorship of *Paleobiology*, and proposed a division of labor in which Schopf would "mainly be concerned with setting up the business and printing arrangements," including "obtaining a printer, designing the format of the Journal, obtaining stationary and printing forms, soliciting advertisers and subscriptions, and handling all communications with the Society," while Johnson would handle "editorial arrangements," including "soliciting manuscripts and reviewers, and processing manuscripts for the printer" (Schopf and Johnson to Valentine, 5 April 1974: PS-*P* 14). A month after accepting the task, Schopf wrote the PS council to update his progress. He reported "fighting the war on three fronts," dealing simultaneously with the tasks of soliciting manuscripts, obtaining subscriptions, and finding a printer (Schopf to Paleontological Society Council, 16 May 1974: PS-*P* 14). Schopf was optimistic, but it is clear from this letter that the editors had a long way to go. With little more than seven months before planned publication of the first issue of *Paleobiology*, many details had yet to be secured. The journal had received no manuscripts, although roughly 70 letters had been sent to individual scientists requesting submissions. The choice of printers had been narrowed to two contenders, but no final decision had been made. And the crucial issue of subscriptions still hung in the balance. It is here that Schopf's personal commitment to the new venture was most evident. As Gould recalled in his 1984 obituary for Schopf, Schopf had been content with doing this "spadework" for the journal:

> I walked into his office one night in Woods Hole, and Tom was in the midst of writing 250 letters *by hand* to libraries that had not subscribed (he knew how much *Paleobiology* needed an institutional basis of support). When I asked why he simply didn't Xerox a single letter, and save himself countless hours of backbreaking, boring, hand-cramping work, he replied that a personal note might get more attention from harried librarians. I do not know whether this ploy worked, but need I say more to illustrate his dedication? (Gould 1984b, 283)

In fact, Schopf reported he had written over 350 such letters, and he promised to follow up with the institutions who did not reply. He also outlined an ambitious plan to advertise *Paleobiology* to the widest audience possible: First, the 1,400 members of the PS would be contacted via the next newsletter. Second, he would ensure that sibling societies like the British Palaeontological Association, the Geological Society of America, and the Paleobotanical Section of the Botanical Society of America printed the announcement in mailings to their memberships. Third, related biological societies would be selectively targeted. This last task was critical, since without support and interest from biologists

Paleobiology would not achieve its mission. Accordingly, Schopf planned to obtain the mailing lists for three societies—the Society of Systematic Zoologists, the American Society of Naturalists, and the Society for the Study of Evolution—which published, respectively, *Systematic Zoology*, *American Naturalist*, and *Evolution*. In a follow-up the next year, he updated the council about his initial success in obtaining subscriptions. He reported having sent more than 600 letters to paleontologists and geologists in academic departments; more than 2,500 advertisements had been sent to biologists; more than 500 solicitations were mailed to members of the Palaeontological Association; and because "we had some envelopes and ads left over," an additional 577 nonsubscribing members of the Society for Vertebrate Paleontology were contacted for a second time (Schopf to William A. Oliver, 8 May 1975: PS-*P* 14). The net haul was over 1,200 initial subscribers, more than a thousand of whom were individuals. Schopf was somewhat disappointed with the number of institutional subscriptions, so he repeated mailings to librarians over the next two years, and by 1977 had reached 2000 total subscribers, including roughly 500 institutions (Schopf and James Hopson, "Solicitation Template," November 1977: PS-*P* 14).

The first issue of *Paleobiology* was published in March 1975, and it included papers by some of the leaders in the new biological approach to paleontology, including Gould, Raup, Stanley, Bambach, Flessa, John Cisne, and Schopf himself. In fact, several of these authors contributed to multiple pieces: Gould coauthored two papers and penned a review, and Raup had a coauthored paper and another paper of his own. The editorial policy Schopf had outlined was also clear in the first issue. In the first place, the selection of papers represented both an orientation towards biologically significant findings in paleontology and a preference for statistical, mathematically sophisticated models. Raup's solo paper, "Taxonomic Survivorship and Van Valen's Law," championed Leigh Van Valen's important mathematical analysis of survivorship "because it is a major step towards a nomothetic paleontology. . . . toward interpreting the evolutionary record in terms of general rules and processes without regard to specific causes . . . [and] it attempts to make generalizations about the fossil record which are not simply enumerations of specific events and causes" (Raup 1975b, 83). This was precisely what Schopf had promised *Paleobiology* would be: a journal that would eschew traditional taxonomic and stratigraphic studies in favor of broadminded analysis of patterns in the fossil record. In fact, Schopf wrote an introductory editorial for the first issue of the journal, which, although never published, is worth quoting at length:

> Rarely does a field of science witness a change in emphasis and direction as severe as has paleontology in the past few years. The general goal remains the same—to determine and to causally explain the patterns of the fossil record. But the scope of approaches now employed involves instruments and methods hardly considered 5 years ago, and in some cases not invented a decade ago. Together with the change in day-to-day activities of paleontologists, the conceptual models used to suggest experiments have lost this special pleading for unique historical conditions. Our practical recognition of this change in paleontology can be seen in our editorial board which includes neontologists and paleontologists. . . . To conclude: biological paleontology has long stood in the shadow of its great counter-part, stratigraphic paleontology, awaiting for the integration of biological principles with the history of life. *Paleobiology* is a vehicle for publication for those seeking to explore the biological implications of the fossil record (Schopf, "Introductory Editorial" [draft]: PS-*P* 14).

The editorial was likely not published because it was recognized as overkill—Schopf had, after all, won his battle—but it demonstrates the degree to which the journal had become part of Schopf's distinctive vision.

Tom Schopf as Editor

The initial success of the journal did not mean that everything went perfectly smoothly for its editors. After the first year's division of labor between Schopf and Johnson, the two shared equally in editorial duties, although each maintained a distinctive editorial style. As Gould later recalled, Schopf "knew that his own personality could not assure smooth sailing, so he wisely chose Ralph Johnson, one of the great gentlemen and diplomats of our profession, for the overt activity" (Gould 1984b, 282). This may have been true, but Schopf quickly inserted himself into the "overt" business of corresponding with authors and handling manuscripts, occasionally with complicated results.

One noteworthy example involved a manuscript Eldredge submitted in 1975 that critiqued a paper by Schopf, Raup, Gould, and Simberloff that appeared in the first issue of the journal ("Genomic versus Morphologic Rates of Evolution"). Eldredge presented his critique as a friendly comment, noting that he found the new journal "so stimulating, in fact, that I have been stimulated to write a commentary of a critical nature on one of the articles," adding, "I take it the editorial policy is in favor of commentary, debate discussion, what-have-you" (Eldredge to Schopf, 1 April 1975: PS-*P* 1). The paper he submitted took Schopf and his co-authors to task for contending "that their results have shed considerable doubt on the generally accepted tenet that 'large differences in rates of genomic evolution among taxa over geologic time' in fact exist" (Eldredge 1976, 174). In his letter to Schopf, Eldredge expressed his expectation that Schopf "will find my remarks moderate, reasonable, and generally to the point, hence probably useful as another way of looking at the problem of differential evolutionary rates."

Apparently, Schopf did not agree. He responded by sending the manuscript back to Eldredge with 30% to 40% of the text boldly crossed out, including most of the passages containing Eldredge's strongest criticisms of Schopf's paper. In fact, Schopf removed nearly everything from the manuscript that could be construed as more than mildly critical: for example, he deleted the entire second half of the paper's opening paragraph (which contained the above-quoted passage challenging the validity of differential rates), so that instead of concluding with an explicit challenge to the thesis of Schopf's paper, the paragraph merely suggested that "their discussions and conclusions merit reexamination" (Eldredge, draft manuscript "Differential Evolutionary Rates: Some Comments on Schopf et al": PS-*P* 1).[2] Schopf made similar suggestions throughout the text, and in numerous places softened Eldredge's statements by replacing words like "would," "has," and "must" with "may," "believe," and "appears." However justified Schopf's alterations may have been from a technical or stylistic point of view, the only possible interpretation is that Schopf systematically took the teeth out of Eldredge's critique.

Unsurprisingly, Eldredge did not take this very well. Schopf returned his changes in early May 1975, and Eldredge apparently waited until late October to reply. In his response to Schopf, he reported that he had essentially "changed nothing" (Eldredge to Schopf, 29 October 1975: PS-*P* 1). This was followed by a point-by-point refutation of Schopf's criticisms, and concluded, "None of this will convince you of anything, but I do think my opinions are (1) legitimately different from yours, (2) within the bounds of orthodoxy . . . hence in general my opinions are 'legiti-

2. This copy of the manuscript contains Schopf's handwritten deletions and comments.

mate' in the sense of publishable, if not (in your opinion) in the sense of 'correct.'" This response was followed quickly by a second, in which Eldredge laid out his feelings more plainly: "I am frankly dismayed by the editorial freedom which you evidently feel you have at your disposal. This is symbolized by the fact that it was the *original* copy of the manuscript you chose to massacre. That takes one hell if a lot of nerve. . . . It is so clear that your job was the product of a wounded author and not the impartial actions of an editor that it only adds insult to injury" (Eldredge to Schopf, 6 November 1975: PS-*P* 1).

Eldredge nonetheless concluded, "I shall *not* withdraw the paper, nor shall I condone publication of the sniveling piece of crap you have turned it into," and he asked Schopf to seek Johnson's guidance in the matter. Johnson immediately sent Eldredge a very conciliatory letter in which he said, "I deeply regret the circumstances that led to your letter of November 6. . . . It is obvious to me that you are perfectly justified in your indignation. . . . In spite of his protestations to the contrary, Tom went beyond normal editorial jurisdiction in my opinion" (Johnson to Eldredge, 12 November 1975: PS-*P* 1). Johnson concluded, "We will be pleased to accept your paper for publication . . . as you originally submitted it." The paper was eventually published the following year. Eldredge seems to have accepted Johnson's assurance that "Tom did not consciously try to suppress your paper or to pressure you into revising it in favor of his viewpoint," and he conceded that "Tom's sincerity was not in doubt" (Eldredge to Johnson, 20 November 1975: PS-*P* 1).

Beyond providing a glimpse of Schopf's personality and the dynamics of the editorial team for the initial years of *Paleobiology*'s existence, there is an additional observation to be drawn from this episode that is important. Eldredge's assumption that Schopf intended the journal to be a forum for debate and discussion of opinion was in fact incorrect. Throughout the episode, Schopf insisted he was acting objectively, hoping, as Johnson put it, "to make our reviews, as well as our papers, as scholarly as is possible and to avoid unnecessary polemics." Schopf's final reply to Eldredge stated his perspective quite clearly: "Too much of paleontology, in particular, has seen the contrasting of personal beliefs about what was and was not the case. . . . What I want to avoid is an opinionated article. That is the way in which I am going to run this journal to the extent that I have anything to do with it" (Schopf to Eldredge, 11 November 1975: Schopf pap. 4, 12). Schopf appears to have been quite concerned not to alienate potential readers by featuring internecine controversies that would reflect poorly on the discipline. He also believed that a crucial step towards making paleobiology a "respectable" or "serious" discipline (truly "nomothetic," disinterested, objective) involved preserving a tone of critical detachment in its publications. As he explained to Eldredge: "I really do think it is far better for scientists (or others for that matter) to avoid the appeal to authority, and ex cathedra pronouncements about what MUST be the case. And I would try, and will in the future try, to remove as much of that as possible from everything that goes into this journal." To Schopf, this meant that "no useful purpose is served by Mayr-like roaring. . . . For he reminds me of the lawyer who if he had the facts, used them, but if not, then he shouted very loudly. I dislike intensely using critical assessments as the place to show off one's ego" (Schopf to Eldredge, 11 November 1975: Schopf pap. 4, 12). Schopf's immediate reaction may have indeed been partially that of a "wounded author," but his general perspective on the matter had (rightly or wrongly) everything to do with his view of the "proper" way to do science.

Maintaining the character of the journal as a fundamentally noncombative, calm outlet for science was a particular point of pride for Schopf over the years. Writing to Earle Kaufman (who was incoming president of the Paleontological Society) in 1980, Schopf noted, "There has been a *minimum* of disputation in *Paleobiology*," and he commented that unlike other journals—particularly *Systematic Zoology*—paleobiology had "*not* gotten bogged down in small details of criticism, and countercriticism." While "this never comes to 'public' notice," he continued, "in a very real sense [it] is one of the things I am most proud of" (Schopf to Kaufman, 4 April 1980: PS-*P* 14). James Hopson, who took over editorship from Schopf in 1981, echoed this sentiment, writing to a prospective author, "Short polemical notes are killing *Systematic Zoology* and I, and Tom before me, wish at all costs to avoid falling into this state with *Paleobiology*" (Hopson to Arthur J. Boucot, 25 March 1981: PS-*P* 13).

Intriguingly, Schopf wrote his letter to Kaufman to answer a "charge" that he "was too involved in *Paleobiology*." In his response, Schopf maintained that "rather than too much, I think I am not enough," and argued that it was his unique editorial vision and stamp that contributed to much of the journal's success:

Now, an editor like myself who has a vision about what science should be like in paleontology, and who aggressively pursues that vision, will always be criticized. The bottom line is the product—*Paleobiology* after 6 years—and in so far as more individuals take it than any other paleontological journal, that it is widely cited, etc. shows, as objectively as can be measured, that the society as been well served in this attempt to provide intellectual leadership. It is on that basis that I would wish to be judged (Schopf to Kaufman, 4 April 1980: PS-*P* 14).

Despite Schopf's prickly temperament, it was unusual for him to trumpet his own accomplishments. In this matter, however, he seems unapologetic: *Paleobiology* was *his* journal.

Paleobiology and Punctuated Equilibria

One area where the editors of *Paleobiology* could not avoid controversy was the debate over punctuated equilibria. Not only were the central protagonists (both pro and con) frequent contributors to the journal, but the authors of the theory also published an important defense in its pages. I have already discussed the formation of the theory, and here I will concentrate on the role *Paleobiology* and its editors played in encouraging discussion of the theory by offering a platform for both support and criticism. I will also examine the actions of Schopf as editor, which (typically) were not intellectually detached from the controversy. Finally, I will consider Gould's use of the theory as a context to promote his paleobiological reinterpretation of evolutionary theory, and the role *Paleobiology* served to give Gould a respectable platform from which to reach his fellow paleontologists.

In the fall of 1976, Gould mailed Schopf a manuscript entitled "Punctuated Equilibria: The Tempo and Mode of Evolution Reconsidered." Gould was its primary author, Eldredge was the official coauthor, and the paper served as the authors' first major reconsideration of their controversial theory. Gould described the "rather exuberant manuscript" as "a labor of love," and as a defense of the theory it served several functions (Gould to Schopf, 6 December 1976; Gould to Schopf, 23 November 1976: PS-*P* 3). First, the new manuscript emphasized "punctuationism" as the center of a philosophical reconceptualization of evolutionary change. Here the authors were more explicit than previously about the exact nature of the conceptual reconfiguration their theory brought to the understanding of macroevolution—in particular, by adapting Stanley's elegant formulation of the asynchrony of micro- and macroevolution published a year earlier in *Proceedings of the National Academy of Sciences* (Stanley 1975).[3] The paper also gave Gould and Eldredge a chance to address some of the criticisms that had been leveled at their theory, and to clarify what they actually intended to say in their first paper. The authors dismissed charges that they were motivated by an *a priori* disdain for gradualism, and that they took a "defeatist attitude" towards the testability of macroevolutionary claims using the fossil record (Gould and Eldredge 1977, 120). They also presented a significant amount of empirical evidence for the theory, by considering a wider sample of taxa than they had in the 1972 essay, and by answering empirical challenges to their earlier conclusions. Finally, Gould and Eldredge extended their model to propose a new and "general philosophy of change" in the natural world. Here they explored the sociocultural basis for gradualism, and proposed punctuationism as a new "metaphysic" which, they suggested, "may prove to map tempos of change in our world better and more often than any of its competitors" (Gould and Eldredge 1977, 146).

Gould and Schopf had developed a very friendly relationship from their collaboration on various projects (which over time developed into a warm friendship), but this submission was potentially complicated by the fact that Schopf was an avowed gradualist who had privately expressed his strong reservations about punctuated equilibria to Gould. While he acknowledged their difference of opinion, Gould's letters did not evidence any expectation of difficulty; on the contrary, he invited Schopf to "collect your thoughts on gradualism into a full-scale paper," since "we would certainly welcome a rebuttal at a higher level than has been directed to us so far." In any case, the paper was accepted with only minor revisions, and Schopf remarked, "We are pleased (and proud) to have your article in *Paleobiology*. . . . I feel this is a most remarkable article, one that will merit and require a lot of careful attention" (Schopf to Gould, undated [early 1977]: PS-*P* 3).

It is somewhat surprising that this 1977 paper did not arouse stronger passions. It was certainly not timid. Nonetheless, not only did Schopf— who would later quite vehemently and publicly attack the theory of punctuated equilibria—receive the paper warmly, but Philip Gingerich,

---

3. Gould acknowledged years later that Stanley "developed the implications that I had been unable to articulate from our original section on evolutionary trends." (Gould 2002, 980).

a staunch opponent of the theory in the 1980s, reviewed the manuscript positively, and characterized it as "an interesting and important paper" (Philip Gingerich, "Review of 'Punctuated Equilibria . . . ,' 31 December 1976: PS-*P* 3). Perhaps Gould was simply right when he later reflected, "The early history of punctuated equilibrium unfolded in a fairly conventional manner for ideas that "catch on" within a field. The debate remained pretty much restricted to paleontology. . . . [and] most discussion, to our delight, arose from empirical and quantitative studies" (Gould 2002, 980).

The turning point in the controversy over punctuated equilibria seems to have come—as Gould himself acknowledged—in 1980 when Schopf asked him to contribute several articles assessing the status of paleontology for the fifth anniversary of the journal. Gould happily obliged, and in a single issue of the journal published two reflective essays on the state of the discipline. In the first, which he titled "The Promise of Paleobiology as a Nomothetic, Evolutionary Discipline," Gould celebrated the recent advances in paleontology and reiterated his call for further progress towards revision of evolutionary theory based on macroevolutionary modeling. One of Gould's major arguments was that macroevolutionary patterns did not follow the same deterministic lines as microevolutionary trends, which, he implied, might challenge the received view of Darwinism (as codified in the modern synthesis).

However, it was the second essay, "Is a New and General Theory of Evolution Emerging?," that stirred up the most controversy. It was here that Gould made his infamous claim that "if Mayr's characterization of the synthetic theory is accurate . . . then that theory, as a general proposition, is effectively dead, despite its persistence as textbook orthodoxy" (Gould 1980b, 120). Years later, while reflecting on the travails of punctuated equilibria, Gould located the origin of much difficulty in this essay: "The received legend about this paper . . . holds that I wrote a propagandistic screed [claiming] . . . first, the impending death of the Modern Synthesis; and second, the identification of punctuated equilibrium as the exterminating angel (or devil)" (Gould 2002, 1002). I will examine these papers in greater detail in chapter 10. The major point here is that it is these 1980 essays, and not the 1972 or 1977 presentations of punctuated equilibria, that initiated the real furor over the theory. But another point worth making is that despite the fallout that resulted, the 1980 papers can be considered a triumph for *Paleobiology*: here, finally, biologists were obviously reading the journal and taking its contents seriously enough to become upset, and it is only really after that point that punctuated equilibria achieved general cultural currency.

This episode also put *Paleobiology* at the center of a very visible controversy, and Schopf made the decision to let the debate play out in the pages of the journal. The issue in which Gould's pieces appeared was mailed to subscribers in March 1980; by April, Schopf had received a manuscript rebutting Gould's claims from Steven Orzack, who was a graduate student in Richard Lewontin's lab at Harvard. Orzack's response—originally titled "Evolution of a Straw Theory," but published as "The Modern Synthesis is Partly Wright"—alleged that Gould had characterized the modern synthesis as far more inflexible and monolithic than it actually was. In particular, he drew attention to the more pluralistic viewpoints of important synthesists, including Sewall Wright, Simpson, and Theodosius Dobzhansky, all of whom, he maintained, eschewed "the blind appeals to natural selection that Gould asserts are characteristic of modern synthetic interpretations of evolutionary phenomena" (Orzack 1981, 130–31). Here Orzack raised an important point: in his 1980 essays and elsewhere, Gould presented a particular interpretation of the modern synthesis that suited his own paleobiological agenda. Gould's response, which was published alongside Orzack's piece, defended his invocation of Mayr's reduction of evolution to micromutation and natural selection as an appropriate general label, and argued that Wright and other pluralists did not have much effect on the articulation of the synthesis. "I didn't, after all, invent a definition to suit my own purposes, but used one common to critics and supporters alike," he claimed (Gould 1981, 131). In presenting Mayr's quite limited definition of the synthetic theory, however, Gould positioned himself perfectly to offer paleobiological, macroevolutionary theory as a correction to, or completion of, the synthesis.

Schopf sympathized with Orzack's argument, as did two rather extraordinary reviewers for the piece: G. G. Simpson and Sewall Wright. Both urged publication, and both agreed substantively with Orzack's correction of Gould's historiography. Simpson tersely commented, "Orzack's brief paper is a reasonable and documented correction of one of the many misrepresentations in Gould's publications. It is fitting that *Paleobiology* publish this partial correction of misrepresentations published in that journal" (Simpson, "Review of 'The Evolution of a Straw Theory,'" 23 September 1980: PS-*P* 7). Wright's evaluation was that he was "not sure that the author fully understands my position, although he does so much more than have Fisher, Mayr or Gould" (Wright, "Review of 'The Evolution of a Straw Theory,'" 29 September 1980: PS-*P* 7). Simpson rated the submission "excellent" while Wright labeled the piece "good." What is particularly interesting is that Schopf appears to have ignored two additional reviews, each of which evaluated the piece as "poor." In the first place, it is highly unusual to have sought four reviews of a short note; secondly, the reviews from Wright and Simpson were commissioned in September of 1980, while the first, negative review—by the biologist Michael Wade—was requested much earlier, in April, and received in August of that year (Michael Wade, "Review of 'The Evolution of a Straw Theory,'" 6 August 1980: PS-*P* 7).

Any question about whether Schopf intentionally suppressed these reviews is answered by a letter he sent to Orzack in which he explained, "We have had your note reviewed, and wish to publish it. I enclose our *two* reviews and your copies" (Schopf to Orzack, 9 October 1980; emphasis added). Needless to say, the correspondence included only Wright's and Simpson's reviews. It is not unprecedented for an editor to mitigate criticisms by reviewers that appear unfairly biased or unqualified, or to request additional reviews. The two negative reviews, however, came from well-respected biologists (the second was by Dobzhansky's former student Francisco Ayala) and had been commissioned well before the second, more positive set of reviews. The question is not whether Orzack's piece should have been published; the positive reviews were sufficient justification, and the final piece itself was well written and measured. The question is why Schopf went about the process in such an underhanded way.

Over the next two years, Schopf finally broke his own silence about punctuated equilibria in a series of highly critical publications. His first step was an item in *Paleobiology*'s spring 1981 Current Happenings section, which was a recently instituted forum in the journal for news and editorial comment about the field. The second was a paper in *Evolution* titled "A Critical Assessment of Punctuated Equilibria," published in late 1982. The third was a 1983 letter to the journal *Science*, coauthored with Polish paleontologist Antoni Hoffman (Schopf 1981; Schopf 1982; Schopf, Hoffman, and Gould 1983). In each of these publications Schopf took a different tactic to attack Gould's ideas—as impartial journal editor, as careful empirical paleontologist, and as concerned scientific citizen—and I will briefly examine and contrast his strategy in each piece. I will also consider a perplexing question: Given Schopf and

Gould's close agreement about the general agenda and goal of paleobiological work, why did Schopf choose to undermine his discipline's most
prominent theorist (and a close personal friend) in a publicly and potentially embarrassing way?

Schopf's "Current Happenings" piece, "Punctuated Equilibrium and
Evolutionary Stasis," was an ostensibly neutral attempt "to place the
paleontological and biological evidence in a 1981 perspective" (Schopf
1981, 156). However, Schopf's personal beliefs very quickly became apparent. He noted, for instance, that punctuated equilibria seemed to demand "some strongly deterministic factors in order to account for patterns of speciation and extinction in the fossil record," and he revealed
that "the major purpose of this 'Current Happenings' is to encourage
the quantitative and qualitative evaluation of these limitations and prediction of the punctuated equilibrium model so that a truer picture of
evolutionary history may be obtained" (Schopf 1981, 158). Schopf then
argued that the case for punctuated equilibria was weakened by several
empirical factors: (1) The incompleteness of the fossil record meant that
the appearance of "suddenness" (or punctuation) of speciation is an artifact "almost guaranteed" by the state of fossil knowledge, and it "indicates nothing of any meaning about the process of evolution which
led to these classes" (Schopf 1981, 160). (2) Because taxa are commonly
assumed to be present for the entire duration of the geologic stages in
which they originate, their "book value" is often exaggerated, meaning that their durations are often exaggerated. (3) Paleontological data
is plagued by insufficient morphological information: "Because only the
most resistant and most numerous of hard parts can ever be studied, paleontologists must recognize species by recourse to only a small part of
an organism's actual evolutionary change" (Schopf 1981, 161). (4) Sample populations are poor, and it is easier to discern species-level evolution in organisms with well-defined hard parts, so we tend to omit organisms with "simple, relatively undifferentiated forms." (5) General
limitations of taphonomy (fossil preservation) mean that short-lived species are much less likely to be preserved or recorded. The majority of
Schopf's complaints thus had to do with "signal" errors—i.e., observations about perceived limitations the reliability of the fossil record. Ironically, Schopf, who was one of paleobiology's most ardent defenders, was
relying on the most traditional line of critique against paleontology.

As a courtesy, Schopf sent Gould a copy of his piece prior to publication, and Gould's response was indignant. He did not object to Schopf's

critique of the theory per se, but rather to Schopf's use of his position as editor to present it:

> I must confess—and I expressed this to Jim [Hopson] when he called me for another reason two weeks ago—that I am not altogether happy with the forum that you have chosen for the piece. If it had been submitted as a regular article, I would have welcomed it entirely (while disagreeing strongly, of course, with its conclusions). In a sense, I am flattered that you consider punctuated equilibrium as a "happening"—and therefore worthy of inclusion in your section. Yet I confess that I do not think it fair for you, as editor, to use this section as a forum for expressing personal viewpoints on issues of the moment (Gould to Schopf, 21 April 1981: Schopf pap. 8, 30).[4]

Gould went on to stress his belief "that accounts of happenings may and indeed even should express a point of view," but added, "Ideas aren't events—and I would argue that the editor of such a section should not use his prerogative as a platform for expressing personal opinions about theoretical issues." It seems that Gould objected particularly to the ostensibly objective way in which Schopf had presented the piece: "I don't think I feel angry about this, but I am not unconcerned either. I just don't see how one man's viewpoint can become a kind of official line in one section of a journal."

Schopf's response was defensive. He noted that there was ample precedent for opinion in the "Current Happenings" section, and moreover he defended the right—and even necessity—for editorial viewpoints to appear in this section of the journal. "The unfairness," he wrote, "could come from a monolithic view in a journal," or "from willful misrepresentation. Neither, I think, is part or parcel of the way I have executed the responsibilities I have had with regard to *Paleobiology*" (Schopf to Gould, 24 April 1981: Schopf pap. 8, 30). He noted that he had the piece carefully reviewed—"In essence I have tried to treat it as a submitted piece"—and that he had been mindful of his critics' reactions. Schopf also contacted Hopson, whom he accused of complicating the dispute by inviting Gould to submit a rebuttal: "With regard to the Current Happenings, it seems to me that you should have consulted me before agreeing to Steve that he could 'reply,' in so far as you wrote me and asked me

---

4. As will be explained shortly, University of Chicago paleontologist James Hopson had by this point succeeded Ralph Johnson as coeditor of *Paleobiology*.

to organize that part of the journal!" (Schopf to Hopson, 23 April 1981: PS-*P* 14). He further chided Hopson "that you should be very, very careful about agreeing to things unless you intend to fully back them up," and faulted his coeditor for not being "able to cool him [Gould] off before he got a full head of steam going." The message was quite clear: in this case, Hopson should defer to Schopf's experience and relationship with Gould, and stay out of it. As Schopf put it, "I have every intention of getting along well with Steve—as I have for years—and as I have defended him in a lot of situations involving 3rd parties."

This disagreement did not cause serious damage to Schopf's friendship with Gould, but it did begin a new, more directly combative, phase of their relationship. For the next several years (until his death in 1984), Schopf devoted considerable energy to attacking punctuated equilibria in print and elsewhere. His next salvo was a paper in *Evolution*, published in late 1982, that added considerably more empirical depth to the arguments he raised in his "Current Happenings" piece. The final published version of this essay took a much more measured tone than the previous one (for example, the paper opened by claiming to "take the position of a devil's advocate" with regard to punctuation), but this was a product of the review process. The manuscript Schopf originally submitted was far more partisan, and often seemed to associate criticism of the theory with that of its author (Gould). One review of the manuscript stated bluntly, "I don't feel that this paper needs to be riddled with *ad hominum* [sic] references to Steve Gould. From a supporter, the term 'Gouldism' would appear laudatory; in this ms it simply seems to be sarcastic" (anonymous, "Review of MS by Schopf": Schopf pap. 2, 5). Another reviewer found it "heavily slanted" and "irritatingly polemical and biased," and recommended extensive revisions before publication anonymous, "Comments on 'A Critical Assessment . . .'": Schopf pap. 2, 5).

It is notable also that in this and other critiques of punctuated equilibria, Schopf focused almost entirely on Gould—it is almost as if he had forgotten Eldridge's involvement entirely. In fact, it is difficult to escape the feeling that Schopf's opposition to punctuated equilibria was mixed with his personal feelings towards Gould. This was even more apparent in the letter to *Science*. The ostensible purpose of the comment was to question whether "a static hierarchy [is] a true and correct view of life" (Schopf, Hoffman, and Gould 1983, 438). Again, however, the published version differs significantly from the original manuscript. In its first draft, the letter began: "S. J. Gould arguably is becoming the most important single force in the shaping of current popular evolutionary thought." It went on to describe him as "a very fine human being who values scholarship," and concluded: "For [his] skill at argumentation and conceptual organization [he] is widely and deservedly admired" (Schopf and Hoffman, "Punctuated Equilibrium and the Fossil Record—Draft A": Schopf pap. 9, 133). The second draft of the letter tried a new (and somewhat toned down) opening, which enthused: "We admire very much [Gould's] efforts and readily acknowledge that no paleontologist has contributed as much to the popularity of evolutionary theory" (Schopf and Hoffman, "Punctuated Equilibrium and the Fossil Record—Draft B": Schopf pap. 9, 133).

All of these statements were ultimately omitted from the published letter, and most readers would have been unaware of Schopf's appreciation for Gould's ideas and character. Nonetheless, they reveal an interesting psychological dynamic between Schopf and Gould. Schopf's feelings about Gould, and punctuated equilibria, were clearly quite complicated and even conflicted. In his editorial capacity at *Paleobiology* and elsewhere (as editor of *Models*, for example), Schopf again and again provided a platform that helped Gould launch and establish his evolutionary views. In Schopf's ideal vision for science, disagreements, though often heated, would always be amicable, and never personal. In his particular agenda for paleobiology, this was necessary in order to establish potentially controversial new methods and ideas on the kind of sound empirical footing that would impress and convince scientists in other fields.

In late 1982 Schopf sent Gould a letter explaining his position towards punctuated equilibria in very candid terms. His major objection, he wrote, was that the theory "got taken too far." He worried that "the many biases of the fossil record that needed carefully, and systematically, to be looked into, never got looked into," and that the effect it would have on future work could be harmful: "Unless those who are brought into the field learn the rigor of testing, it will be for naught" (Schopf to Gould, 19 September 1982: Schopf pap. 9, 106). Still, he explained, "I don't have a campaign against P. E."; rather, "I do have a campaign *for* rigorous testing if these ideas." Part of Schopf's justification for his critical stance—and for his continued warm regard for Gould—was that he felt Gould was not entirely responsible for the excesses connected with the theory:

I don't think it's entirely your fault that P. E. got out of hand. As I see it, the "press" (Roger Lewin et al.) discovered Steve Gould, and what Steve Gould *happened* to be on was P. E.. If it had been some other issue, then *that* issue would be well known. The press recognizes personalities. It publicizes what those personalities are saying and doing. Sometimes, those particular sayings, and doings, are beyond their worth. I *think* this is what happened to P. E.. The press didn't discover P. E. It discovered S. J. G.—and S. J. G. happened to discover P. E. Pure accident. Five years later, or earlier, it would have been different.

Schopf was very clear, however, that his feelings on the matter did not jeopardize his affection for Gould: "So, Steve, I have felt and do feel, very close to you. But, I have to (had to) go my own way on P. E.. But I'll defend you as a person as long as I can write."

Paleobiology's Success

During the initial period of the journal's existence, its editors received a number of testimonials from colleagues, which reflected the widely-held view that *Paleobiology* was an unqualified success. In 1975, at the end of *Paleobiology*'s first year, Valentine sent Schopf a letter reporting on the journal's reception in England (where Valentine was spending a sabbatical). He noted that while some paleontologists found the content "over their heads," students had "reacted quite favorably, especially in departments where they are exposed to some biological aspects of the [fossil] record." He concluded, "The journal is pumping badly needed biological light into paleo," and he was "tickled" by its success (Valentine to Schopf, 18 November 1975: PS-*P* 14). The response from others was similar, as a sampling of reactions shows: A.J. Rowell admitted to initially harboring doubts about the advisability of a new journal, but added that the success of the endeavor had "proved me wrong" (Rowell to Schopf, 27 November 1976: PS-*P* 14). Anna K. Behrensmeyer called it "one of the best journals in my realm of interest," and labeled it "both readable and informative" (Behrensmeyer to Schopf, 20 December 1976: PS-*P* 13). Warren Addicot of the US Geological Survey lauded Schopf's stewardship as "remarkable and truly superb," adding that he was "deeply impressed and very pleased" by the "rather overwhelming response to your new journal" (Addicott to Schopf, 1 November 1977: PS-*P* 13). Everett

Olson was sufficiently impressed to send a contribution to the journal's Patron's Fund, and called *Paleobiology* "a very fine journal . . . much the best of its type and in many ways unique" (Olson to Schopf, 4 January 1979. *PS-P* 14). And Stephen Wainright, a prominent biologist at Duke University, was unequivocal in his praise for the journal, surprising the editors with the following note:

> Because I am stimulated by the directions that are being taken by many bi-
> ologists today in weaving ideas and information of the extant and the extinct
> into evolutionary cloth; and because the best of this material has/is appeared/
> ing in *Paleobiology*; and because I believe the next 20 years in biology will be
> dominated in the matter of synthesizing information and ideas by paleobiolo-
> gists and their "like"; I wish to make the enclosed contribution to the *Paleo-
> biology* Patron's Fund. You have my joy and respect (Wainright to editors,
> 6 May 1981: PS-*P* 14).

From this anecdotal evidence, Schopf could clearly take heart that his message was reaching its desired audience, and that this audience greatly appreciated his efforts.

Financially, the journal was also a success. The goal for the first year was to obtain 1,700 subscribers, and while the initial response fell just short, by 1976 *Paleobiology* had roughly 1,850 subscriptions. More important, the journal was in the black for its first two years. The 1976 report to the Paleontological Society Council recorded $23,700 in subscription income, balanced against $24,000 in editorial costs. Additional income—including page charges—made up the remaining deficit (Schopf, "Semi-Annual Report," 12 October 1976: PS-*P* 14). The next year the journal again broke even, and the number of subscriptions reached 2,000; in 1978, this number rose to 2,100, and the journal actually made a profit. At the end of his second three-year term as editor of the journal, Schopf cowrote a report (with Hopson) to the council summarizing the success of the venture during its first six years. They reported, first, that the journal was continuing to receive a healthy volume of high-quality manuscripts—an average of 80 to 85 per year, with an acceptance rate of 40–50% (Schopf and Hopson, "Six Year Report to the Council of the Paleontological Society Regarding *Paleobiology*: 1974–1980": PS-*P* 14). They also drew attention to the yearly increases in subscriptions, and optimistically estimated that the current total could be increased by at least 25% during the next three years. Additionally, they noted that the journal "has operated in the black every year of its existence," and proposed that the growing Patron's Fund become an endowment with income used to support non-operating expenses, such as reprinting back issues. Finally, they made the case that *Paleobiology* had genuinely contributed to shaping the field of paleontology: "It is no accident that the success of *Paleobiology* as a journal has paralleled the increasing attention which paleontology in general has received during the past few years."

In 1985, *Paleobiology* celebrated its tenth anniversary with a special issue devoted to assessing the success and significance of recent developments in the field. In a sad twist, neither of the founding editors was alive to see its publication, both having died short of their fiftieth birthdays. Ralph Johnson passed away in 1976 at the age of 49, having helped steer the journal through its first year of existence. Schopf's obituary for him praised "his dedication to integrity and responsibility in both professional and personal matters" as "complete and unswerving," and cited the "enormous trust in his judgment" he inspired in colleagues (Schopf 1976b, 391). After Johnson's death, Jim Hopson took his place as coeditor, where he joined Schopf until 1981, when Schopf's second three-year term as editor expired. In 1984, while leading students on a field trip in Texas, Schopf died suddenly of an apparent heart attack at the age of 44. Gould's extraordinarily personal eulogy appeared in the journal in the Spring issue, and the next year coeditors Jack Sepkoski and Peter Crane dedicated the anniversary issue to the memories of both Schopf and Johnson.

According to the editorial introduction, the tenth anniversary issue was conceived as an opportunity to review "many of the topics that have been the journal's mainstay over the past ten years," and to assess "what progress has been made in the various areas of inquiry over the past several years and what problems are most likely to be of concern in future years" (Sepkoski and Crane 1985, 1). The contents were dominated by authors who had most frequently contributed to the journal, and who had become the first generation of workers in the new paleobiology: Gould on hierarchical levels of selection; Stanley on rates of evolution; Raup on mathematical modeling; Eldredge on systematics; Valentine on biogeography; and Daniel Fisher on morphology. Younger, up-and-coming paleobiologists including David Jablonski, Jennifer Kitchell, Susan Kidwell, and Andrew Knoll also contributed essays. Overall, the issue was a celebration of the success of the journal in publishing papers

uniting "paleontology with modern biology," and in promoting methodologies in theoretical paleontology, morphology, taphonomy, paleoecology, paleobiogeography, "and theoretical aspects of population biology,
systematics, and stratigraphic paleontology—subjects united by a prevalent attempt at a biological interpretation of the fossil record" (Sepkoski
and Crane 1985, 1).

The only obvious omission in this group was Schopf, who before his
death had been slated to have the first essay in the issue, titled "The History and Influence of *Paleobiology*" ("Draft Table of Contents for *Paleobiology* Anniversary Issue," undated [circa June 1983]: PS-*P* 11).
Schopf had taken an active role in planning this issue, despite having
stepped down as editor in 1980. In 1982 he wrote to Hopson suggesting
that Gould be commissioned to write an essay on the subject "Reflections on the last 20 years of paleontology" (Schopf to Hopson, 13 December 1982: PS-*P* 13). At this time, Gould was battling cancer, and the
prognosis was fairly grim. Schopf suggested that Gould write the essay
then—well in advance of the anniversary issue—and "if, most unfortunately, something should happen to him, *Paleobiology* would run it." If
not, Gould could be invited later to compose a new essay, as a substitute. Hopson decided not to act on Schopf's proposal for several reasons.
First, Gould had recently written about the state of the field (in 1980).
Second, Gould was ill, and "it is difficult enough for Steve to maintain a
positive psychological state without being asked to write something that
only has value because he might die very soon." Finally, Hopson reasoned, "If Steve wanted to write such an overview, I think he would be
doing it." He added, "For an editor of the journal which stands to benefit should he die," to ask was "in bad taste." Hopson closed by advis-
ing, "As his friend you should drop it" (Hopson to Schopf, 17 December
1982: PS-*P* 14).

Schopf's historical essay was never published, but it was preserved
in the journal's files. The text offers a somewhat sentimentalized assessment of his personal achievements and regrets as editor. Among his
proudest accomplishments, Schopf listed minimizing "the extent of vituperation" published in the journal and the "zero backlog for every issue" (e.g., the fact that returned papers had gone directly into the next
issue with no delay). Interestingly, he seemed to have learned very little
from those occasions on which he was taken to task by authors for abus-
ing his editorial position, writing: "Dealing with authors (each of whom
has his or her own various degree of paranoia) takes no special talent—

only a certain shaking of the head as the worst among them fire off letters to the President of the Society complaining of evil doings—without sending a copy to the editor, of course. . . . Seeing such individuals in action is one of the sadder aspects of being an editor." Schopf didn't spare himself criticism, although even in describing his failings he managed to sound defensive: "My major failure has been in efforts to try to get those whose business it is to know the systematics of a group to think of that group chiefly as an experimental system, and thereby to use that group as a means to address paleobiological issues of broad interest, and to submit that work to *Paleobiology*." In other words, his "failure" was that of expecting "that this group of middle-aged systematists, having put in years of homework to learn about a group, would then contribute a large number of relevant papers." In the end, though, Schopf seems to have been unable to take undiluted pleasure in the accomplishments of his six years as editor, ruefully concluding that while "few would have predicted that *Paleobiology* would have done as it has in its 6 year history," nonetheless "some may believe, I among them, that it should have done better" (Schopf, "*Paleobiology*," undated [1980]: PS-*P* 14).

It is difficult to imagine, however, that *Paleobiology* would have had the same character without Schopf's leadership.; For better or for worse, his personality, professional agenda, and goals for the field made an indelible stamp on the journal, and consequently on the emerging field it promoted. In 1979, having served two consecutive terms as editor, Schopf evidently felt that his work was not yet finished, and he wrote to Paleontological Society President Frank Stehli to ask that his position be extended by an additional three-year term. As he explained,

> The chief reason for this is that I believe the journal is not quite yet in a position where it can take care of itself. Many of the most extensive and important papers which we have published have been ones which I personally have gone after. . . . There are many things that can be done with *Paleobiology* but which require intense dedication—dedication which is not likely to come from most journal editors (Schopf to Stehli, 4 January 1979: Schopf pap. 5, 42).

Schopf realized (correctly, as it turns out), that the council might not like this plan, so he spelled out two prerequisites for anyone chosen to succeed him:

If this suggestion is not acceptable, I would urge that new co-editors be young persons. In the hands of an older person whose reputation has been made *Paleobiology* does not mean much. In the hands of a younger person *Paleobiology* offers the opportunity to develop over a 5 year period a national or international reputation. Thus *Paleobiology* is likely to fare much better in the hands of a younger person. The second requirement which I have in mind is that the person must know zoology and zoologists. This is a stringent requirement, but I think for this journal it is an absolutely necessary one. The journal is interdisciplinary, and its editors must reflect that.

Schopf handed over the reins in 1980, and his life ended only a few years later. But in the short period of his activity in the paleobiological revolution between 1971 and 1980, he had a tremendous influence on the institutionalization of paleobiology and the establishment of theoretical, evolutionary paleobiology as a legitimate discipline.

CHAPTER SEVEN

"Towards a Nomothetic Paleontology": The MBL Model and Stochastic Paleontology

The Roots of Nomotheticism

By the early 1970s, the paleobiology movement had begun to acquire considerable momentum. A number of paleobiologists began actively building programs of paleobiological research and teaching at major universities—Stephen Jay Gould at Harvard, Tom Schopf at the University of Chicago, David Raup at the University of Rochester, James Valentine at UC Davis, Steven Stanley at Johns Hopkins—which would flourish as centers of the movement over the next decade. Schopf had organized a well-received book (*Models in Paleobiology*) that presented exciting new theoretical approaches to the study of fossils. The movement even had its own textbook—*Principles of Paleontology*, published in 1971 by Raup and Stanley—that would provide an essential pedagogical foundation for paleobiology for the next two decades. And the establishment, in 1975, of the journal *Paleobiology* ensured that the "new paleobiology" would have a friendly outlet for publication and a platform from which to promote its agenda.

Nonetheless, not all of the elements of a paleobiological revolution were yet in place. While punctuated equilibria would eventually acquire iconic status as one of the movement's signature theories, there was as yet no central theoretical and methodological commitment among paleobiologists. Paleobiologists could agree on some basic tenets: that the fossil record provided genuine insight into evolutionary rates and processes, that quantitative analysis and modeling using computers promised exciting new avenues for research, that insights from biology and ecology could more profitably be applied to paleontology, and that the future lay in assembling large databases as a foundation for analysis of broad-scale patterns of evolution over geological history. But in comparison to other expanding young disciplines—like theoretical ecology— paleobiology lacked a cohesive theoretical and methodological agenda. However, over the next ten years this would change dramatically.

One particular ecological/evolutionary issue emerged as the central unifying problem for paleobiology: the study and modeling of the history of diversity over time. This, in turn, motivated a methodological question: how reliable is the fossil record, and how can that reliability be tested? These problems became the core of analytical paleobiology, and represented a continuation and a consolidation of the themes we have examined thus far in the history of paleobiology. Ultimately, this focus led paleobiologists to groundbreaking quantitative studies of the interplay of rates of origination and extinction of taxa through time, the role of background and mass extinctions in the history of life, the survivorship of individual taxa, and the modeling of historical patterns of diversity. These questions became the central components of an emerging paleobiological theory of macroevolution, and by the mid 1980s formed the basis for paleobiologists' claim to a seat at the "high table" of evolutionary theory.

This chapter will explore the beginnings of this development in paleobiology by examining the emergence of a new, clearly-articulated agenda: the often-expressed desire to construct a "nomothetic paleontology." This phrase originated in 1973 in the first of an important sequence of papers that sought to develop a stochastic, equilibrial simulation of evolutionary dynamics over time—what came to be known as the MBL model—and quickly became a rallying cry (most often repeated by Gould, its inventor) for the new approach to paleobiology. The term "nomothetic" essentially means "law-producing," and John Huss has outlined some of the essential features of this approach:

> "Nomothetic" paleontology would: (1) work from the top down, deducing consequences from general models (i.e., models applicable across times and taxa); (2) predict the behavior of evolutionary events as statistical ensembles, rather than individually, (3) use equilibrium models as comparative baselines to single out elements requiring explanation in terms of specific, non-

recurring causes; and (4) attempt to replicate empirical phenomena using the minimum necessary departure from a simple random model (Huss 2004, 53).

This goal would manifest itself in a number of ways throughout the period that will be considered in this and succeeding chapters, and while its ultimate accomplishment may have been elusive, it nonetheless both helped focus paleobiologists towards a set of more achievable aims, and also gave the movement a greater sense of purpose.

The 1970s also saw a decade of debates among a core group of paleobiologists that illuminate some of the central themes and tensions in the new paleobiology. These themes encompass some of the basic methodological and philosophical issues confronting paleobiology, which ultimately resolve to between distinct and competing interpretations (or metaphors) for how best to "read" the fossil record, which entailed divergent responses to how to solve "Darwin's dilemma" for paleontology. Darwin himself recognized that if the fossil record were to be taken at face value—or "read literally"—it would challenge many of his assumptions about the tempo and mode of the evolutionary process. However, a small but influential group of paleontologists, from John Phillips to George Gaylord Simpson, argued forcefully that the fossil record was a less imperfect document that Darwin had suspected. It was precisely on this basis that Eldredge and Gould's 1972 paper on punctuated equilibria endorsed a more prominent role for paleontology:

> Many breaks in the fossil record are real; they express the way in which evolution occurs, not the fragments of an imperfect record. The sharp break in a local column accurately records what happened in that area through time. Acceptance of this point would release us from a self-imposed status of inferiority among the evolutionary sciences. Our collective gut-reaction leads us to view almost any anomaly as an artifact imposed by our collective millstone—an imperfect fossil record (Eldredge and Gould 1972, 96).

From the earliest drafts onward, Eldredge and Gould invoked the metaphor of "reading," calling for "a more literal interpretation of the fossil record" or, as elsewhere stated, "the concept that phylogeny can be read directly from the rocks" (Eldredge, draft 1, 14 and 17: Eldredge pap.).

However, developments to be discussed over the next two chapters would introduce an alternative, approach in which the fossil record was not to be read literally, but rather idealized, abstracted, and generalized from. In various ways, paleobiologists attempted to use analytical techniques to identify and circumvent limitations in the fossil record, a document they acknowledged could never be "read literally." Almost from the very start, then, the revolutionary phase of paleobiology endorsed two, potentially conflicting metaphors for rereading the fossil record: a "literal" versus an "idealized" approach (to complicate matters even further, Gould was involved as a primary advocate of *both* interpretations). Reconciling these two metaphors was a central obstacle for the success of the paleobiological movement. I will argue in the next chapter that the emergence of a "taxic approach" to paleobiology in the late 1970s brought a partial reconciliation of these two metaphors, though lingering disagreements still remained.

Reading as Idealization: The MBL Model and "Stochastic Paleontology"

The Origin of the MBL Project

The simulation project that came to be widely known in the paleobiological community as the MBL model was the result of a collaboration, instigated by Tom Schopf in 1972, between Schopf, Gould, Raup, and Dan Simberloff, who met informally for discussions at the Marine Biological Laboratory in Woods Hole (hence the name MBL). The model itself was presented in a series of jointly authored papers between 1973 and 1977, after which point the collaboration effectively disintegrated. In its most basic terms, the MBL model was a computer simulation of the evolution of hypothetical, randomly-generated lineages or "phylogenies" that tested whether random phylogenetic patterns could be produced that matched actual patterns in the fossil record. The general idea was that these simulations provided a null hypothesis against which the assumption that evolutionary patterns are deterministic could be tested; more broadly, this can be seen as part of paleobiologists' efforts to determine whether the fossil record is a reliable document. As with the *Models* project and the launching of *Paleobiology*, Schopf's participation was formative and essential. However, whereas Schopf's role in those other endeavors was primarily organizational and administrative, in the MBL project Schopf made an important creative and theoretical contribution. Indeed, Schopf would come to regard the MBL model as his primary intellectual contribution to paleobiology, and identified it very closely with his grand personal vision to build a "stochastic" or "particle paleontology" (Schopf's term) in which the history of life could be idealized as a series of "gas laws." Schopf's energy and enthusiasm for this vision initially bound the group together and inspired its successes; as time went on, however, it also became a source of internal discord that ultimately led to the collaboration's dissolution.

Like many of the major events in the growth of paleobiology, chance and contingency played significant roles in the origin of the MBL model. In early 1972, even as the *Models* volume was being finalized, Schopf was looking ahead to the next opportunity to advance the agenda of paleobiology. As discussed in chapter 6, he had already begun to explore the possibility of starting a new journal. But he also wanted to capitalize on the intellectual momentum of the *Models* symposium and book. Schopf usually spent his summers at Woods Hole, and in 1972 Gould was also planning to spend time there working on the draft of what would eventually become his first book, *Ontogeny and Phylogeny* (Schopf to Gould, 3 February 1972: Schopf pap. 5, 14; Gould 1977). Schopf decided to take advantage of this coincidence by organizing a small brainstorming session involving Raup and Simberloff as well. He described the meeting in his invitation to Raup as an opportunity "to get together for about 3 days to discuss the way in which theory can be more directly introduced into invertebrate paleontology." Clearly, Schopf envisioned the session as an explicit continuation of the *Models* project, although there is little indication he had any distinct ideas about what form the collaboration might take. As he continued in his letter to Raup,

> Of course, one can never "program" good research, and in any event research is always done by individuals and not teams, yet the self-conscious attempt to introduce more theory into our mass of facts might be a very useful thing to do. I have heard about similar affairs working out well; perhaps the most spectacular were the various outcomes when Lewontin, Ed Wilson, McArthur [*sic*], Levins and Egbert Leigh met for a week or so over a couple of summers at McArthur's [*sic*] New Hampshire farm in the early '60's (Schopf to Raup, 5 March 1972: Schopf pap. 3, 30).

At first blush it might seem odd that Simberloff—a newly minted PhD in ecology with very little experience in paleontology—would be included in this group. Schopf's only previous interaction with Simberloff had been in producing the *Models* volume, where Simberloff had, after all, been a second-choice replacement for E. O. Wilson. It appears, however, that from the very start Schopf imagined that equilibrial island biogeography would have a prominent place in the brainstorming sessions, and in that light Simberloff was a natural choice. Not only had he trained with Wilson, but as Schopf went on to explain, "Simberloff has the mathematical tools which might be required, is hungry to advance a field different from his dissertation area, and during a 2–3 hour conversation over dinner [at the GSA meeting in 1971] appeared to be a very perceptive, responsible and reasonable fellow" (Schopf to Raup, 5 March 1972: Schopf pap. 3, 30).

In retrospect, MacArthur and Wilson's theory of island biogeography had a formative influence on inspiring paleobiology to become more models-oriented. Schopf was a central conduit for this influence. He had come to know Wilson casually during summers spent at Woods Hole, and the two developed a cordial relationship. When, for example, several years later Wilson became embroiled in his infamous controversy over sociobiology with Gould, Lewontin, and others, Schopf wrote Wilson to express his "greatest personal sympathy" and "great deal of affection" for Wilson. He also noted, "You have been of great help to me at Woods Hole, and I have, as you know, a very high regard for you scientifically, and to the extent that I know you and your family, also personally" (Schopf to Wilson, 2 March 1976: Schopf pap. 5, 35). Wilson wrote back to thank Schopf and to "reciprocate the feelings of friendship that motivated it" (Wilson to Schopf, 4 March 1976: Schopf pap. 5, 35).

Schopf's friendship and admiration for Wilson had a significant intellectual effect on Schopf, and from fairly early in his career Schopf gave serious consideration to Wilson's equilibrium model. In an autobiographical sketch written in 1976, Schopf noted that his first use of equilibrium models was in a paper that was "helpfully reviewed by E. O Wilson" (Schopf 1972a; Schopf, "Notes on 1956–1976," 5 July 1976: Schopf pap. 7, 1). Additionally, while still completing the *Models* volume, Schopf wrote a paper, entitled "Ergonomics of Polymorphism: Its Relation to the Colony as the Unit of Natural Selection in Species of the Phylum Ectoprocta," that paid explicit homage to Wilson's work on the caste system in ants. In that paper, Schopf examined polymorphism in bryozoans, which he likened to the caste structure Wilson had observed in ant colonies and had described as a form of "kin selection" (Schopf 1973, 259). In the acknowledgements, Schopf thanked Wilson for conversations that had led to the paper's conception and for commenting "extensively" on the manuscript. Schopf noted in his autobiographical sketch that this paper foreshadowed a "general model of speciation" in Bryozoa, published in 1976, that presented "an equilibrium model using stochastic processes" (Schopf 1976a).

Schopf's most explicit early theoretical engagement with equilibrium models came in his introduction to the *Models* volume, "Varieties of Paleobiologic Experience." In this essay Schopf examined a variety of theoretical approaches to paleontology, but focused especially on equilibrium models, which he likened to "gas law[s] in which the state of any particular molecule is immaterial to the general description of the behavior of the volume as a whole." The value of such an approach, he argued, would be that "where the particular history of species may be immaterial. . . . in some important senses, every species is 'equally good'" (Schopf 1972b, 12–13). Schopf drew an analogy between this "particle paleontology" and the MacArthur-Wilson equilibrium model in that the evolutionary history of an individual species might be ignored, just as the life history of an individual island colonizer was merely part of the statistical dynamic steady-state of arrival and extinction. In either case, the behavior of individuals in the system is essentially stochastic, but the behavior of the whole is predictable. Ultimately, this perspective would lead Schopf to determine—as he explained in his 1976 personal narrative—that "the basis for any general theory of the history of life is, I feel, in investigating equilibrium models in which stochastic processes are the important processes" (Schopf, "Notes on 1959–1976": Schopf pap. 7, 1). Schopf, therefore, can be regarded as one of the earliest and most significant conduits for the transfer of MacArthur-Wilson equilibrial biogeographical theory to paleobiology.

When Schopf set a date for the first Woods Hole meeting on the weekend of 26–27 August 1972, he wrote to Simberloff with details and a reading assignment (the newly published textbook *Principles of Paleontology*), and reminded him, "It is extremely critical that you be able to attend, since you have the mathematical expertise, and the first hand experience with developing simple mathematical models" (Schopf to Simberloff, 27 March 1972: Schopf pap. 5, 34). Schopf also sent a short agenda to the three other participants, outlining the questions he hoped might be addressed during the meeting. In particular, he suggested attacking the conventional emphasis in invertebrate paleontology "on the facts of history, often stressing unique events" that "has led to historical models where the ideas are closely tied to empirical summaries." As examples of this approach Schopf cited Valentine's 1969 paper on increasing marine diversity over time, A. Lee McAlester's correlation of historical variations in oxygen with major periods of extinction, and Frank Stehli's correlation of global diversity patterns with temperature fluctuations (Schopf to Raup, Gould, and Simberloff, 20 April 1972: Schopf pap. 3, 30). Schopf suggested that "one way we could proceed would be to examine these and other patterns in terms of equilibrium models," an approach he noted which "worked extremely well in a field very similar to ours, i.e., biogeography, with the species equilibrium." Schopf then presented the four initial problems to be explained—the investigation of "organismal diversity," "morphological themes," "chemical themes," and "phylogeny" through time—and suggested, "We want to ask what are the processes underlying these patterns, and what are their long-term equilibrium consequences" (Schopf to Raup, Gould, and Simberloff, 20 April 1972: Schopf pap. 3, 30).

It seems that Schopf hoped to develop some kind of approach to analyzing fossil data using sources like the *Treatise on Invertebrate Paleontology* or the 1967 compilation *The Fossil Record* to produce generalizations similar to the MacArthur-Wilson island model. Raup recalls that "the original meeting in Woods Hole was stimulated by the success of island biogeography (MacArthur-Wilson)," and that "Tom hoped to apply the same thinking to the Phanerozoic record" (Raup, e-mail communication with John Huss, 13 March 2002). Raup also remembers that Schopf brought the entire multivolume *Treatise* to the meeting, "and we put it on the table" while Simberloff brought one of the earliest programmable hand calculators. "The question was: 'What can we do that's different?'" (Raup, quoted in Sepkoski 2009, 463). Unfortunately, Schopf's hopes were dashed. Simberloff recalls that "right at the start they presented me with a whole raft of paleo data and patterns," but that he immediately had "problems with analyzing all of them in ecological terms, largely centering on the fragmentary nature of the data" (Daniel S. Simberloff, e-mail communication with John Huss, 15 March 2002). In other words, the plan seemed to have optimistically hinged on Simbrloff's ability to work his mathematical magic on the data, and fallback options had apparently not been considered. Or, as Raup succinctly summarizes, "We got nowhere. Dead zero" (Raup, quoted in Sepkoski 2009, 463).

However, this initial failure had the effect of producing an unexpected new direction. Frustrated by three days with little success, and in some desperation on the final afternoon of the meeting, Raup posed a radical idea. He effectively asked, "What if we take natural selection out of the equation?" More specifically, he proposed simulating the history of life *as if* it were random—"that is, if extinction or survival of lineages was merely chance"—to serve as a null hypothesis against which to test the traditional evolutionary assumption that phylogenetic patterns observed in the fossil record are solely the result of deterministic and directional processes. As Raup explained, "The idea was not meant to suggest that things like the extinction of species occur without cause. Rather, that there are so many different causes of extinction operating in any complex ecosystem that ensembles of extinctions may behave *as if* governed by chance alone" (Raup, quoted in Sepkoski 2009, 463). Since the meeting was over, Raup volunteered to write a computer simulation program and report back to the group.

This, in essence, was the genesis of the MBL model. However, before moving on to consideration of the MBL papers themselves, there are two additional influences that need to be addressed. The first concerns the source of the introduction of stochastic or random models. Despite Schopf's later contention that he was led directly to consider random processes by his interest in equilibrium models, there is no evidence in his work prior to the MBL meetings that he had given stochastic models any thought. Gould also would have been an unlikely source for this idea, given that he has explicitly described his theory of punctuated equilibria (at least in its original inception) as a deterministic, even "Darwinian" model. This leaves Raup, who, in 1969, published a paper (with Adolph Seilacher) on a "grazing track model" for analyzing trace fossils that described the grazing patterns of individual animals as "partially stochastic" (Raup and Seilacher 1969). Although the paper did not involve detailed analysis of stochastic factors, it did indicate that Raup had previously developed some familiarity with mathematical treatments of randomness. Since Raup confirms that by 1972 he was "far enough along with random number generators" that writing the original MBL program posed no great difficulty, it is entirely possible that this was the inspiration for generating random phylogenetic trees (Raup, quoted in Sepkoski 2009, 464).

There is other evidence that Raup came to the meeting prepared to discuss random models. In a letter to Schopf prior to the first MBL meeting, Simberloff made the suggestion that "another model type" (i.e., another theorist) be invited to the meeting, "someone especially at home with stochastic models (which may be what turns out to be appropriate for paleontology)" (Simberloff to Schopf, 19 April 1972: Schopf pap. 5, 34). Schopf mentioned in the agenda sent to all participants that "Dan felt that we should consider the role of stochastic models, and in particular that someone with this background be involved." However, Schopf also concluded "that unless the issues are fairly clearly delineated, that just the four of us should meet this first time" (Schopf to Raup, Gould, and Simberloff, 20 April 1972: Schopf pap. 3, 30). In response to this letter, Raup replied "I am a strong advocate of stochastic models, as you know," but also cautioned "they are, as you know, dangerous as hell," and he endorsed Schopf's cautious stance that "stochastic models will come up where appropriate—but that they should not be treated as ends in themselves" (Raup to Schopf, 27 April 1972: Schopf pap. 3, 30).

The second influence was University of Chicago evolutionary theorist Leigh Van Valen's so-called Red Queen's hypothesis. Although the idea was first proposed in a paper (entitled "A New Evolutionary Law") published a year after the initial MBL meeting, as Van Valen noted, his ideas had "been circulating in *samizdat* since December, 1972," and he thanked both Schopf and Raup in the acknowledgements of the paper for previous discussions (Van Valen 1973, 21–2). It is clear, then that members of the MBL group were aware of Van Valen's paper while they were developing their own initial model, and Raup has stressed that any discussion of "important historical elements [in the MBL story] must include Van Valen's original [1973] Red Queen paper" (Raup, e-mail correspondence with John Huss, 13 March 2002).

The Red Queen's Hypothesis would have an important influence not just on the MBL models, but also on the debates over taxonomic diversity that will be discussed in the next chapter. Van Valen's argument was essentially a reexamination of the application of survivorship curves used in population ecology to paleontology. From the time of Simpson's *Tempo and Mode*, survivorship curves—expressed as a plot showing the probability of an individual's extinction with age—were a standard method in paleontology for depicting the longevity of taxa. While Simpson and other paleontologists had plotted survivorship arithmetically, one of Van Valen's innovations was to plot his own survivorship curves with a logarithmic ordinate—as was standard in population ecology— which meant "that the slope of the curve at any age is proportional to the probability of extinction at that age" (Van Valen 1973, 1). Van Valen also plotted survivorship curves for much larger samples than Simpson had used: his initial study analyzed the first and last appearances in the fossil record of more than 25,000 subtaxa of vertebrates, invertebrates, and plants. Conventional wisdom was that a taxon's probability of extinction increased with its longevity. However, Van Valen's results confounded this expectation: his survivorship curves were linear (plotted on a logarithmic ordinate), which meant that the probability of extinction for a subtaxon within a given group remained *constant* over time. In other words, according to Van Valen's analysis, there is no correlation between the age of a taxon and its probability of extinction; within a higher taxonomic group, extinction rates are effectively constant, and all subtaxa have an equal probability of extinction at any given time— regardless of age.

Van Valen's paper explored the meaning of this result, which he interpreted in an ecological context as an evolutionary 'law': "The effective environment of the members of any homogeneous group of organisms deteriorates at a stochastically constant rate," or "extinction in any adaptive zone occurs at a stochastically constant rate" (Van Valen 1973, 16). Van Valen explained this result with a metaphor drawn from Lewis Carroll's *Through the Looking Glass*, where the character of the Red Queen proclaimed, "Now, here, you see, it takes all the running you can do, to keep in the same place. If you want to get somewhere else, you must run at least twice as fast as that!" In the face of ecological pressures, Van Valen argued, "each species is part of a zero-sum game against other species," in which "no species can ever win, and new adversaries grinningly replace the losers" (Van Valen 1973, 21). Because of this, each species must keep "running" just to maintain its fitness from one moment to the next. Since the age of a given taxon neither detracts nor adds to its momentary fitness, "the probability of extinction of a taxon is then effectively independent of its age" (Van Valen 1973, 17). Van Valen's hypothesis was potentially revolutionary because it was unsettling to conventional assumptions about evolution, as Van Valen himself acknowledged at the end of his paper:

> From this overlook we see dynamic equilibria on an immense scale, determining much of the course of evolution by their self-perpetuating fluctuations. This is a novel way of looking at the world, one with which I am not yet comfortable. But I have not yet found evidence against it, and it does make visible new paths and it may even approach reality (Van Valen 1973, 21).

The major influence this idea had on the MBL model was that it justified the assumption that the evolutionary properties of species (or other
subtaxa) are essentially "equivalent," which in turn endorsed the construction of a model in which species could be treated as identical "particles" in space and time (Huss 2004, 58–60). This was part of the "novel
way of looking at the world" Van Valen referred to, and as we will see,
the MBL model would eventually test the comfort each of the participants felt in employing it.

The Initial MBL Model

In the fall of 1973, Schopf addressed a letter to Raup, Gould, and Simberloff that began "Dear Fellows of the 'Radical Fringe of Paleontology'"
(Schopf to Raup, Gould, and Simberloff, undated [fall 1973]: Schopf
pap. 3, 30). This was just one of several nicknames that the group would
acquire; they would be variously labeled elsewhere as the "gang of four"
and "the four horsemen of the MBL."[1] There was indeed something quite
radical about the MBL model, and in many ways the very first paper,
"Stochastic Models of Phylogeny and the Evolution of Diversity," which
appeared in late 1973, was the most radical of the entire sequence.

One of the ironies of this story is that such a bold, theoretical paper—a
paper that declared itself to be "the first in a projected series of papers
that might bear the general title 'nomothetic paleontology'"—was published in the very conservative *Journal of Geology*, which had a traditional emphasis on hard-rock geology and a fairly limited circulation
(Raup et al. 1973, 526). However, it turns out the choice of venue had less
to do with exposure than with expediency: the *Journal of Geology* was
owned by the University of Chicago, and Schopf, whose office was in the
"same building" as the journal's, was able to arrange for quick, uncom-
plicated approval and publication (Raup interview). This is an important reminder that conceptual developments cannot be separated from
institutional resources and context: in the days before *Paleobiology*, paleontologists faced limited options and sometimes daunting obstacles to
getting theoretical papers published. In fact, Van Valen became so frustrated with the policies of established journals that he founded his own
journal, *Evolutionary Theory* (where the original Red Queen paper and

1. The "four horsemen" reference was Phil Signor's, and was not a compliment (Signor
to Jack Sepkoski, 7 June 1981. Sepkoski pap. '1981').

many of his subsequent papers appeared), which for many years he produced on a mimeograph machine in his own office.

From its opening lines, the first MBL paper announced its debt to the MacArthur-Wilson equilibrium model, and also the hopes that its authors had to imitate the success of that research program:

> The application of equilibrium models to population biology and theoretical ecology has yielded important generalizations and redirected field exploration. Most striking has been the direction of animal biogeography away from a purely descriptive summation of distributions toward the prediction that species diversity is in equilibrium dependent upon immigration rates, extinction rates, and the area available for colonization. . . . This result raises major questions about the guiding principles behind evolution—in particular, can the processes resulting in local equilibria in ecologic time also be used to predict events in evolutionary time? (Raup et al. 1973, 525)

The paper began with the basic observation that equilibrial diversity models, developed to describe populations of organisms in ecological time, might have a fruitful translation to the fossil record and geologic time. This observation would be at the heart of both the MBL model and the subsequent debates over Phanerozoic diversity. Here it was expressed as "a continuation of the conscious application of equilibrium models to paleontological data," and an attempt to determine "whether an equilibrium pattern of phyletic radiation can exist, and if so, what its characteristics would be" (Raup et al. 1973, 525).

This initial paper also promoted a new, idealized approach to reading the fossil record. As the authors noted, "The essential assumption of this strategy is that the true complexity of events in the real world can be adequately rendered by models using relatively few generating factors." They pointed to the success of this program in theoretical ecology, a discipline they characterized as "nomothetic," in contrast to paleontology, which "has traditionally focused on the idiographic (Why did *this* crinoid become extinct at *that* time)." For their own part, Raup et al. expressed their determination to "try to abstract the common elements hidden beneath the bewildering and all but impenetrable verbiage of our nomenclature and stratigraphy by constructing a model that makes no reference to individual taxa" (Raup et al. 1973, 526). One of the unstated advantages of this nomothetic approach to paleontology was that it rather conveniently freed its adherents from their dependency on the notoriously fragmentary and unreliable fossil record. However, this point was not belabored in the first paper, nor was it mentioned that the paper's central idea originated in response to the explicit failure of the group to make headway with actual fossil data.

The basic MBL model itself was quite simple: Raup had designed a computer program that simulated the evolution of phylogenies by generating an initial "lineage" and then randomly subjecting it and its "offspring" to one of three possible outcomes—extinction, persistence with branching, or persistence unchanged—over a predetermined number of steps. This was an application of a randomization process known as a Monte Carlo simulation. The computer was used to randomly draw numbers to determine outcomes with prespecified possibilities, much as a dealer might randomly draw cards from a deck and arrange the outcomes into hands. At the end of the run, the program output the results graphically in the form of a branching phylogenetic tree. The program also automatically grouped lineages into "clades" when lineages had accumulated a preset number of branches. These clades were graphically depicted in "spindle diagrams" representing clade diversity, in which the width of the spindle varied in proportion to the number of lineages it represented (fig. 7.1). Both kinds of graphical representations would have been "instantly recognizable" to paleontologists (Huss 2004, 63–64).

In addition to these basic parameters, a few additional controls were built into the program. Most significant was the adjustment of the probabilities set for extinction and branching at each step, in which the initial probability of extinction was damped so that branching was favored. As the program approached a predetermined "equilibrium diversity," the probabilities of extinction and branching were equalized so that diversity would oscillate around a mean equilibrium value. Raup does not remember this constraint having been motivated by any explicit biological assumptions, but rather that it addressed the limitations of early 1970s computer technology. Because of limited computer memory, if the initial probability of extinction were too high, most runs would simply terminate after only a few steps. On the other hand, if no equilibrium value were set, the program would quickly generate too many lineages for the computer to handle (only 500 lineages could be stored in the computer's memory at any time). In the final paper, the authors recast this technical limitation as an explicit endorsement of the MacArthur-Wilson model, explaining, "We did not choose to limit diversity simply to restrict a

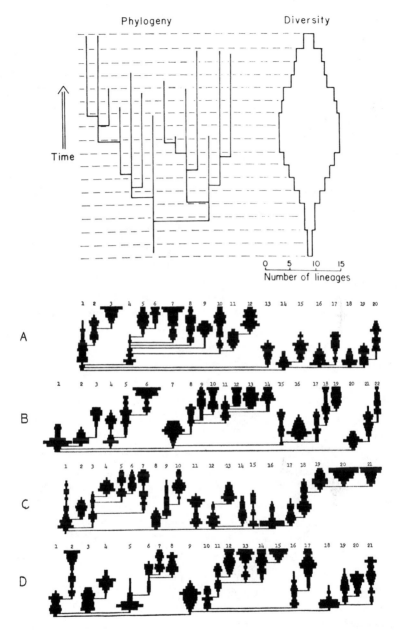

FIGURE 7.1. Graphical outputs from the original MBL phylogeny simulations. David M. Raup et al., "Stochastic Models of Phylogeny and the Evolution of Diversity," *Journal of Geology* 81 (1973): 527, 533.

potentially boundless phylogeny to a tractable size." Rather, "the maintenance of an equilibrium diversity in the present work implies that an
adaptive zone or a geographic area becomes saturated with taxa and remains in a dynamic equilibrium determined by the opposing forces of
branching (speciation) and extinction." This they directly attributed to
the influence of the MacArthur-Wilson model (Raup et al. 1973, 529).

"I grant that the equilibrium aspect came to be important but I certainly did not think this way while formulating the MBL program,"
Raup recalls. "I used an equilibrium constraint in the MBL algorithm
only as a protection against saturating the computer's memory. Given
my wish to make MBL constraint-free, I would not have included a control on population unless forced to." Nonetheless, reflecting on this fact
nearly 40 years later, Raup concedes, "I [recently] opened the original paper and was astonished to see Equilibrium as the first word of
the abstract and to read the emphasis on equilibrium throughout the
paper. . . . Was our thinking recast as the project developed, or is my
memory playing tricks?" The most likely explanation is that Raup's solution to a technical problem was seized on by the more equilibrial-minded
collaborators (Schopf and Simberloff) as a serendipitous endorsement of
their own theoretical predilections. "I submit that most of our interesting results could have been achieved with [an] unconstrained algorithm,"
says Raup. "I suppose my colleagues would have eventually insisted on
adding equilibrium" (Raup, personal communication, 29 April 2010).

The MacArthur-Wilson model was only one of three major assumptions about the dynamics of evolution that the MBL program accomo-
dated. The second had to do with the way evolution was reflected in the
model. Since lineages could change only by branching and persistence of
the original lineage, the phenomenon of "pseudo-extinction," in which
lineages gradually transformed from one to another, was excluded from
the model. This was an explicit accommodation of punctuated equilibria; as the authors explained, the MBL model "describes a situation in
which phyletic transformation is absent and where new taxa arise only
through speciation," pointing out that "Eldredge and Gould (1972) have
argued that it corresponds closely to biological reality" (Raup et al. 1973,
528). The third assumption related to extinction. As Huss observes, Van
Valen's Red Queen's hypothesis provided two important justifications
for the MBL model. First, it validated treating individual taxa as essentially identical units, since Van Valen had shown that despite individual
differences average rates of extinction should be the same for all taxa.

Second, the Red Queen justified the assumption of stochastically constant rates of extinction, and endorsed choosing a resolution level that ignored individual selection. However, Huss also notes, "In applying Van Valen's ideas, MBL excised his causal language and his distinctions between causal mechanisms and their resultant patterns," while retaining the "mathematical behavior" of the phenomenon being described. Huss sees this as "inevitable" and "perhaps the whole point of a mathematical model" (Huss 2004).

There are other significant features of the MBL model worth discussing. First, it helped to introduce stochastic null hypotheses into paleobiological modeling. The model was not assumed to depict any actual mechanisms or processes in the real world; rather, in excluding traditional deterministic evolutionary assumptions, it implicitly tested those very assumptions. As the authors explained, "We do not suggest that evolution be viewed as a haphazard process, independent of basic relations of cause and effect." Nonetheless, "We wished to predict what phylogeny would look like if it were determined by random processes and then to compare this with the real world, to enable us to separate random elements from those that require interpretation in terms of specific and perhaps nonrecurring causes" (Raup et al. 1973, 526–27). In other words, the null hypothesis of randomness should be examined and dismissed before assumptions of determinism are accepted. If the null model could not be easily dismissed, than traditional assumptions would have to be re-examined.

Second, the results of the initial study confirmed the importance of statistical random walks in the interpretation of patterns of evolution and extinction. The authors tested their simulations against actual diversity data in 178 higher taxa of reptiles drawn from *The Fossil Record* (Harland 1967). In doing so, they found a number of striking similarities between the real and simulated clades. To take one example, the characteristic shape of the simulated spindles—a tendency to taper on both ends and bulge in the middle—was similar to about two-thirds of the reptile clades (Raup et al. 1973, 538). For another, the pattern of initial rapid diversification with low extinction values, tapering off at an equilibrium level where speciation and extinction were roughly equal, was clearly evident in both real and simulated clades (Raup et al. 1973, 539). How could this similarity be explained? Another way of putting it is to ask how an apparently random process produced patterns of apparent directionality and biological "meaning." The initial 1973 paper did not explore these apparent similarities in detail, but follow-up publications (as we will see) applied the concept of the random walk as an explanation for how random factors can produce apparently directional patterns. This conception invoked a special kind of statistical process known as a Markov chain, which became an important explanatory tool in analytic paleobiology.

Finally, the simulations performed in the initial paper represented a new and important current in the application of computer technology to paleontological data, where computers were used not just as calculating tools, but as sites for the construction and testing of novel hypotheses. This is similar to Raup's simulation and analysis of coiled shell morphology. In both cases, the computer performed two very important functions in addition to crunching numbers: First, computers allowed for a kind of idealized "experimentation" that would not have otherwise been possible in a historical discipline like paleontology. In the late 1960s, Raup was able to manipulate and produce hypothetical shell morphologies that are not necessarily found in nature, which in turn allowed him to reach conclusions about the constraints that govern the structure of actual organisms. Similarly, the MBL model allowed paleontologists to experiment with patterns in the fossil record—it gave paleobiologists the ability to "rewind the tape of life," in Gould's later memorable phrase, and to produce counterfactual or alternative histories that allowed paleontologists to interrogate the actual record in new and interesting ways. Huss discusses the importance of this kind of "numerical experimentation," which he defines as "the practice of creating mathematically modeled reality within which experiment is conducted via simulation" (Huss 2004, 150). Second, computers introduced an important and new kind of visual imagery into paleobiology. Raup's computer-generated shells are one obvious example of this; the graphical output of the MBL program is another. In both cases, the visual imagery itself was as important in the arguments being put forward as was the text presenting the data, and are examples of what Jan Golinski calls a "visual hermeneutics" (Golinski 2005, ch. 4). It was the striking visual representation of simulated clades that was often most compelling to paleontologists who read these papers; as Huss puts it, "The most salient argument remained a visual argument based on an overall gestalt, the likes of which had been doing the heavy lifting" since the first MBL paper (Huss 2009, 336).

The 1973 paper certainly raised more questions than it attempted to provide answers for. The authors quite readily acknowledged that their analysis "has been primitive so far," and pointed out a number of potential problems with analogies they had made between real and simulated data (Raup et al. 1973, 539). What they did propose was that the simulation experiments potentially challenged many traditional assumptions in interpreting—or "reading"—the pattern of the fossil record:

> When faced with such variation in evolutionary patterns, paleontologists are inclined to suspect or even to postulate that the organisms involved are inherently different—that the various taxonomic groups differ from one another because they differ in population structure, reproductive systems, mutation rates, dispersal systems, and so on.

In other words, paleontologists traditionally assumed that the fossil record looks the way it does because individual and specific ecological and biological constraints have determined that the history of life must look this way, and no other.

> But the simulation modeling shows that two or more groups operating under identical constraints—that is, having the same evolutionary potential—can behave very differently. We do not rule out that such differences in the real world may be due to inherent biological differences. We only contend that the observed variation does not in itself demand such an explanation (Raup et al. 1973, 534).

The salient point here was not that traditional paleontological explanations were necessarily wrong, only that they might be underdetermined by available evidence. Raup et al. warned "above all . . . against using patterns of diversity as the *major evidence* for differences in evolutionary potential." The question that would be pursued in succeeding MBL papers—and which would surface behind the scenes in internal arguments among the collaborators—was the extent to which this potentially radical, even non-Darwinian picture of the history of life actually applied to reality.

Extensions of the MBL Model

Even as the first MBL paper was in preparation, Schopf pressed for the group to move forward with plans to expand the initial effort into a genuine research program, and proposed additional meetings and papers.

However, at the same time, cracks had already begun to form in the nascent collaboration. Just months after the first meeting at Woods Hole, in January of 1973, Schopf wrote to Gould to thank him for "your gracious remarks to Dave [Raup] about my overblown writing on the importance of our meetings, and what has come of it" (Schopf to Gould, 27 January 1973: Schopf pap. 5, 14). It is unclear to what Schopf was referring, or if he even sent the letter (the version that appears on Schopf's departmental letterhead has a handwritten note reading "not sent"), but this is an early indication that Schopf, who would continue to press for a more "radical" interpretation of the MBL model, was already butting horns with Raup, who would favor a more conservative one. Over the next several years this tension would grow, and Gould would find himself in the role of mediator between the two.

Schopf had originally hoped to have the MBL group convene during the 1973 GSA meeting in Dallas, but schedule conflicts made a Christmastime gathering back in Woods Hole more feasible. In September of that year Schopf wrote to Gould, Raup, and Simberloff to propose this option, but also to express "varying degrees of uneasyness [sic] about trying to hold our loose arrangement together" (Schopf to Gould, Raup, and Simberloff, 28 September 1973: Schopf pap. 3, 30). This comment was motivated by the fact that, following the first paper, the collaboration had begun to unravel. Shortly after finishing the original paper, Raup and Gould struck out on their own with a paper extending the original MBL program to the evolution of morphology, while Schopf and Simberloff produced a linked pair of single-authored papers applying the MacArthur-Wilson species-area effect to analysis of the great Permian extinctions (Raup and Gould 1974; Schopf 1974; Simberloff 1974). These papers were each very different in methodology and emphasis, and highlighted some of the inherent differences that had already begun to appear in the collaboration.

Schopf's letter began by characterizing his initial motivation for bringing the group together as "the romantic, idealistic notion that Paleontology, rich in evidence, weak in theory, strong on history, nearly devoid of equilibrium models . . . could be redirected by the self-conscious application of a way of doing science not found previously in paleontology." He explained that while he maintained his hope that "our relative strengths could be dedicated to a higher purpose which would result in a quantum increase in understanding the biological meaning of the fossil record," it was becoming clear that "there is some difference of opinion" over the best course to take. Apparently some in the group were advocating continuing to publish papers authored by all four members, despite the fact that "in each case, one or two persons [would be] doing 90% of the work," while others felt that "honorary authors should be avoided." Schopf therefore considered the collaboration to be at a crossroads.

> While we were close enough together on the first paper that this was not a serious problem, the situation has obviously fallen apart with the [Raup and Gould] morphology paper, and the odd situation of my ending up writing a geological paper which called for a short explicit test that Dan was then able to provide. Accordingly we really have gone our separate ways, and without a face-to-face meeting of the 4 of us again to reestablish possibilities for mutual input, I think we will continue to go our various ways (Schopf to Gould, Raup, and Simberloff, 28 September 1973: Schopf pap. 5, 30).

Even so, Schopf was not ultimately pessimistic. "I think that our base intellectual input is so unique and powerful that it will have the desired effect," he wrote. "I cannot help but think that this approach will revolutionize the fundamental questions that are asked in paleontology." He closed the letter by suggesting that the group consider sponsoring a "joint course or workshop stressing stochastic processes, and equilibrium theory, in the context of the fossil record in particular, and evolutionary biology in general," which could be aimed at a broad audience of paleontologists and offered at some point in the next year or two.

In his response, Gould (perhaps surprisingly) conceded many of Schopf's points, though he remained unequivocally convinced of the importance of the research program. He went on to reassure Schopf:

> You are right in arguing that we are among the first group of scientists to use a consciously nomothetic approach to paleontology (there's that word again, but it's the right one and I don't apologize). Previous explanatory schemes have been either particularistic (functional morphology of this or that group), general theoretic applied to particulars (biogenetic law) or representative of that curious technique of theorizing that tries to establish general laws by induction from historical facts (Cope's law etc.). . . . The more you think about it, the more you realize how heretical the search for a kind of timeless generality appears within a science so deeply committed to historicity. . . . The only reason we may not cause the stir we should is that paleontologists are so no-

toriously unaware of the philosophical implications of their methodologies (Gould to Schopf, 5 October 1973: Schopf pap. 5, 14).

He therefore encouraged continued meetings "if we are to reestablish mutual input—and I think we owe such a meeting to what we have already done," and endorsed the idea of a public workshop to spread the new ideas. This was apparently enough to keep the collaboration going; the next letter from Schopf is the one which bears the cheeky heading "Radical Fringe in Paleontology," and it set out the arrangements for the next meeting.

Simulated Morphologies, Directional Evolution, and Markov Chains

Gould's reassurances aside, Schopf's concern had not been unfounded. Raup and Gould's morphology paper—"Stochastic Simulation and Evolution of Morphology: Towards a Nomothetic Paleontology"—was perhaps the most important and original of the entire sequence. It would not be surprising if Schopf felt left out, and even hurt, at being excluded. Whereas the original MBL paper simulated evolution and extinction in lineages that had no individual characteristics, Raup and Gould modified the MBL program to simulate a randomly changing "morphology" consisting of a set of hypothetical "characters" that randomly "evolved" at each step in the program. Whereas the first study had asked, "What would diversity look like in the absence of selection?," Raup and Gould now asked, "What would trends in morphology look like without unidirectional selection?" The importance of this modification was that it allowed a more sophisticated interrogation of the assumption that directional trends were the product of "directed causes": as the authors explained,

> Modern paleontology has retained a vestige of idealistic morphology in its traditional argument for the role of directional causes in macroevolution. The presence of order in the results of evolution is taken uncritically as definite evidence for the production of such order by directed causes. . . . The marks of order include evolutionary trends, correlation between characters, and morphological coherence of taxonomic groups. The postulated cause, for the last thirty years at least, has been uni-directional selection (Raup and Gould 1974, 305).

This question was a logical extension of the discussion Eldredge had appended to the first draft of the 1972 punctuated equilibria paper, which had essentially asked, "How do we explain trends?" Raup and Gould responded in this paper that "we have come to doubt that the formal pattern of change is an adequate argument for directed causes," and suggested that apparent order "can arise in random systems of change, bounded only by the conventional assumptions of monophyly, continuity, and equilibrium" (Raup and Gould 1974, 306). The surprising result they announced was that directional selection was not needed to produce apparent trends in morphology.

Raup and Gould defended the biological assumptions in their model by arguing that random morphological changes could be imagined to have been produced either as adaptive responses to random fluctuations in environment, or by genome alteration produced by random drift. Thus, the model was neutral towards "Darwinian" or "non-Darwinian" evolutionary interpretations. The null hypothesis it set out to test was the assumption that directional selection was required to produce evolutionary order: if apparent order could be produced without directional causes, then paleontologists would be justified in looking beyond orthoselection for other mechanisms that could produce directional trends (Raup and Gould 1974, 306). In many ways this question was a natural outgrowth of previously established commitments of both authors. Gould had primarily explored the role of mathematical allometric relationships as constraints on morphologic evolution, while Raup had similarly investigated geometric constraints in his simulations of theoretical morphospace (Gould 1966).

Raup and Gould made a number of runs of their simulation, which assigned a set of hypothetical "characters" to an original ancestral "lineage," and the tracked the changes in those characters over successive branching points. They then examined the results by comparing them to expectations for what they termed the "general orderliness" in an evolutionary tree. In the first place, they argued, in a nonrandom evolutionary tree we would expect to see an incomplete filling of morphospace; the traditional assumption is that morphological similarities along an ancestral line are a function of recency of ancestry, which constrains the degree of departure from the initial state. This incomplete filling might also be taken as evidence of selection, wherein morphospace would be "colonized" in a nonrandom way as selection pursued certain advantageous regions and rejected others (as Raup had modeled in his 1966 paper on theoretical morphospace) (Raup and Gould 1974, 308–10). In the second place, selective evolution would be expected to show a close correspondence between phenetics and cladistics. In other words, morphology should "unfold" in an orderly way with progressively greater likelihood of morphological change occurring the further the tree moved away from its ancestral state. Phenetics assigns groups to taxonomic categories based on degree of shared morphological similarity, while cladistics infers recency of common ancestry by measuring the degree of morphological (and hence genetic) divergence in ancestor-descendant relationships. While cladistics and phenetics have different methods and assumptions, there should be a basic correspondence between phenetic and phylogenetic classifications (Raup and Gould 1974, 313).

In both cases, the simulation confounded expectations for the behavior of a randomly changing morphology. On one hand, a random simulation would be intuitively expected to fill morphospace in a chaotic pattern: nothing should constrain a randomly changing set of characters from evolving rapid and quite different states from its original setting, and all regions of morphospace should theoretically have equal probability of being filled. This, however, did not occur: when analyzed for two particular characters, which the authors imagined might represent adult body length and width, the frequency distribution of characters was "clearly non-random." Values were strongly skewed in a positive direction, while negative regions of the morphospace remained unoccupied (Raup and Gould 1974, 310). On the other hand, there should be no necessary correlation between the phenetic and phylogenetic aspects of the simulation: since the original branching simulation of the MBL program was kept entirely separate from the morphological simulation, phenetic and cladistic patterns should be entirely independent of one another. However, when Raup and Gould performed cluster analysis on the morphological results of the simulation (e.g., they reconstructed taxonomic groups based on morphology alone), they produced a second classification that was "remarkably similar" to the cladograms produced by the program. In other words, for some reason the independent simulations of evolution via branching and random morphological change very closely corresponded with one another (Raup and Gould 1974, 313).

The problem was to figure out what this result meant. Naturally, the fact that these results could be obtained by random simulation did not mean that actual directional processes normally assumed to produce evolutionary order did not exist. However, Raup and Gould argued that the simulation cast doubt on certain traditional assumptions in paleontology. First, they stressed that the assumption that apparent directional changes in morphology are necessarily produced by deterministic causes—whether "internal factors in the bad old days of orthogenesis and vitalism" or "directional selection today"—was misplaced. As the simulation showed, trends "of outstanding duration and unreversed direction" could be produced by random factors alone (Raup and Gould 1974, 314). Second, while an inductive approach to functional morphology assumes some causal basis for correlations among characteristics (e.g., shell volume and number of whorls in a chambered nautilus), strong correlations between characters were produced in the simulation in the absence of any causal relationship between characters. For the 200 lineages simulated, correlation coefficients for the 45 possible pairs of character states showed that 75% of pairs yielded statistically significant correlations. As Raup and Gould put it, if this "were encountered in the real world, we could usually assume either that the characters are not genetically independent, or that there is a functional or structural relationship between them, such that the expression of one is effectively dependent on the other." Of course, in the simulation no such relationship existed (Raup and Gould 1974, 314–15).

Third, significant variation in rates of morphological change is usually interpreted as evidence of "real biological differences among structures subjected to unidirectional selection." Put simply, different kinds of organisms (e.g., clams versus mammals) should evolve at different rates because of inherent biological/genetic differences. In the simulation, all characters should have evolved—on average—at the same rate, because there were no inherent differences between 'organisms.' However, the simulation produced different rates of morphological evolution in different lineages, suggesting that "the operation of chance" is enough to account for differences between rates of evolution among different taxa (Raup and Gould 1974, 316). Fourth, Raup and Gould argued that the simulation made a positive case for the significance of episodes of evolutionary convergence, which the authors noted are "widely regarded as crucial tests for the understanding of natural selection," since they act as 'natural experiments' for testing hypotheses about the efficiency with which natural selection responds to adaptive problems. In the simulation, convergence was represented by similar character states in widely separated lineages, and we would expect phenetic reclassification to group convergent lineages more closely than their phylogenetic relationships warrant. However, this did not happen: convergence was not a phenomenon in the simulations. Raup and Gould explained that this suggested convergence may be a genuine result of "unidirectional selection for (presumably) functional adaptation" (Raup and Gould 1974, 316–17). Finally, the simulation shed light on the consequences of evolutionary specialization. It had often been assumed that "terminal overspecialization"—that is, the tendency of some lineages to develop such overspecialized characteristics that no longer had adaptive value (e.g., the antlers of the Irish elk)—was the result of some intrinsic law that defined ideal adaptive parameters (e.g., Cope's law). However, in the simulation extreme morphological values (high positive or negative numbers for individual characters) were found at the terminal end of lineages simply because they had had greater opportunity to diverge. Overspecialization in nature, then, might just be a product of chance over time (Raup and Gould 1974, 317–18).

Raup and Gould summarized their results with a striking visual example: in order to make the simulation results more concrete, they arbitrarily assigned each of five of the simulated characters a morphological 'value' to create an imaginary organism they named a "triloboid." These triloboids were then represented along 11 of the simulated clades, so that the evolution of their morphology could be graphically depicted (fig. 7.2). The graphical representation alone made a strong visual argument: the evolution of each triloboid lineage looked realistic in the sense that while "they display considerable morphological variation," they nonetheless shared characteristics that appear common to the clade as a whole. As the authors put it, if this imaginary example "depicted an actual case, it would probably not stand out as unusual: the relation between cladistics and phenetics presents a plausible picture" (Raup and Gould 1974, 319). The results of this simulation, then, seemed to lend support to the argument that many paleontological assumptions about trends were underdetermined by evidence. If apparently directional trends could be produced by random generation, then many of those assumptions would need to be reexamined.

However, this did not explain *why* these random simulations had the ability to mimic directional patterns. Here Raup and Gould leaned heavily on the statistical concept of the random walk, a particular kind of result produced by a special statistical process known as a Markov chain, in which the probability of a future event is independent of the

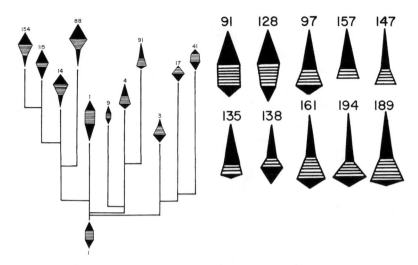

FIGURE 7.2. Raup and Gould's simulation of hypothetical "triloboid" phylogeny. David M. Raup and Stephen Jay Gould, "Stochastic Simulation and Evolution of Morphology: Towards a Nomothetic Paleontology," *Systematic Zoology* 23 (1974): 319. Reproduced by permission of Oxford University Press.

configuration of past events. Markovian processes are notable because they often produce apparently directional patterns even though the overall system is stochastic. Take, for example, a series of coin flips. This is a Markovian process, since the results of successive flips are in no way conditioned by prior throws. Whether or not the next flip will be heads or tails has nothing to do with previous results. A Markov chain reflects no "memory" of past events, and future outcomes are dependent only on the states that immediately preceded them. Imagine a series of coins are flipped, but that after each toss the result is plotted on a graph, with each flip moving one step forward on the abscissa (*x*-axis), each "heads" indicating a positive step on the ordinate (*y*-axis), and "tails" a negative one. Each flip has equal probability of being heads or tails, but the *pattern* of flips will likely show strong directionality, because each successive step can only be one unit above or below the previous one. Furthermore, the probability of heads or tails is reset after each flip; the coin does not "know" that after several successive heads it is due for a tails. Thus, as the number of flips increases and the graph moves forward along the abscissa, there is a strong likelihood that directional patterns will be produced—for example, that the points will "walk" on the positive side of the ordinate for some or all of the run. Markov processes therefore have the ability to generate directional patterns from otherwise random events; in fact, when tested with conventional statistics most random walks produce statistically significant trends (so-called nonsense correlations).

The morphological version of the MBL model presented by Raup and Gould is a Markov process. As Huss puts it, "A lineage's morphology in large part reflects the morphology of its ancestor, since each character of a given lineage in a given interval will vary at most by one increment or decrement of one unit from the state of its most immediate ancestor" (Huss 2004, 78). The randomization function of the program was counterbalanced by the provision that each successive state is based only on the immediately preceding one, therefore constraining possible changes from one time-unit to the next. For example, a character state could not jump from "1,3,-2,1,2" to "5,5,5,5,5" in a single step; in practice, lineages tended to be quickly constrained in a particular direction, and change occured very slowly and gradually.

Raup and Gould recognized this feature of their simulation, from which they extrapolated a broader conclusion: that "the basic order of morphology on an evolutionary tree need not reflect any special biological process, but arises inevitably . . . from topological properties of the abstract form of the tree itself" (Raup and Gould 1974, 320). In other words, that there are certain predictions that can be made about the properties of an evolutionary tree based solely on the laws of probability, irrespective of any particular biological processes. One example is what they termed the "spreading effect," which predicts that lineages which by chance quickly produce a large number of descendents will establish a "crowded" region of morphospace filled with "near replicates." Since morphologic change is gradual, this region will be the center from which subsequent morphological variations will slowly 'spread,' leaving large regions of morphospace unoccupied. This will tend to happen whether or not particular morphologies are assigned adaptive 'values.' Another example relates to the "irreversibility" of morphological evolution: if we consider the initial ancestral character state to be "unspecialized" (i.e., set to values of zero), it is inevitable that successive branchings will produce ever more "specialized" states as the simulation walks from one step to the next. With each step, due to the number of characters involved, it becomes increasingly unlikely that a previous state can ever be recovered. Again, this takes place without respect to any selective value in the character states. In both of these examples, a result which had been normally attributed to deterministic biological factors could be demonstrated to follow logically from the basic rules of probability (Raup and Gould 1974, 321).

Despite this potentially profound challenge to the logic of traditional paleontological (and Darwinian) approaches to selection and evolution, Raup and Gould insisted that their paper was "not an attack upon the concept of uni-directional selection." Nonetheless, they did offer two conclusions: their "methodological" conclusion related to biologists' tendency to infer evolutionary insight from the "abstract geometry of form," a tradition stretching well back before Darwin. Since the simulation showed that "random processes can produce most of the patterns generally associated with directional causes," it "constitutes a challenge to the formalist position." Raup and Gould hoped that as a cautionary lesson, their paper would encourage "more attention" to the study of functional morphology, and to the principle that "adaptation is better demonstrated by the mechanics of form in relation to environment than by evidences of directional change through time." The second, "substantive" conclusion was somewhat more iconoclastic: "The fact that examples of steadily changing characters do not occur with much higher frequency in the fossil record than in the computer simulations suggests that, over long stretches of time, undirected selection may be the rule rather than the exception in nature" (Raup and Gould 1974, 321).

More broadly, the results of the paper appeared to invite to two possible interpretations of the role of randomness in the history of life: a conservative interpretation was that the stochastic model was simply a useful heuristic for rethinking certain traditional assumptions in paleontology, but that it shed no light on the actual processes that govern evolution. A more radical interpretation drew the conclusion that evolution really is, in some ontological sense, a stochastic process. It seems that at the time of the paper's publication, Raup and Gould were not in complete agreement. A short aside stated that whereas "one of us (SJG) remains an unrepentant Darwinian, the other (DMR) has his doubts" (Raup and Gould 1974, 321). However, the second, "substantive" conclusion suggests that both authors were seriously entertaining the radical interpretation of stochastic paleontology. The next phase in the development of the MBL model would bring this potential opposition into clearer focus, but it would also highlight divergent interpretations within the group, as each member considered, and reconsidered, the implications of a "stochastic view" of life.

Implications of Stochastic Paleontology

Although the entire MBL group would produce two more jointly authored papers, Raup and Gould's morphology paper might be considered the high point of the collaboration (Schopf et al. 1975; Gould et al. 1977). Raup himself recalls, "Looking back over the MBL history, I think the (only?) really good paper is that with Steve Gould in 1974. And that was a wonderful experience: I did the analysis but Steve saw what it meant" (Raup, e-mail communication with John Huss, 14 March 2002). While publications and other activities would continue through 1978, the group's final meeting at Woods Hole was in December, 1973, shortly after the Raup-Gould paper was submitted to *Systematic Zoology*. After this point the MBL group would never regain its initial intensity and momentum, and internal tensions within the group would become more pronounced.

The final MBL meeting took place the weekend of 21–23 December, and discussion centered around how the original MBL program could be modified to address additional questions about rates and patterns of evolution. Gould's graduate student Jack Sepkoski had been experimenting with different parameters and permutations of the program, and he was invited for the first two days to give a report of his findings. As Schopf put it in his letter to the group, "Modification of the program of course presupposes a purpose, and that will then lead to questions of what additionally we want the program to do for us," and suggested that this might be "the time to bring out additional topics such as rates of evolution." He closed the letter by reminding the group of his ambition to produce a book which would serve as a manual or primer on stochastic methods, for which Raup had agreed to write an initial outline (Schopf to Raup, Gould, and Simberloff, undated [circa November 1973]: Schopf pap. 3, 30).

The meeting appears to have been a success. By February of 1974 Sepkoski reported to Schopf that he had developed "a package of about 20 subroutines" based on Raup's original code that included routines for analyzing numerical and cladistic taxonomy, survivorship curves, and geological sampling. He noted that the new version of the program would be "more robust so that alternative hypotheses can be tested," and promised a working version within the next several months (Sepkoski to Schopf, 21 February 1974: Schopf pap. 4, 60). In April, Sepkoski followed up with Schopf to report that the completed revision of the program was in Raup's hands (Sepkoski to Schopf, 7 April 1974 (draft): Sepkoski pap. "1974"). Sepkoski's involvement in the project was becoming increasingly important, although up to this point his role had been only as a technician. Gould had already employed him to write the program that performed the comparison between the phenetic and cladistic phylogenies for the morphology paper with Raup, and had mentioned to his colleague Richard Wassersug that Sepkoski was "the person most knowledgable about the computer program used in the Raup-Gould study" (Wassersug to Sepkoski, 10 April 1975: Sepkoski pap. "1975"). In addition to writing the MBL subroutines for Schopf, Sepkoski also provided him with a program he had written (QUAJAC, for Quantified Jaccard's Coefficient) for performing cluster analysis on the MBL results. At the same time, Sepkoski was busy working on an improved database of first and last appearances of orders, families, and genera in the fossil record that Gould had commissioned to use for tests against the output of the simulated clades of the MBL program (Sepkoski 1994, 135).

Differential Evolutionary Rates

The result of the 1973 MBL meeting was the second "quadripartite-authored" MBL paper, "Genomic Versus Morphologic Rates of Evolution: Influence on Morphologic Complexity" (Schopf et al. 1975). This paper appeared in early 1975, in the very first issue of the new journal *Paleobiology*, alongside Raup's important review of Van Valen's Red Queen hypothesis (which will be discussed in the next chapter). The paper itself built on the analysis of morphological trends in Raup and Gould (1974) by directly addressing whether different groups of organisms have inherently different evolutionary clocks, or whether rather inherent rates of evolutionary change are effectively constant for all groups. Simpson had famously concluded in *Major Features of Evolution* that mammals, for example, had evolved more rapidly than bivalves by up to an order of magnitude, and this assumption was often taken for granted in paleontological literature. Schopf et al. argued, in contrast, that all taxa evolved at about the same rate, and that apparent discrepancies in the "genomic rate" of evolution among different taxa were artifacts produced by different levels of morphological complexity in different groups. Put simply, the paper argued that organisms with greater morphological complexity only *look* like they are evolving more quickly than simpler ones, because genetic changes are more readily expressed phenotypically in organisms with complex morphologies. Underneath, at the genomic level, "there may be a rather small range in rates of genomic evolution for different taxa over geologic time" (Schopf et al. 1975, 63).

Part of the reason previous analyses had missed this fact, the authors argued, was that it is extremely difficult to quantify and correlate degrees of "morphologic complexity" and "genomic change" in fossil organisms. In the first place, morphologic complexity is a subjective measure, and must be inferred as best as possible from imperfect fossil evidence. Secondly, rates of genomic evolution in fossils can only be estimated indirectly by guessing at how closely phenotypic change reflects underlying genetic variation. This presented a potentially circular argument: taxa that display slower rates of morphological change experience slower rates of genomic evolution because morphological change is an indication of rate of genomic change. Schopf et al. proposed to probe this relationship by employing the modified, morphological version of the MBL program used in Raup and Gould (1974) to produce cladistic lineages with 20 morphologic "characters." They then applied Sepkoski's QUA-JAC cluster analysis program to group the lineages phenetically using four "stages" of morphologic complexity: 3, 5, 10, and 20 characters, respectively. This produced four dendrograms, each of which associated the simulated lineages based on perceived degree of phenetic relatedness using a different number of characters (Schopf et al. 1975, 65).

The results of this simulation indicated that there was a direct relationship between the number of characters attributed to a group and its apparent rate of evolution. The four dendrograms were analyzed for survivorship, where the authors assumed that "short duration implies rapid turnover and rapid evolution," and vice versa. They found that, plotted against an arbitrary time scale, the survivorship curve produced when clustering was based on 3 characters was twice as long as when it was based on 20. In other words, when greater morphologic complexity was taken into account, the lineages appeared to evolve "faster."

Ultimately, then, the paper argued that assumptions about inherent rates of genomic change needed to be recalibrated: organisms with less complex morphologies weren't necessarily less evolved than more complex ones. In fact, Schopf et al. suggested that average rates of genetic change (e.g., those involved in speciation) may be significantly higher for all taxa than had been commonly assumed, and that those rates were more or less constant throughout the lifetime of a lineage. One problem with this conclusion is that the fossil record appears to record frequent instances of rapid change are followed by long periods of stagnation (e.g. punctuation and stasis). The traditional explanation for this pattern was that rates of genomic evolution are variable. However, the paper argued that this pattern could be explained via one of two alternative assumptions: in the first possibility, the rate of change is uniform, but phenotypic expression of genomic change decreases over time because of internal structural constraints. For example, "the structural limitations of frog morphology" allows less morphologic variation than mammalian morphology, which "allows a similar degree of genetic change to be translated into morphological adaptations in widely different environments" (Schopf et al. 1975, 67–68). In the second possibility, rates of evolution are proportionate to area available for colonization, and as environment fluctuates, so do opportunities for advantageous colonization by peripheral isolates with new morphological adaptations. In other words, populations are always experiencing the same rate of genomic change, but environment strongly determines whether those changes will be preserved as successful adaptations (Schopf et al. 1975, 68).

While the MBL group argued that the results of the simulation were important for reevaluating taxonomic practice, they concluded by connecting the paper to the broader project of investigating "equilibrium models in place of strictly historical explanations," or taking a deductive approach to the fossil record: "Our emphasis on the role of stochastic processes *over geologic time* bears upon a much larger issue—what are the general implications for evolutionary theory if we are correct in claiming that genomic rates of evolution differ far less among major taxa than do rates of morphological change" (Schopf et al. 1975, 69). Although they admitted to being "not unanimous among ourselves regarding our favored guesses," they presented two possible interpretations: the first, a "neutralist" view, held that the constant genomic rate of evolution was simply the result of "random fixation of mutations," while the variation in rates of morphological expression was "merely the differential use of this genetic resource by the epiphenomenon of natural selection." Though not directly attributed, this view would seem to be consistent with Schopf's growing conviction that, at its root, evolution is primarily a stochastic process in which selective (individualistic, historical) factors play only a secondary role. On the other hand, the second interpretation spoke for Gould's more traditional view:

> (1) that the genetic rate is primarily a function of the rate of speciation (*viz.* Eldredge and Gould, 1972), (2) that true rates of speciation among taxa differ far less than the differential propensities in various groups for parlaying large series of speciations into sequences that we recognize as evolutionary "trends" in morphology, (3) that genetic differences appearing at speciation are (nearly) completely controlled by natural selection (Schopf et al. 1975, 70).

Here, in short, were the two developing rival interpretations of the broader implications of the MBL simulation studies: Schopf's vision of a "particle paleontology" which was ontologically stochastic and functionally non-Darwinian, versus the more restrained "heuristic" interpretation that preserved some measure of both historicity and of Darwinian evolutionary assumptions.

Real versus Random Clades

As the MBL group continued to explore more even more radical interpretations of the history of life, the community of evolutionary biologists' reactions were mixed. E. O. Wilson, one of the referees for *Paleobiology* for the 1975 Schopf et al. paper, gave the manuscript an unequivocally positive review. In rating the paper "excellent" and "acceptable with little or no revision," he described it as "an absolutely first-rate paper, of fundamental importance," and he complimented the authors for having "fashioned an essentially new basic idea fully" (Wilson, review of "Genomic versus Morphological Rates of Evolution," 29 July 1974: PS-*P* 4). The MBL group also received friendly publicity in the form of a "Research News" column in a 1975 issue of *Science*, which gave a detailed summary of the papers up to that point and described the work as "arousing a great deal of interest among paleobiologists" (Kolata 1975, 625). While the essay did note paleontologist Arthur Boucot's assessment that the model was "clever, polished, but of limited use," it undoubtedly brought the project welcome attention from a broader audience (Kolata 1975, 660).

A less friendly response, however, was published the following year in *Paleobiology*, when Niles Eldredge challenged the conclusions of the paper on genomic rates (Schopf's frosty editorial handling of this piece was discussed in the last chapter). In arguing that the simulation approach "obscure[d] some probably real biological phenomena of great theoretical interest," Eldredge questioned the basic logic of the 1975 paper's conclusion that genomic rates are constant (Eldredge 1976, 174). If, he argued, we assume that most genetic change occurs through geographic variation, genome sampling in isolated populations, and character displacement, then the better question would be "Is there variation in rates of speciation among different kinds of organisms?" He then argued—using as examples both the "living fossil" *Limulus* and a comparison between clams and mammals—that disparities in evolutionary rates are real and that Schopf et al. were wrong to minimize the importance of geographic and environmental influence on local populations, and he admonished the authors to restore their discussion of evolution to "the language of evolutionary and ecological theory" in which "'population' and 'species' are the core words" (Eldredge 1976, 177). Gould's immediate response in the same issue of *Paleobiology* confirmed that Eldredge had scored a direct hit: while Gould defended what he called "the Lyellian tradition of probing 'behind appearances,'" he nonetheless conceded that Eldredge's argument that questions about genomic evolution ultimately reduced to questions about speciation was one with which he "can scarcely disagree" (Gould 1976a, 177). That there were pronounced differences in rates of speciation among different taxa was a foregone conclusion. The question was how to interpret those differences.

Gould then summarized the strong and weak interpretations of the MBL model as he saw them (which correspond to what I have described as the "ontological" and "heuristic" interpretations of the stochastic null hypothesis) and admitted that in his view the 1975 paper supported only the weak version, although "some of its authors (not including myself) believe or at least prefer" the strong one. Then, in a particularly candid and revealing passage, Gould articulated for the first time (at least in print) what would become one of the central metaphors in his own, emerging view of the history of life:

> Are the profound differences [in rates of speciation between Coelacanthini and Coleoptera] . . . a determined result, predictable *a priori* from the intrinsic biology of the two groups? Or is the difference a complex result of stochastic opportunity and a different number of founding lineages. . . . In other

words, if we could replay the tape of life, might coelacanths become the te-
losts and beetles the monoplacophorans? In short, empirical differences in
rate of speciation exist—how could it be otherwise in a non-typological world.
But we must know whether these differences are essentially random scattering about a mean rate (the [genetic] metronome hypothesis) or a determined
and predictable feature of certain morphologies and environments.

Here, in short, was the central question: Is evolutionary change essentially a historically determined, contingent process, or is it rather a random walk? Gould confessed that while his "own intuition leads me to Eldredge's conclusion—that the differences are predictable consequences
of functional anatomy and environmental space," he nonetheless concluded, "I cannot vindicate this intuition with any satisfactory data"
(Gould 1976a, 178). Thus, in his response to Eldredge, Gould found himself torn between conflicting loyalties: on the one hand, to his co-author
Eldredge and their "literal" reading of the fossil record as a sequence of
historically contingent events, and on the other to his coauthor Schopf,
and to their reading of the record as an idealized statistical record of stochastic fluctuations around an equilibrium.

Despite mounting disagreements, the MBL collaboration produced
one last major paper, "The Shape of Evolution: A Comparison of Real
and Random Clades" (Gould et al. 1977). This paper revisited the original 1973 *Journal of Geology* study but provided an expanded comparison with clade diagrams derived from real taxa. As such, the technical
aspects of the paper followed a by now familiar procedure: First, clade
diagrams were generated using a modified version of the original MBL
program. Next, the size and shape of the simulated diagrams were measured and tested to determine frequency distributions, or "clade statistics" for the simulated clades. These clade statistics measured the size,
maximum diversity, duration, center of gravity, and extent of fluctuation
for each clade. As in earlier studies, a "damped-equilibrium" function
was built into the program, which was justified as a representation of the
MacArthur-Wilson insular equilibrial model. In the final stage, the paper computed clade statistics for 144 orders, 206 families, and 1442 genera of actual fossil organisms, derived primarily from the *Treatise on Invertebrate Paleontology* (Gould et al. 1977, 30). Clade statistics for real
and simulated clades were then compared in order to identify features
of real clades not found in random ones. The computations were mostly
performed by Sepkoski, who wrote a FORTRAN program that calcu-

lated and plotted clade statistics. Sepkoski was therefore added as a fifth author, in a departure from the "quadripartite" scheme debated by the MBL group.

The paper itself was written primarily by Gould, who prefaced the technical sections with a fairly extensive meditation on the meaning of the stochastic simulation "experiments." He explained that while the authors did not necessarily assume "that life's history is ontologically random," the stochastic null hypothesis was an effective "'criterion of subtraction' for ascertaining what amount of apparent order requires no deterministic cause" (Gould et al. 1977, 23–24). As Gould explained in the introduction, though, the paper also had a "larger, ulterior motive":

> We believe that paleontology—the most inductive and historical of the sciences—might profit by applying some deductive methods commonly used in the non-historical sciences (without sacrificing its important documentary role for the history of life). We may seek an abstract, timeless generality behind the manifest and undeniable uniqueness of life and its history (Gould et al. 1977, 25).

Gould acknowledged the debt this approach owed to "the other branch of natural history most celebrated for the complexity and uniqueness of its subject—ecology," and also defended a version of the species-as-particles idea where "untimebounded" and "untaxonbounded" models "treat all times and taxa alike" (Gould et al. 1977, 25).

Nonetheless, in important ways this paper did not pursue the more radical, ontological interpretation of randomness in the history of life. Despite the fact that the comparison between real and random clades appeared to indicate "the outstanding feature of real and random clades is their basic similarity," the paper concluded by highlighting three significant departures from randomness in the real clades. In the first case, the study found that living clades tend, in the "real world," to be significantly larger than extinct ones. While the paper discussed possible sampling biases which might produce this phenomenon, the authors nonetheless concluded that "a plausible story in the deterministic mode" may offer the best explanation: "The real world is 'taxonbound'—superior designs tend to persist, diversify, and survive" (Gould et al. 1977, 34). In the second instance, the extent of fluctuation of diversity—measured as the "uniformity" (UNI) of a clade—was much higher in certain real clades than in any of the simulated ones. Low values of UNI—meaning fairly significant fluctuations of diversity in a taxon—were found especially in amphibians and mammals, and the authors were unsuccessful in adjusting the parameters of the simulation to produce random clades with similar values. Furthermore, fluctuations in diversity among different groups appeared to be correlated in time. Again, the authors concluded that this may "represent a non-random effect," as a result of "real, biological interaction" between groups. Thus, "the real world may be, in this respect, 'timebounded,'" meaning that "some times really are 'good' for certain groups" (Gould et al. 1977, 37). Finally, the comparison also found that the average "center of gravity"—"a measure of the relative position in time of the mean diversity"—was lower for extinct clades among the real sample than in the simulation. This effect was especially pronounced for early clades of the Cambrian-Ordovician period of rapid evolutionary expansion, where clade shapes in groups such as trilobites, brachiopods, and nautiloids tended to have clade shapes that were wide at the base, well before the midway point in the groups' chronology. The "deterministic" explanation for this effect was that "an average real clade tends to be wide at the bottom because it radiates rapidly following the invasion of new ecological space or the evolution of new morphological designs by ancestral lineages," then achieves optimal diversity before tapering off very slowly towards extinction. While the authors explored a variety of ways in which such a pattern could be accounted for in the simulation, they also frankly admitted it posed a potential "violation" of the stochastic model by suggesting that "all times are not alike to members of a clade if chances for diversification are characteristically greater during their early history" (Gould et al. 1977, 38).

All in all, these limitations in the random model can be interpreted as an endorsement of the heuristic value of the stochastic null hypothesis. In this paper, the "criterion of subtraction" appeared to work very well for separating those patterns in the history of life that did not require deterministic explanations from those which did. One such case, it would emerge, was mass extinction, which was explicitly acknowledged to be outside of the testable parameters of the MBL model. And despite the confident spin with which the paper concluded—calling for a revival of Charles Lyell's dynamic steady-state model of the history of life that rejected directionalism and progress—it is no coincidence that this paper would be the final entry in the MBL collaboration. One could argue that, with this study, the MBL model had simply run its course.

As we will explore below, it may also be that this final paper exacerbated the growing divisions within the group over how to interpret the model. As primary author of the 1977 paper, Gould presented his own more moderate heuristic interpretation, but Schopf, in particular, was becoming ever more attached to a radical, ontological interpretation of stochasticity. It is also the case that the 1977 paper brought more uncomfortable critical scrutiny to the model than had any of the previous efforts. The manuscript was initially submitted to *Science*, where it was rejected after several lukewarm reviews. One referee commented that the paper was "not important enough to warrant publication as a lead article in *Science*," and felt it was overly long and wordy. This view was echoed in other reviews as well, and while the most positive found the paper "generally excellent and stimulating," the most damaging report contended that the paper's "results have no explanatory significance." This last review, which was unsigned, was the most strongly worded, and Gould speculated that its author was "GGS himself—I'd know his style (& typewriter) anywhere" (PS-*P* "1977"). This reviewer—George Gaylord Simpson or not—chastised the authors for confusing the terms "random" and "stochastic," and criticized the ambivalent stance they took on the implications of randomness for evolution, noting that while "there is some implication that the 'real' distributions studied are acausal. . . . [which] would be a revolutionary doctrine not only for evolution but for science in general . . . no such conclusion is really inherent in their results."

This paper also generated the most damaging published criticism of the MBL model. These criticisms centered on improper scaling in comparisons of real and simulated clades in both the original 1973 paper and in the 1977 follow-up, and were first presented by Steven Stanley in his 1979 book *Macroevolution: Pattern and Process* (Stanley 1979, 279). Stanley's basic argument was that the similarity in shapes between real and random clades was artificially enhanced by the comparatively low numbers of taxa composing each of the random clades. In other words, the random clades consisted of far fewer units than average real higher taxa, which gave them artificial instability and made them much more likely to rapidly diversify via random branching and extinction. In 1981, Stanley coauthored a paper in *Paleobiology* that reiterated this point and argued that species, rather than genera or higher taxa, were the appropriate units for simulation (Stanley et al. 1981). While the reaction from the MBL group was initially quite defensive—leading to an acrimonious exchange between Stanley and Schopf over the paper's publication—Stanley's criticisms have for the most part been vindicated. As Raup acknowledged years later, "They were bang on—they were dead right" (Raup interview).

The Demise of the MBL Group

Schopf versus Raup: Determinism and Extinction

Up through the 1977 paper on clade shape, the MBL group had successfully, if sometimes uneasily, juggled its various allegiances and interpretations. However, internal disagreements would soon lead to tensions that ultimately ended the collaboration. A particular locus of conflict was the proposed stochastic methods primer, on which Schopf began to fixate as a vehicle for proselytizing his view of stochastic paleontology. Raup had been tasked with bringing an outline to the December 1973 meeting, an arrangement that made sense given the fact that Raup was most intimately knowledgeable about the workings of the simulation models. However, even as he accepted this task he worried that "it [the book] might turn out to be more of a swell foop than a fell swoop" (Raup to Schopf, 20 November 1973: Schopf pap. 3, 30). Apparently Raup was not encouraged by conversations that took place at Woods Hole, because in his next communication with Schopf on the subject—more than a year later—he announced, "I have been doing a lot of thinking about my role in the enterprise and have finally decided to opt out" (Raup to Schopf, Gould, and Simberloff, 7 April 1975: Schopf pap. 3, 30). Raup explained that his "negative feelings about the venture have been aired at length" during the 1973 Woods Hole meeting; nonetheless, he left the door open for future collaboration, and gave his "complete blessing and cooperation" to Schopf and Gould to continue the manuscript without him. Although the letter did not elaborate what those "negative feelings" were, Raup recalls experiencing a growing concern that Schopf's aggressive promotion of stochasticity and "gas laws" might have an alienating effect on more traditional paleontologists, and a fear that "Tom's messiah approach to the whole thing" might undermine the positive potential of the MBL collaboration (Raup interview).

Raup's concern must have gotten through to Schopf, because the next time Schopf broached the book with Raup and Gould, his tone was considerably more restrained. He explained that in "considering why it is the MBL work does not receive the credit it is due in the profession," he realized that the group had not sufficiently articulated to the profession how their approach contributes to the "day to day activity" of traditional paleontologists and geologists. After all, Schopf admitted, since "the primary concern of geologists is the explanation of particular events," and "what we have done does not (generally) relate to events," it was hardly surprising that "our colleagues look at what we have done and say 'so what!'" (Schopf to Raup and Gould, 30 August 1977 [draft]: Schopf pap. 3, 30). "It matters not one iota that species may be particles like atoms" Schopf continued. "After all, the blastoids went extinct didn't they . . . and *that* is what needs explaining—not some fancy footwork about how some group went extinct, and it 'happened' to be blastoids." The solution Schopf proposed was to find a way to "provide bridges" between traditional paleontological questions and random models, in which case it was incumbent on the MBL group to "be able to provide a prospectus of what *are* the interesting questions which is both attractive *and within the capabilities* of the guy who was previously worried about why the blastoids were done in." For that reason, he said, "a book-length treatment is desperately needed" because "only in that way will the full panoply and inherent richness of the stochastic models approach become apparent."

Schopf then outlined an approach in which questions about particular historical events could be redefined so as to preserve their legitimacy, but "in such a way that we also provide an answer—nontraditional to be sure" for why stochastic models were relevant to those questions. One strategy was to argue that "specific causes are so multifaceted and unable to be sorted out" that continuing to search for individual, deterministic answers to questions like 'why did a particular group become extinct' was futile. In other words, here Schopf was advocating the weaker, heuristic MBL interpretation, in which stochasticity was most effective as a null model, and proven statistical techniques like survivorship analysis gave insight into specific paleontological problems. "By far our best strategy," Schopf concluded, is "to encompass and conquer. But if we do so, we need to treat the traditional problems of paleontologists. Thus our taking-off points start with a real world problem. Every chapter in the book should have as its first sentence the statement of a specific problem the day-to-day paleontologist meets with—& can relate to. We need to do the extra work of going more than half way, or so I feel" (Schopf to Raup and Gould, 30 August 1977 [draft]: Schopf pap. 3, 30).

The final suggestion Schopf made was that the book project be put on hold—temporarily—in favor of offering a workshop on stochastic models in order to provide "feedback to see how things can best be done." Schopf's change of heart encouraged Raup to put aside his reservations and participate, and the two began collaboration on a short primer for the workshop (funded by the National Science Foundation), which they envisioned as a rough first version of the eventual book. Raup wrote an initial draft, and plans went ahead to present a workshop on stochastic models at the National Museum of Natural History in early June 1978. A few days before the workshop began, Schopf proudly sent a revised copy of the primer to Gould. "I am rather pleased with the way this is developing," he commented, asking Gould to consider writing a "major Preface" for the published version. Schopf suggested that "such a forward [*sic*] might be historical, might be 'idiographic' and 'nomothetic' with regard to paleontology, and would be acknowledged on both the dust cover and the title page as 'Preface by Stephen Jay Gould'" (Schopf to Gould, 2 June 1978: Schopf pap. 5, 14).

Unfortunately, Schopf badly misjudged Raup's reaction to the draft. A month later—and after the workshop concluded—Raup sent Schopf what must have been a deflating letter in which he spelled out reasons why he was "not optimistic about the primer at this point." Chief among Raup's concerns were "technical faults and ommissions [*sic*] which will require a *lot* of work to remedy," and an impression that Schopf had deviated from the more moderate approach outlined in his earlier letter. As Raup explained,

> Our basic objective has been to produce a 'how to' treatment of certain *techniques* of paleontological importance. We have *not* had in mind the presentation of a definitive theory of macroevolution. . . . We wanted to produce a book that would be used by everyday paleontologists on everyday problems—not necessarily confined to those in evolutionary theory. I am afraid we have misfired: the techniques answering the above requirements are few and far between and there is (I think) much too much theory mixed in (Raup to Schopf, 17 July 1978: Schopf pap. 3, 30).

Following a chapter-by-chapter enumeration of technical and conceptual faults in the draft, Raup concluded that Schopf's revision "has indeed moved you away from the primer objective" and warned "if we publish a book in something close to its present form, we will be absolutely crucified by friend and foe alike!"

Despite declaring "little faith in the present attempt," Raup closed his letter with an attempt at finding a constructive solution. He agreed that "we would all like to put what we have learned from the MBL project in some sort of hard-hitting book form," and suggested "a different tack":

> We have written a lot of MBL papers. They have gained considerable attention and have started a lot of argument. Two of the papers were in semi-popular journals and there was the commentary in *Science*. . . . As a result of all of this, I suspect that there are relatively few of our paleontological colleagues who remain unaware of the MBL work. This means to me that a book version must really be stronger than the sum of its parts. If it isn't, it is a waste of the reader's time. A collected reprint volume would do as well (Raup to Schopf, 17 July 1978: Schopf pap. 3, 30).

The solution he proposed was for Schopf to take the project in his own direction, since "what you are trying to do calls for a totally different format." Raup imagined that the book would present "a clear statement of the species-as-particles idea" through rigorous testing and documentation of a specific set of models. However, he concluded that this would be "a tough job," and stressed that he did not see a role for himself in such a project.

Gould had been on the sidelines during much of this discussion, but after receiving both Schopf's invitation and Raup's response he sent Schopf a letter in which he attempted to play peacemaker. On the one hand, he reasoned, Raup's objections put further progress on the project in jeopardy, and suggested "I'd do best by just holding on to the ms. for a while until the dust clears." On the other, he tried to reassure Schopf of his ultimate support: "let me reiterate my agreement with you that such a document represents a consistent, logical, and forceful next step and that our major task in pushing our view" (Gould to Schopf, 7 July 1978: Schopf pap. 5, 14). Essentially, Gould left the decision about how to proceed to Schopf, but stressed his willingness to contribute the invited preface "once we all clarify and agree." Schopf replied a week later to express 'appreciation' for Gould's support and 'puzzlement' at "the Raup situation." He rather breezily waived aside Raup's hesitation as the result of Raup's "very high aspirations for the National Academy," and advanced a "theory" that Raup might feel threatened by a co-authored project which would "steal his thunder and dilute the impact of his work." "Fearing that," Schopf continued, "and wanting the NAS recognition, he may feel very uneasy about doing anything as major as this with coauthors. If so, we'd better get him elected as soon as possible so that the business of remaking paleontology can progress as rapidly as possible" (Schopf to Gould, 30 July 1978: Schopf pap. 5, 14).[2] Schopf made no comment about any of the substantive criticisms or suggestions Raup had raised, but promised to revisit the matter after he had spent time with Raup in Chicago.

At least initially, Schopf seems to have believed that his differences with Raup could be patched up. In a letter to Raup in late December 1978, he reiterated his continuing "belief of the importance of the stochastic paleontology" and stressed his view of the importance "of a primer which has this theory at its core." He even optimistically closed his letter by mentioning that he "look[ed] forward to reading a revised draft of either the book or the primer when I return [from a semester in Germany] at Christmas!" (Schopf to Raup, November 1978: Schopf pap. 3, 30). However, this cheerful optimism could not prevent further disintegration of the collaboration, and in early 1979 Raup sent Schopf a long, philosophical letter that highlighted the widening gulf between the two.

At the time of this letter, January 1979, Raup was experiencing a radical change in his views about randomness and determinism. Over the several years that had passed between the publication of the first MBL paper, in 1973, and the paper on clade shape in 1977, Raup's interpretation of the role of stochastic processes in nature appears to have remained fairly consistent. Raup's own recollection about his original attitude in 1973 is fuzzy, but Simberloff recalls that "of the group, I felt at the time that Dave was almost, but not quite, committed to the view that many things that were interpreted as having a very specific cause were in fact random, at least with respect to that putative causal factor" (Simberloff, e-mail communication with John Huss, 11 March 2002). Raup acknowledges that his 1974 morphology paper with Gould reflects an interpretation in which "random processes often produce patterns that APPEAR to be worthy of deterministic interpretation," and it seems likely that he was one of the authors to whom Gould alluded, in 1976, as supporters of the "strong version" of MBL (Raup, e-mail communication with John Huss, 13 March 2002). In the same year that the final collaborative MBL paper was published, Raup also wrote an essay for *American Scientist* entitled "Probabilistic Models in Evolutionary Paleobiology," which advocated a version of the more radical, ontological in-

---

2. Raup was indeed elected to the National Academy in 1979.

terpretation of MBL. In summarizing the advances of the MBL model for a semi-popular readership, Raup explained that specific causes for changes such as faunal succession may be viewed as the result of random fluctuations in a Markov chain, and he drew explicit attention to the possible conclusion that "natural selection may behave (mathematically) as a random variable." While he acknowledged that describing "such changes as 'random' does not deny cause and effect," he stressed that "the distribution of these causes in geologic time may be essentially random," and concluded that "evolution certainly should be viewed in a Markovian framework" (Raup 1977, 51).

Nonetheless, as Raup recalls, by the time of the 1978 workshop, he was ready to concede that "the external rare event was deterministic . . . [and] I became more and more convinced that the externalities were important" (Raup interview). This brings us back to Raup's letter to Schopf, which began with a meditation on the consequences for the profession "if you and I and Steve really succeed in selling our current brand of nomothetic paleontology." He imagined that "scores of young paleontologists will be plotting survivorship curves (or whatever) in a slavish and unthinking manner and much of the work will have to be thrown out ultimately" since "the scientific work that is done will probably be more wrong than right" (Raup to Schopf, 28 January 1979: Schopf pap. 3, 30). After all, he reminded Schopf, "Every generation of paleontologists has had its nomotheticists." Nonetheless, taking "a more positive stance," Raup asserted his firm conviction "that the current stochastic models are right—at least at some scale." The problem, he argued, was determining the appropriate scale.

Quickly, however, Raup revealed that a subject that would come to occupy much of the remainder of his professional career—the interpretation of mass extinctions—had largely precipitated the shift in his thinking about random processes. As he continued, Raup explained, "I am becoming more and more convinced that the key gap in our thinking for the last 125 years is the nature of extinction." He went on to argue:

> If we take neo-darwinian theory at face value, the fossil record makes no sense. That is, if we have (a) adaptation through natural selection and/or species selection and (b) extinction through competitive replacement or displacement, then we ought to see a variety of features in the fossil record that we do not such as: (a) clear evidence of progress, (b) decrease in evolutionary rates (both morphologic and taxonomic), (c) possibly a decrease in diversity

(at least within an adaptive zone). Now we do not see these things because: (a) we are too dumb, *or* (b) the record is lousy, *or* there are features of the evolutionary mechanism that prevent the approach to a steady-state.

Raup explained that his "candidate explanation is, of course, that extinction is random with respect to fitness. By this scenario,"

> the neo-Darwinian system is at work all the time—producing trilobite eyes and pterosaur flight—but never really gets anywhere in the long run because the trilobites and pterosaurs get bumped off (through no fault of their own!). . . . The system is always heading toward a steady state but never gets there (Raup to Schopf, 28 January 1979: Schopf pap. 3, 30).

So far, Raup's view was consistent with Schopf's vision of an evolutionary process held in check by stochastically varying fluctuations in extinction rates around an equilibrium. Extinction is not a matter of bad genes, in a Darwinian, selective sense, but rather bad luck.

The problem with this more radical view, Raup confessed, is that there was simply too little "convincing documentation" to support it, making it "nothing more than a just-so story." What was required were studies of actual groups in the fossil record in which extinction could be shown to be nonselective. Take, for example, the trilobites, whose abrupt departure in the Permian is one of the most spectacular examples of extinction in the fossil record. "By my grand scenario," Raup explained, "the demise of the trilobites was just bad luck on their part": in other words, the extinction rates of trilobite taxa were no different from those of other groups, but trilobite diversity simply had the 'misfortune' to "wander down to zero over the course of a couple of hundred million years." Unfortunately, Raup admitted, he no longer believed such a scenario was mathematically plausible. According to standard equilibrium assumptions from island biogeography, speciation rates (p) and extinction rates (q) should be roughly equal for all species over the Phanerozoic (in order to prevent either total extinction or exponential expansion of life). Using accepted values for p and q of 0.1 per million years, and assuming that approximately 1,000 trilobite species existed at the height of the group's diversity, then the probability that any one of the 1,000 species would become extinct over the 200 million year existence of the group is 0.95. To put it in terms of survivorship, roughly 1 in 20 trilobite species should have had living descendents at the end of the Permian.

However, Raup pointed out, the real question was the probability that *all* of the 1,000 species would become extinct. This can be expressed as the extinction probability (95%) multiplied exponentially by the number of species ($0.95^{1,000}$), which yields the astronomically low probability 5×10^{-23}, or 0.00000000000000000000005. "In other words," Raup concluded, "the chances are nil that the trilobites could have drifted to extinction and one must conclude that there must have been something different about their extinction rates" (Raup to Schopf, 28 January 1979: Schopf pap. 3, 30). Or, to put it another way, trilobites would have to have had extinction rates an order of magnitude greater than other groups' in order for their total extinction to have been plausible. In either case, the assumption that extinction is non-selective was in serious jeopardy: there must have been something special that made the trilobites candidates for complete extinction. If this was true, Raup reasoned, the "Paleozoic extinctions [appear to have been] actually selective . . . meaning in turn that trilobites were inferior beings and deserved to die." And he closed the letter with the candid admission that "if I were to be completely objective . . . I would have to conclude that the stochastic model does not apply to the distribution of species extinctions between and among classes or phyla," which led to "a purely Darwinian conclusion." His final caution to Schopf was, "I don't think it is wise or fair to present [the stochastic model] as the final solution until it can be proven rigorously" (Raup to Schopf, 28 January 1979: Schopf pap. 3, 30).

Schopf responded with a long letter in which he attempted to sway Raup with an argument about survivorship curves, claiming "the Raup paradox" was "fallacious" since it implied "that NO GROUP the size of trilobites could ever go extinct" (Schopf to Raup, 8 February 1979: Schopf pap. 3, 30). Of course, the obvious rejoinder was that Raup's proposition was a paradox only if one insisted on viewing extinctions as nonselective and randomly distributed through time. If, on the other hand, one rejects either of those assumptions—admitting either that some groups have bad genes or that historical events like mass extinctions have determinate causes—then Raup's calculation appears to be fairly damning evidence against the stochastic, species-as-particles view. But by this point Schopf had become even more committed to an ontological interpretation of the MBL model; as he explained to Raup, "In my view, all of paleontology, i.e., all of those fossils, is (are) simply a metaphore [sic] for what is really the statistical mechanics of a series of interacting hollow curves."

Schopf's statement can be interpreted as the most extreme application of the idealized metaphor for reading the fossil record: in his view, *the fossils themselves* are a "metaphor" for the model, implying that the statistical idealization itself which is the true "text." Schopf's commitment to this extraordinary interpretation was, by this late stage, inflexible and dogmatic: his commitment to nondeterminism prevented him from seeing Raup's probability calculation as a legitimate obstacle to the stochastic view, because he refused a priori to admit alternative mechanistic explanations (such as inherent maladaptiveness or historically contingent and extrinsic causes of mass extinction). Raup obviously recognized this, and in his next letter he confided to Schopf,

> On a somewhat serious note, your recent letters have given me some concern. Could it be that you are getting a bit carried away by stochastic approaches to paleontology? I fear that you will lose your credibility if you press the gospel too hard. My concern is difficult to express. I am sure you are more inhibited in 'public' than in personal letters but the message that comes through the letters is that you have found the Holy Grail and that anyone who does not recognize this (and join in the feast) is a poor slob in desperate need of salvation (Raup to Schopf, 22 February 1979: Schopf pap. 3, 30).

He went on to warn Schopf that this attitude might well end with Schopf as "the ultimate loser," and even raised the specter that Schopf could "join the ranks of T. Y. H. Ma, Petrunkovitch, Meyerhoff, Goldschmidt, et al. . . . [all of whom] lost credibility and thus lost the ball game." "All I am asking," Raup concluded, "is that you (1) think hard about your modus operandi, and (2) think hard about just how compelling the case is for our particular breed of stochastic paleontology."

Schopf versus Gould: Replaying the Tape of Life

At around the same time, Gould finally voiced his own concerns about Schopf's increasing attachment to the 'species as particles' view of paleontology. At the time of his last exchange with Raup, Schopf was putting the final touches on an essay that would appear in *Paleobiology* in late 1979 entitled "Evolving Paleontological Views on Deterministic and Stochastic Approaches." Despite the seemingly neutral title, the paper was a polemical statement of Schopf's grand vision for stochastic paleontology, and opened with the declaration that chance and stochastic processes form "the first order pattern of organization of the history of life" (Schopf 1979, 337). As an introductory analogy, Schopf drew on historian Herbert Butterfield's 1931 classic *The Whig Interpretation of History*, which he extrapolated to an argument "that Whig values are not limited to human history but are found in a pervasive way in interpretations of the history of life," in which evolutionary "progress" is interpreted as the result of a causally determined, directional process (Schopf 1979, 338). In place of "the extraordinarily strong reliance on determinism in paleontology," Schopf explained that

> the proper type of theory to apply to large statistical summaries is some form or another of stochastic theory, such as occurs in chemistry (the gas laws), population biology (demography), and physics (the Heisenberg uncertainty principle). The fate of any given molecule, or individual animal, or atom is of no concern per se. Rather, the ensemble statistical properties of the particles and the types of predictions which those properties allow are what is of interest.

While he stressed that such a view did not ignore particular historical events, he nonetheless asserted that "the incorporation of stochastic thinking into paleontology does say that there was nothing inevitable either in evolution or in history that a priori determined either the present state of affairs, or any specific past configuration" (Schopf 1979, 343).

Schopf then applied this insight to two major topics of paleontological interest: rates of speciation and extinction. In the first case, he essentially recapitulated the argument of the 1975 MBL paper that different taxa have inherently different rates of genetic evolution. However, he departed from the 1975 paper by steering the discussion to a critique of the "philosophical assumption" that "a literal reading of the objects which comprise our fossil record is assured to reflect a 'true' image of a previous reality" (Schopf 1979, 344). In fact, Schopf maintained, such a "literal reading" gives "a grossly distorted view of rates of evolution," and paleontologists would be better served by taking a more idealized, statistical view of the record (Schopf 1979, 345). In the second case, Schopf reiterated the criticism he had expressed privately to Raup about fallacious attribution of selective causes to extinction, which practice he termed "tautological" since it depended (he argued) on a post hoc determination of 'fitness' based merely on the record of which species did and did not survive. "'Fitness' or 'adaptation' may not have been a factor" in extinctions, Schopf contended, "unless one *defines* them to have been a factor. In a world of fixed total resources that fluctuates from time to time and place to place, where individuals cannot live forever, it may be no more than bad luck as to which species persist, or die" (Schopf 1979, 346). Schopf concluded the essay with a seven-point outline of a "stochastic view" that mixed elements of biogeographic equilibrium theory with precepts of the MBL model. In point 5—that rates of extinction and origination are random with respect to membership in a particular taxonomic group—Schopf paraphrased (without attribution) Gould's "tape of life" metaphor, observing that if we were to "replay geologic history within the same general ecologic constraints as have prevailed . . . it could be that (for example!) blastoids live through the Permian, and crinoids die out" (Schopf 1979, 348).

When Schopf sent Gould a draft of this essay just prior to publication, Gould made a comment that, in many ways, epitomizes the growing distance between his own and Schopf's understanding of the role of chance and determinism in historical processes:

> You continually conflate (though they are not unrelated of course) the notions of predictability and stochasticity. Stochastic models can, of course, lead to a high degree of predictability, at least for general patterns of events. Of course I agree that the most fascinating aspect of life on earth is that it would probably play itself out in a totally different way if we started again from the same initial conditions—but this metaphysic (which I share with you) is not the essence of maintaining a stochastic perspective in paleontological theory (Gould to Schopf, 25 June 1979: Schopf pap. 5, 59).

As Gould was implying, sensitivity to initial conditions in fact presupposes determinism, since while the initial state of a system may be unpredictable, subsequently unfolding events can, as Gould noted, be quite predictable. Gould was also drawing attention to two alternative interpretations of the metaphor "replaying the tape of life." By 1979, Gould's interests were in the midst of a slow turn back to one of his original preoccupations: the role of physical developmental constraints in evolution. This was heralded by the publication of his first book, *Ontogeny and Phylogeny*, in 1977, which combined a historical and scientific reexamination of Ernst Haeckel's intuition about parallels between embryonic development and evolutionary history (Gould 1977). The same issue of

Paleobiology in which Schopf's essay appeared also contained a paper, coauthored by Gould (along with David Wake), entitled "Size and Shape in Ontogeny and Phylogeny," which drew attention to the relationship between the timing of developmental stages and morphological evolution, and which is now considered one of the early contributions in the field of evolution and development, or evo-devo (Alberch et al. 1979). Finally, 1979 was the year that Gould and Richard Lewontin published their famous critique of adaptationism, "The Spandrels of San Marco," in which they emphasized the role of nonselective or accidental by-products of evolution ("spandrels") in contributing to—and constraining—the evolution of organic forms (Gould and Lewontin 1979).

In other words, Gould's own distinctive view of life was experiencing significant change at exactly the time when Schopf showed him his essay on stochastic paleontology. In Gould's developing view, stochastic processes were balanced by three important kinds of constraints. The first was intrinsic structural factors like allometric relationships that limit or direct morphology along particular pathways. This notion of constraint harkened back to Gould's early interest in D'Arcy Thompson's "laws" of form, and while it was essentially nonselective, it did consider form to be in an important sense determined by the laws of geometry. The second kind of constraint was the influence of developmental pathways in ontogeny and phylogeny. The essential idea was that certain directions, once taken, were irreversible: organisms with bilaterally symmetrical appendages radiating from an axial notochord may not have been in any sense evolutionarily "necessary," but once that particular "choice" was taken by natural selection, certain morphological options were closed off to further exploration. The final kind of constraints Gould recognized were geological and environmental factors: he argued that the history of life was shaped by particular, distinctive historical events that are independent of normal Darwinian processes. For example, Gould would become quite interested in the role of extraterrestrial mechanisms for mass extinctions, which periodically reset the adaptive conditions of the physical environment in unpredictable ways. Each of these senses of constraint contributed to what Gould would famously define as "contingency" in the history of life: the interaction between unpredictable or random events, and the consequences of those events, which were in an important sense deterministic. He would explore this idea at length in his 1989 book *Wonderful Life*, and also in his work with Elizabeth

Vrba on macroevolutionary hierarchy and "exaptation." The crucial idea in both cases was that we live in "a world built by irrevocable history" (Gould 1989b; Vrba and Gould 1986, 226).

The message Gould took away from the MBL model, then, was subtly but importantly different from Schopf's. Gould was impressed by the way random processes could generate apparent order, but was not ready to give up a "literal" reading of the fossil record. The "tape of life" metaphor was a lesson about sensitivity to slight differences in initial conditions, and Gould realized that determining *which* particular outcome resulted in the end was a matter of "irrevocable history." The fact that it was precisely these contingent factors that gave history its particularity—such as major mass extinctions—that the MBL simulation did not take into account was something Gould frequently acknowledged (Gould 1978, 279). Schopf, on the other hand, saw the MBL model as a lesson that the history of life could be interpreted in such a way that all particularity was ignored—in a sense, it allowed history without historicity. This helps to explain Schopf's adamant objection to two ideas Gould championed: punctuated equilibria and catastrophic mass extinctions. Schopf consistently rejected the notion that mass extinctions have deterministic, identifiable causes out of hand. He argued, for example, that the great Permian extinctions were a consequence of the slow shifting of continents and the MacArthur-Wilson species-area effect, and he campaigned against the theory that dinosaurs became extinct as a result of a bolide impact up until his death in 1984 (Schopf 1974).

As for punctuated equilibria, Schopf best expressed his view in a letter to Gould in 1981, in which he related his social views to his philosophy of science:

> I was struck again—and not for the first time . . . that the social level was *far* more important than the talents of any individual. And if so for *Homo sapiens* . . . then it must be the case for *all* species. The luck of the draw is the dominant factor in how we live our lives—mostly by how it constrains the opportunities before us.
>
> Now, if one has this sort of a social view, it seems to me that "stochastic" paleontology flows effortlessly—it agrees with a luck of the draw view of life. . . . This being so, it is now completely unacceptable to me that any *species* is really any different from any other species. They are *all* out there trying like hell to do well. Sure, God helps those that help themselves. But what does God do when *all* species are helping themselves?? The traditional view

is that, under these circumstances, God is picking and choosing! But surely that is nonsense.

So where does that leave one? Well, it leaves me thinking there *must* be something wrong with Raup's conclusion that trilobites are morally degenerate (or at least that they couldn't have gone to extinction by mere accident!). ... If all species *through time* are equally "successful"—and if all species *at any given moment in time* are equally successful—then the notion of "success" (*sensu latu*) has no place in evolutionary theory (Schopf to Gould, 4 December 1981: Schopf pap. 8, 31).

This view conflicts in one important way with Gould's: While Gould might have agreed that "success" was a problematic term when applied broadly to the history of a lineage, his contingent view of history rejected the argument that all organisms were, at any particular time, equally prepared for certain environmental conditions. If one accepts that unique events—a bolide impact at the Cretaceous-Tertiary boundary, for instance—play a role in the history of life, as Gould did, then one accepts that, even if for no fault of their own, some organisms (dinosaurs) failed while others (mammals) succeeded in meeting the new conditions that resulted.

However, Schopf's "species-as-particles" philosophy ultimately rejected historicity (e.g., causal contingency) because it assumed that, like molecules in a volume of gas, individuals change place with respect to one another so frequently that their individual positions are of no importance to the description of the greater whole. As Schopf explained in his letter to Gould, evolution happens too fast to establish preferences for particular organisms: "Sure some animals or plants are more abundant than some other animals or plants at some moment in time. But the shuffle is so fast one can't even learn the players." And given that assumption, Schopf revealed that his overwhelming objection to punctuated equilibria was based not on its assumption of rapid change, but rather on its assumption of stasis: "The notion of stasis—that the mean duration of a species is millions and millions of years, that then becomes the MAIN SUPPORT for a deterministic view of life!" In other words, in a world where change is extremely rare, one must conclude that it is also "special"; this implies, as Schopf put it, that "punctuated equilibrium becomes nearly the main argument for biological determinism" (Schopf to Gould, 4 December 1981: Schopf pap. 8, 31). As Schopf explained in another letter to Gould,

I hope you will see it not as a campaign *against* something (PE), but rather as a campaign *for* something (a view of the world where change is easy and continuous). I think you hit the nail-on-the-center when you said it is a question of change 'difficult' vs. change 'easy'. . . . I am as convinced as I can be (&, possibly, as wrong as can be) that with 10^6 to 10^7 living species, & $\approx 10^{10}$ over geologic time, that species are particles in a never-ending biological world. Thus, in order to *avoid* Raup's determinism . . . I am forced to a view that species durations must be quite short ($\approx 10^5$ years). If so, change must be easy. Or so it goes, in this (my) view of life (Schopf to Gould, 22 November 1982: Schopf pap. 9, 106).

Conclusion

One irony is that, while the very basis for Schopf's view appears to be its purportedly literal reading of the fossil record, Schopf ultimately rejected that approach to reading on first principles in favor of his idealized, statistical vision. In the end, despite their initial excitement about the prospect of stochastic paleontology for producing a radically revised interpretation of the history of life, Raup and Gould each concluded that particular, historical, and deterministic factors could not be entirely abandoned. This did not mean, however, that the idealized approach to reading the fossil record was rejected as a consequence. As we will see in the next chapter, it was preserved in the approach to taxic paleobiology that negotiated many of the dichotomies—general versus particular, determinism versus randomness, selectivity versus stochasticity, models versus empirics—that lay at the heart of disagreements surrounding the MBL model. However, Schopf would not participate in this compromise. For the next several years—up until his death—Schopf clung to his species-as-particles view even as his onetime collaborators moved farther and farther away, and as a result he became increasingly bitter and isolated from his onetime compatriots. When, in 1984, Gould memorialized his friend in an obituary in the journal Schopf had founded, he acknowledged the painfulness of this fact:

> Tom was a prickly, often difficult colleague, so driven by his unconventional vision, so committed to its fundamental truth, so brave (or foolhardy) that he would sacrifice friendship and human relations to its zealous advance. I can-

not psychoanalyze him. . . . But I do know one thing: I know that it led him to much personal misery. I often cried inside that he could not break out of it and spare himself the pain, but I and his other friends could do nothing. Tom was so committed to a unity of vision that he hopelessly conflated his sense of the factual with his belief about the ethical. He saw what he deeply believed as not only true but just, right, and moral. . . . But my love and admiration for him were never compromised (though I was, of course, often annoyed), for I hope we can all understand that a pure and burning commitment to knowledge and understanding drove him, often (sadly) in destructive ways (Gould 1984b, 282).

Nonetheless, Schopf's hardheaded determination to press the case for stochastic paleontology had an undeniably salutary effect on the field, and pushed paleobiology in directions that would ultimately help it to secure its place at the "high table" of evolutionary theory. Huss identifies three ways in which the MBL model innovated paleobiology, despite its "inadequacy as a theoretical model of evolution." First, it presented a new tool—simulation—for investigating evolution that "changed paleontologic practice" by making it possible "to generate patterns from known starting assumptions (i.e., a model) and compare them with observed natural patterns." Secondly, MBL-style computer simulations made it possible to conduct paleontological "thought experiments" that had "tremendous heuristic value." And third, "The MBL group changed the kinds of questions paleontologists are able to ask and answer, and even the ways they are able to resolve disputes" (Huss 2009, 340). Raup somewhat more modestly concludes that while "stochastic models were developing in many fields and someone was sure to apply those methods to the fossil record," in the end the MBL model "increase[d] awareness of the power of stochastic processes and [provided] a general warning against jumping to deterministic interpretations of pattern" (Raup, e-mail communication with John Huss, 13 March 2002).

I would add that the history of the MBL group also sheds important light on the negotiation that took place between the heterogeneous theoretical commitments of individual paleobiologists and the institutional agenda they shared. From the very start, what bound the group together was a shared commitment to creating a "nomothetic paleobiology" that would advance the status of the discipline within evolutionary biology and distance it from more traditional modes of explanation in paleontology. This commitment never faltered, despite disagreements over exactly what that nomothetic science should look like. In the end, however, Gould and Raup came to feel that a nomothetic approach could comfortably exist alongside more traditional empirical methods that emphasized historicity and even determinism, while Schopf never wavered in his belief that such factors must be rigorously excluded. This led to the demise of the MBL group, but not of the paleobiological movement itself.

CHAPTER EIGHT

A "Natural History of Data": The Rise of Taxic Paleobiology

In the last chapter, I examined the development of the MBL model of simulated phylogeny as an example of "idealized rereading" of the fossil record. This approach, I argued, attempted to circumvent Darwin's dilemma for paleontology by departing from a literal reading of the empirical record in favor of a model which abstracted and simulated evolutionary patterns in a search for generalizable statements or even "gas laws" that could be applied to the history of life. Ultimately, despite stimulating valuable insights, the MBL model faltered in part because the dynamic processes being simulated were simply too complex to be fully or adequately represented by a simple computer program run on 1970s-era technology. One of the most obvious questions the MBL simulations raised was whether the patterns of evolutionary rates and diversity generated could be identified in fossil data. While several of the MBL papers made tentative steps to answer this question, paleobiologists realized that much more substantial efforts were required using far more complete sources of data than had been available previously. The MBL model flirted with the notion that evolution was at some level an essentially random process, but this did not sit well with all of its authors. George Gaylord Simpson, himself no enemy of "nomothetic" paleontology, wrote in 1960 that the presence of evolutionary trends, such as directional morphological changes in a single lineage, or parallel trends in across many different lineages "are so common and so thoroughly established by concrete evidence" that they effectively "rule out any theory of purely random evolution" (Simpson 1960, 167). Nonetheless, he recognized that "what directional forces the data do demand, or permit, is one of the most important questions to be asked of the fossil record."

For all of their usefulness as heuristic tools, the MBL studies showed that fairly clear discrepancies emerge when predictions based on a purely stochastic model are compared with patterns identified in actual fossil data. For this reason, the MBL simulations were only one element in the development of a more comprehensive paleobiological approach to large-scale evolutionary patterns. An equally important source was the analysis of marine faunal diversification begun by Valentine (1969) and David Raup (1972) and carried through the 1970s. The debate between James Valentine and Raup was eventually joined by Richard Bambach and Jack Sepkoski, and culminated ultimately in a consensus solution to the problem of marine diversity in the early 1980s (Sepkoski et al. 1981). This consensus helped to establish what would be labeled the "taxic approach" as a dominant perspective in paleobiology, and contributed to the solidification of what came to be known as the Chicago School of analytical paleontology. It also represented the culmination of a long tradition in paleobiology—dating back to studies by Simpson, Newell, and Imbrie in the 1950s—of drawing statistical interpretations of patterns in the history of life from an otherwise imperfect fossil record. This approach, in many ways, epitomizes the paleobiological tradition in the second half of the 20th century.

The term "taxic paleontology" was coined by Niles Eldredge in 1979, in a paper where he described the following conundrum: "Paleontologists have always had the option of looking at the fossil record, in either or both of two ways—(1) distributions in space and time of discrete taxa, which differ among themselves to a greater or lesser extent, and (2) distributions in space and time of different states of morphological characters assumed to be evolving" (Eldredge 1979, 9). The modern synthesis reified option number two, but Eldredge pointed out that the previous decade (the 1970s) had seen a renaissance of the first approach on the part of paleobiologists. What Eldredge commented on in his 1979 paper was in fact the convergence of several distinct threads in paleobiology that I have been tracing over the last several chapters. Both punctuated equilibria (with its assumption of a stasis-plus-rapid change model) and the MBL model (incorporating Schopf's "species as particles" concept) helped prepare the path for a taxic view. In each case, species (and even higher taxa) are considered to be discrete entities with clearly demarcated "births" and "deaths." This was not, however, the essence of taxic paleobiology. As Eldredge noted, the taxic approach is also implicitly an *ecological* view, since it understands evolution to consist "essentially of the origin, maintenance, and degradation of diversity" (Eldredge 1979, 10). This view was inspired by the mathematical modeling approach of the MacArthur-Wilson insular model of biogeography, but developed its own uniquely paleobiological perspective with the advent of massive fossil databases in the late 1970s and 1980s.

Patterns of Phanerozoic Diversification

The basic methodology of taxic paleobiology is to identify patterns in the evolution of diversity by counting the first and last appearances of taxa in the fossil record. This is not as simple as it sounds for a variety of reasons, the most basic of which is that the fossil record is biased in a number of ways that makes a genuine estimate of diversity at any particular interval difficult to obtain. We will recall from earlier chapters that the English geologist John Phillips first confronted this problem in his 1860 book *Life on Earth*, where he attempted an analysis of Phanerozoic diversity based on a study of strata in Britain. Phillips recognized that his fossil evidence could not be taken at face value: not all organisms would have been preserved with equal frequency, and not all stratigraphic units were of equal thickness. For this reason, he calibrated his diversity estimate for the number of species per unit thickness, and was able to obtain a remarkably accurate estimate (by modern standards) of broad global diversity trends. Phillips's approach is a classic example of the taxic view in paleontology.

Raup versus Valentine

A hundred years later, paleontologists took up the problem of Phanerozoic global diversity once again, but the basic methodology and assumptions were largely the same as those used by Phillips. As I described in chapter 4, James Valentine addressed the problem of global marine diversity as part of his project to recast paleoecology in the idiom of contemporary theoretical ecology. In his 1969 paper "Patterns of Taxonomic and Ecological Structure of the Shelf Benthos during Phanerozoic Time," Valentine drew on the most up-to-date taxonomic references available, such as the *Treatise of Invertebrate Paleontology*. But despite having access to data of much greater scope and finer resolution than Phillips, Valentine's conclusion that standing species diversity increased over the Phanerozoic by about one order of magnitude agreed roughly with Phillips's initial estimate (Valentine 1969). The superiority of Valentine's data (which was based both on a century of continued collecting and on refined taxonomic procedures) did allow Valentine to identify patterns that Phillips had missed. For example, Valentine determined that different diversity patterns existed for different levels of taxonomic hierarchy: taxonomic diversity at the highest levels (phylum, class, order) was greatest early in the Phanerozoic and declined over time, while diversity at the lower levels (family and genera) began low but rose steadily over the Phanerozoic. Another way of putting this is to say that while diversity (measured by the total number of different taxa) increased, disparity (a measure of the amount of taxonomic difference between groups) declined. Valentine explained this pattern as a consequence of the greater specialization required for individual species as global marine ecospace became divided into smaller and smaller units (Valentine 1969, 706).

Valentine's interpretation of the history of marine diversity also initiated a controversy that lasted well over a decade. The most direct initial response to Valentine's proposal was Raup's 1972 paper "Taxonomic Diversity during the Phanerozoic," published in *Science*. Raup's central concern was similar to Phillips's: Is the fossil record of diversity biased because of the uneven volumes of preserved stratigraphic intervals? This question, he argued, was important because trends in diversity can help clarify "general models of organic evolution." In particular, diversity trends can indicate whether "the evolutionary process [is] one that leads to an equilibrium or steady-state number of taxa, or should diversification be expected to continue almost indefinitely?" (Raup 1972, 1065). This question is quite similar to the one MacArthur and Wilson posed in their island biogeography studies: Are diversity, adaptation, and evolution indefinitely expanding or steady-state processes?

In his analysis, Raup used many of the same sources for fossil data as had Valentine, including the *Treatise on Invertebrate Paleontology* and its Russian counterpart, *Osnovy Paleontologii*. He described Valentine's model of increasing diversity as the "traditional view," noting that it had also been supported by Norman Newell and A. G. Müller (Raup 1972, 1065). Raup's main worry was estimating the amount that diversity estimates were biased. It is vitally important to accurately estimate the ranges of taxa in order to get a valid measure of diversity over time. However, since fossils in older rocks are more likely to suffer from "nonexposure or destruction by erosion or metamporphism," it is likely that the dates of first occurrence for older taxa may be artificially truncated. In other words, some taxa may appear to be younger than they actually are, because their earlier histories have been lost; the impression that many new taxa emerged in later geological periods may be a false one, produced simply because younger rocks provide better samples (Raup 1972, 1068). Raup then showed that the likelihood of this kind of bias was high: since higher taxa invariably have many subtaxa, it is likely that representatives of the individual higher taxa will be identified long before all of their constituent subtaxa (e.g., genera) will be found. Additionally, higher taxa with fewer subtaxa will be completely identified much sooner than those with many subgroups. Moreover, as geological time progresses, the likelihood that subtaxa will be more completely preserved increases, thus giving the appearance that more recent higher taxonomic groups have a greater number of subtaxa, or in other words, that taxonomic diversity among lower taxa increases over time (Raup 1972, 1067). This is one aspect of the effect Raup would later term "the pull of the Recent" (Raup 1979).

From this elegant thought experiment, Raup reached several conclusions about resulting biases in models of taxonomic diversity. Most importantly, he argued that "time-dependent biases" likely "shift any diversity peak toward the Recent . . . and the amount of shift should be greatest at the lowest taxonomic levels." In other words, what Valentine and others had interpreted as a genuine biological signal "may actually be due to the effects of biases" (Raup 1972, 1070). Raup then offered an alternative model of Phanerozoic diversity, which he represented graphically by superimposing a curve on Valentine's projection, in which diversity increased to a maximum point somewhere in the mid-Paleozoic, and then declined to an "equilibrium level." In what was becoming a hallmark of Raup's methodology, he used a simulation program to generate a large number of hypothetical lineages and then "destroyed" part of the record (fig. 8.1). Instead of peaking and falling to equilibrium, the graph now rose steadily from the middle section of the range chart towards its endpoint, much like Valentine's curves for actual taxa. Raup concluded that the simulation provided a basis for "a plausible [model] for the Phanerozoic record of marine invertebrates" (Raup 1972, 1071).

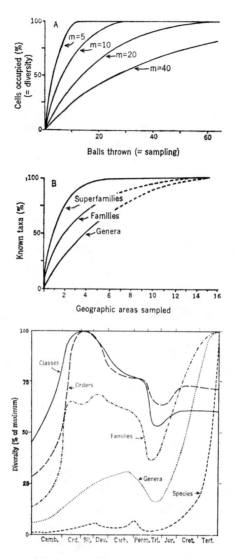

FIGURE 8.1. Raup's analysis of diversity as a function of sampling. *Top and center*, an illustration of the cell occupancy problem as it relates to diversity sampling. *Bottom*, variation in apparent taxonomic diversity for Phanerozoic marine invertebrates. David M. Raup, "Taxonomic Diversity During the Phanerozoic," *Science* 177, no. 4054 (1972): 1067, 1070. Reprinted with permission from AAAS.

Factor Analysis and Diversity Data

Another important message in Raup's and Valentine's studies was that complex ecological-evolutionary relationships could be profitably examined using simplifying models and analysis. One of the most commonly used techniques in quantitative paleobiology during this period was factor analysis, a branch of multivariate statistics used to describe complex relationships among phenomena as a function of the smallest number of variables, or factors. As described in chapter 3, one of the pioneers in the application of multivariate statistics to paleontology was John Imbrie, who developed many of the standard statistical techniques used in paleontology. Imbrie had drawn specific attention to the value of factor analysis in solving paleontological problems involving complex ecological interrelationships. While Imbrie and other authors had noted the potential for factor analysis to produce insight into paleontological problems, little headway was made in actually applying the technique to large data sets (Imbrie 1964; Rohlf and Sokal 1962; Sokal and Sneath 1963; Sokal 1969). This was likely due to both the technical demands of the statistical procedures as well as the limited availability of adequate sources of comprehensive data sets.

However, by the 1970s the combination of powerful digital computers and new compilations of fossil data such as Harland et al.'s *The Fossil Record* and the updated *Treatise on Invertebrate Paleontology*, made such studies more feasible (Harland 1967; Moore 1969). In 1973, Imbrie (who had moved from Columbia to Brown University) published a paper with Karl Flessa that applied factor analysis to the problem of Phanerozoic diversity. The question that Flessa and Imbrie set out to examine was whether changes in rates of diversification among Phanerozoic taxa are correlated. In other words, did multiple groups diversify at the same time? The primary mode of analysis used in the paper was Q-mode factor analysis, which the authors described as a method "that attempts to reduce many variables (taxa) into a smaller number of groups or variables, each group consisting of taxa which share rates and times of diversification and decline." This approach is an excellent candidate for what I have described as "rereading as generalization": Flessa and Imbrie asserted that patterns such as diversification and decline "are most accurately described and objectively discerned" by using "simplifying models" that "are valuable in clarifying known patterns of evolution; re-

vealing previously unsuspected patterns; and providing a quantitative index of biotic change" (Flessa and Imbrie 1973, 248).

Flessa and Imbrie noted that the correlation they were testing had
been explained in Norman Newell's 1952 paper "Periodicity in Invertebrate Evolution," where Newell concluded that "the rise and fall in apparently evolutionary activity is not at random" (Newell 1952, 385). To
address this question, Flessa and Imbrie drew primarily on Harland's
The Fossil Record to construct matrices representing the diversity of 59
marine taxa across 71 geologic stages, and 20 terrestrial taxa in 60 stages.
They estimated 'diversity' as a measure of the relative number of families within each higher taxon (Flessa and Imbrie 1973, 250). To give taxa
equal weight, the statistical procedure expressed the diversity of each
higher taxon as a percentage of its total observed diversity range. These
data were then subjected to a factor analysis that attempted to resolve
the n variables (individual marine and terrestrial taxa) into a smaller
group of m independent parameters (hypothetical assemblages of taxa);
if m was found to be smaller than n, then it could be concluded that "certain taxa share rates and times of diversification and decline."

The results of the factor analysis showed unequivocally that fluctuations in diversity of the marine taxa were correlated, and Flessa and Imbrie determined that these correlations could be grouped into 10 "diversity associations" that explained 96% of these fluctuations. Diversity
associations (or factors) are groups of taxa whose individual diversity
fluctuations are found to be strongly positively correlated. For example, "marine factor 1" consisted of *Trilobita*, *Archaeocyatha*, and *Inar-
ticulata*, which shared a diversity pattern that peaked in the Cambrian
and fell sharply toward extinction by the mid-Ordovician. Based on their
analysis of these diversity associations, Flessa and Imbrie determined
that Newell's conclusion was confirmed, reasoning "that many ecologi-
cally and phylogenetically unrelated taxa have common modes of diversification and decline is clearly a phenomenon of substantial evolutionary significance" (Flessa and Imbrie 1973, 264).[1]

However, the factor analysis did not explain *why* "certain taxa radiate and decline more or less in concert"; it merely expressed the pattern
in quantitative terms. The authors acknowledged that their quantitative

---

1. Flessa and Imbrie found four diversity associations for terrestrial taxa, which they
noted were somewhat less strongly correlated than marine taxa. I will focus on marine diversity in the discussion that follows.

analysis of diversity could perhaps be used to test a number of hypotheses which had been presented for a mechanism that could explain patterns of global taxonomic change. So in addition to resolving the extent of "coincident evolutionary activity" through factor analysis, Flessa and Imbrie also hoped to determine the total (or cumulative) amount of taxonomic change experienced over time in the marine and terrestrial biotas, respectively, and to compare that pattern to the expectations of proposed mechanistic hypotheses. "Thus," the authors contended, "this *quantitative* model of the patterns of Phanerozoic diversity provides an objective framework within which *conceptual* models of fluctuations in diversity and trophic characteristics may be tested" (Flessa and Imbrie 1973, 266).

Flessa and Imbrie reported that while the cumulative rate of taxonomic change for Phanerozoic marine and terrestrial taxa has remained roughly steady since the Cambrian, the steady increase was composed of several distinct 'taxonomic turnovers' or "evolutionary pulsations" in which one or more diversity associations gave way to another (fig. 8.2). For marine fauna, moderate spikes appeared throughout the Cambrian, Ordovician, and Devonian; the two largest were at the Permian-Triassic and Triassic-Jurassic boundaries; significant events also occurred in the mid-Cretaceous and at the Cretaceous-Tertiary boundary (Flessa and Imbrie 1973, 270). Flessa and Imbrie found that marine and terrestrial patterns corresponded quite well (although intensity of change was more dramatic for terrestrial taxa), and also that their combined pattern showed a "reassuring similarity" with Newell's 1967 estimate of extinctions and radiations (Newell 1967, 79–80).

Flessa and Imbrie concluded by identifying four hypotheses for explaining global changes in diversity: change in the area of shallow seas, fluctuations in atmospheric oxygen, the rate of reversal in the earth's magnetism, and global tectonic changes. Here the value of the quantitative approach is striking: each of these mechanisms involved well-understood data that could be graphically superimposed on the pattern of taxonomic change, in many cases providing an instantaneous visual indication of the adequacy of the hypothesis. The authors were particularly drawn to tectonic changes, which coincided with a well-established ecological mechanism. As they explained, changes in plate tectonics, which matched remarkably well with major changes in diversity, would have had a direct impact on the area available to organisms for colonization, according to the MacArthur and Wilson species area effect

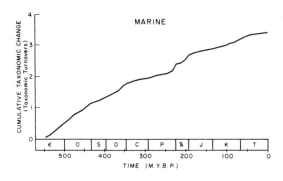

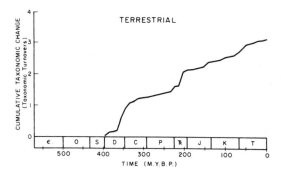

FIGURE 8.2a. Flessa and Imbrie's analysis of patterns and rates of change in Phanerozoic marine and terrestrial diversity. Cumulative change. Karl W. Flessa and John Imbrie,
"Evolutionary Pulsations: Evidence from Phanerozoic Diversity Patterns," in *Implications
of Continental Drift to the Earth Sciences*, vol. 1 (Academic Press, London: 1973), 268.

($S = CA^z$) (Flessa and Imbrie 1973, 273–77; fig. 8.3). Although they characterized their exploration of this mechanism as "speculative," Flessa
and Imbrie nonetheless concluded that the MacArthur-Wilson model of
insular biogeography provided the most promising future direction for
quantitative analysis of fossil data on diversity:

> If dispersed islands and clumped islands may be considered analogs of frag-
> mented and assembled continents respectively, and if immigration rates and
> local extinction rates are the ecological analogs of speciation and extinction
> rates respectively, MacArthur and Wilson's model will prove to be a powerful
> tool in understanding the biological consequences of continental drift (Flessa
> and Imbrie 1973, 281).

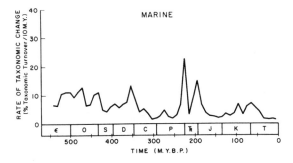

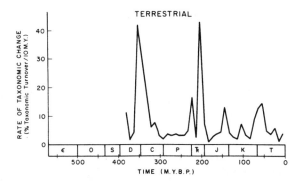

FIGURE 8.2b. Rates of change. Karl W. Flessa and John Imbrie, "Evolutionary Pulsations; Evidence from Phanerozoic Diversity Patterns," in *Implications of Continental Drift to the Earth Sciences*, vol. I (Academic Press, London: 1973), 269.

Survivorship and Taxonomic Diversity

Thus far, and despite their individual interpretive differences, studies of Phanerozoic diversity were converging on a central conclusion: in some important respect, the mathematics of equilibrial biogeography offered a compelling heuristic for understanding patterns of evolutionary change as an explicitly ecological problem. As methods for studying the history of global marine diversity evolved during the 1970s, an important additional ingredient came from Leigh Van Valen's work on survivorship and his "Red Queen's" hypothesis. As previously discussed, Van Valen's "new evolutionary law" proposed that "extinction in any adaptive zone occurs at a stochastically constant rate," or in other words that for any given time span, all taxa present have a statistically equivalent probability of extinction, meaning that "the probability of extinction

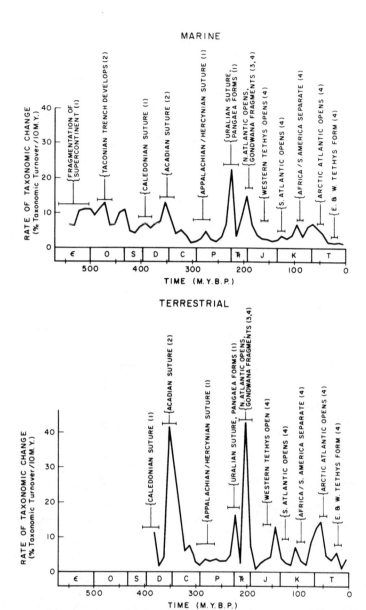

FIGURE 8.3. Correspondence of rates of taxonomic change with tectonic events during the Phanerozoic. Karl W. Flessa and John Imbrie, "Evolutionary Pulsations; Evidence from Phanerozoic Diversity Patterns," in *Implications of Continental Drift to the Earth Sciences*, vol. I. (Academic Press, London: 1973), 274.

of a taxon is then effectively independent of its age" (Van Valen 1973, 16–17).

In the last chapter, we saw how the members of the MBL group interpreted this generalization as endorsing a "particle view" of taxonomic groups; Van Valen's "law" seems to imply that, in some sense, all taxa in a given adaptive zone can be treated as statistically equivalent units. In 1975—in the first issue of the new journal *Paleobiology*—Raup drew out the broader implications of this view for a "taxic" approach to paleobiology in a paper entitled "Taxonomic Survivorship Curves and Van Valen's Law." Calling Van Valen's work "a major step towards a nomothetic paleontology," Raup began the discussion by describing his paper as

> a step (anticipated by Van Valen, 1972) toward interpreting the evolutionary record in terms of general rules and processes without regard to specific causes operating on specific taxa: it attempts to make generalizations about the fossil record which are not simply enumerations of specific events and causes. In this sense, Van Valen's Law is analogous to the equilibrium approach taken by MacArthur and Wilson (1967) so successfully in biogeography and perhaps is even comparable to the approach which led to the gas laws of physics (Raup 1975b, 83).

But Raup's purpose in examining the law had less to do with exploring its "ecological and evolutionary interpretation" than with probing its assumptions about survivorship. For this reason, I argue that Raup's analysis is more an extension of his 1972 critique of Valentine's diversity interpretation than it is a continuation of the theoretical MBL work, and that it is therefore an important step in constructing the "rereading as generalization" methodology implicit in the taxic approach to paleobiology.

A central question Raup addressed was whether the fossil record used as the basis for Van Valen's survivorship analysis was a reliable source. In particular, he worried that the range data Van Valen had drawn on, taken from existing sources such as *The Fossil Record* and the *Treatise on Invertebrate Paleontology*, might have been corrupted by "error and uncertainty, especially at the higher taxonomic levels," as well as by "monographic confusion" and "lack of adequate stratigraphic control" (Raup 1975b, 83). Ultimately, Raup concluded that Van Valen's approach had promise, but that the use of survivorship curves for generalizing about the patterns and mechanics of extinction should proceed cautiously, especially "with respect for the uncertainties inherent in the stratigraphic and taxonomic data contributing to the analysis" (Raup 1975b, 96). Raup also worried, though, that taxonomic practices for identifying higher taxa, as well as time-dependent biases (especially the underrepresentation of long-lived taxa) might have produced the curves' "linearity" which Van Valen claimed supported his argument that extinction occurs at a stochastically constant rate.

The next year, Raup's colleague Jack Sepkoski, who had recently taken his first job under Raup's wing at the University of Rochester, published a paper extending Raup's analysis of stratigraphic bias in survivorship. Sepkoski examined two sources of nonbiological bias in the fossil record—variation in the length of stratigraphic intervals and incompleteness in fossil sampling—that might have produced the apparent linearity in Van Valen's survivorship curves (Sepkoski 1975, 343). In the first instance, Sepkoski noted that taxonomic durations are determined by persistence of taxa across stratigraphic intervals, not by "direct measurement." Because stratigraphic intervals in the geological record have unequal lengths, these estimates are biased, especially for taxa of short duration (i.e., which appear in one or a few intervals) which are "positively skewed," resulting in overestimated durations. This could produce Van Valen's characteristic flat-topped, lognormal (linear) survivorship curves (Sepkoski 1975, 344–47). Since this kind of bias "can be merely minimized by considering only paleontological data with the greatest possible stratigraphic resolution," Sepkoski's unstated implication was that the only way of properly confronting this bias would be to improve the resolution of the fossil record beyond what current available sources allowed (Sepkoski 1975, 351).

The other source of bias Sepkoski considered was incomplete fossil sampling. This was potentially the greatest obstacle to the taxic approach, since taxon counts are useless unless they can be relied on to make generalizations. However, Sepkoski concluded that incompleteness can be easily corrected for using statistical techniques. He simulated the effect of systematic temporal sampling bias by producing 200 hypothetical lineages, and used Monte Carlo simulation to randomly destroy range information in proportion to the age of the intervals sampled from. Survivorship curves produced from the "biased" data were not substantially different from the original ones, which he argued indicated that "incomplete sampling of the fossil record may not severely alter the shape of the survivorship curve of a taxonomic group" (Sepkoski 1975, 355).

Equilibrium Interpretations of Diversity

In late 1975, Raup published a paper in which he returned to the issues he had raised in his 1972 critique of Valentine's Phanerozoic diversity estimate. He began this paper with a question: how can paleontologists estimate the number of taxa in a given stratigraphic interval?
This issue related directly to Valentine's proposal that disparity (order
or phylum-level diversity) decreased after the Cambrian, while diversity (the number of species and genera within higher taxa) increased by
up to an order of magnitude. Valentine's source for his data on taxonomic diversity was the *Treatise on Invertebrate Paleontology*, which did
not compile species-level data. Estimates for diversity at the lower taxonomic levels therefore had to be calculated indirectly, by extrapolating
what *would* be found if species level data existed. Valentine's 1970 paper
"How Many Marine Invertebrate Fossil Species?" attempted to approximate this number by comparing fossil and living phyla, extrapolating
the ratios of taxa empirically observed at different levels, and inferring
changes in paleoenvironments based on assumptions about shifting tectonic plates throughout the Phanerozoic (Valentine 1970). His estimate
of an order of magnitude increase in species-level diversity was based on
the results of this extrapolation.

But Raup's misgivings about time-dependent biases in this estimate
had not been resolved, and his 1975 paper explored the statistical technique of rarefaction to address this concern. Rarefaction is an interpo-
lation method that takes a taxonomic sample of a given size and asks
how many species would be found if the sample was smaller. This allows samples of unequal sizes to be compared in such a way that their
diversity, or "richness," can be compared. Raup applied rarefaction to
samples of post-Paleozoic echinoids (urchins, sand dollars) consisting of
nearly 8,000 species, and produced estimates for the number of species
within families, genera within families, and species within orders. Based
on the results of the rarefaction, Raup concluded that the apparent rise
in the number of echinoid families was likely a real biological signal, and
not just an artifact of sample size. However, he also found that this increase was likely not as great as was suggested by the raw, unrarified
data, and more significantly that rarefaction indicated little change in
both the number of families since the Cretaceous and the number of
orders since the Jurassic (Raup 1975a, 341). If Raup's analysis of echi-

noids could be generalized, this was an implicit critique of Valentine's estimate: Raup's species/order rarefaction curves indicated that equilibrium, rather than exponential growth, was the proper model for post-Paleozoic species richness.

The next year, Raup published two articles in the same issue of *Paleobiology* that presented his most compelling evidence against a general increase in species diversity during the Phanerozoic. In the first of these papers, "Species Diversity in the Phanerozoic," Raup outlined the basic problem: If the fossil record is taken at face value, there appears to have been a marked increase in the number of species since the Paleozoic, a fact which "has broad implications for large-scale interpretations of community evolution, the development of provinciality, and other aspects of the evolution of life." However, two factors complicated this interpretation. In the first place, no reliable estimates of species diversity existed, since previous efforts—such as Valentine's—were based "largely on inference" drawn from compilations of higher taxa. Second, even if a better survey of fossil data at the species level could be obtained, the specter of sampling bias still loomed. As Raup himself had pointed out in earlier papers, "It is thus possible that much or even all of the increase in diversity through time is an artifact" (Raup 1976a, 279).

Rather than relying on compilations of paleontological literature on the fossil record (which, as noted, did not include species-level data), Raup turned to the *Zoological Record*, a bibliographic database of species descriptions stretching back to 1864. In this way, Raup hoped to improve on Valentine's estimates by directly sampling the best available species-level data (Raup 1976a, 280–81). From the *Zoological Record*, Raup compiled a list of more than 71,000 fossil invertebrate species, which he extrapolated upwards (to 144,251) to correct for biases in collection practice, and encoded on a computer tape to subject to a variety of analyses. He then arranged this species data according to geological time intervals, and produced a graph depicting apparent variation in species diversity through the Phanerozoic (fig. 8.4). Raup observed that the shape of this graph does indicate an accelerating increase in species diversity in the Cretaceous and Cenozoic, but noted that the Cenozoic peak is only roughly 3.9 times the Paleozoic average (and 2.4 times the Devonian peak), which is much less than Valentine's proposed tenfold increase (Raup 1976a, 286).

Raup's first 1976 paper stopped short of interpreting this pattern, concluding only that "a much more important question raised by the

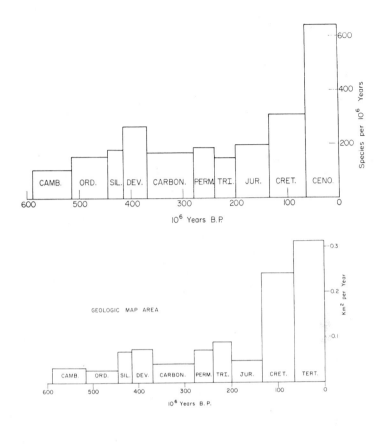

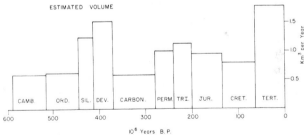

FIGURE 8.4. Figures from Raup's two species diversity papers of 1976. *Top*, Raup's estimate of apparent species diversity through the Phanerozoic. *Center and bottom*, surviving
area and rock volume for sedimentary rocks. Note the correspondence between the figures
in several intervals. David M. Raup, "Species Diversity in the Phanerozoic: A Tabulation,"
Paleobiology 2 (1976): 286. David M. Raup, "Species Diversity in the Phanerozoic: An Interpretation," *Paleobiology* 2 (1976): 291. Courtesy of the Paleontological Society.

Zoological Record data is the meaning of the Phanerozoic diversity distribution" (Raup 1976a, 288). That question was taken up in the second paper, "Species Diversity in the Phanerozoic: An Interpretation" (Raup 1976b). The basic question Raup asked in this paper was whether any identifiable biasing factors can be put forward to account for the apparent increase in species diversity towards the end of the Phanerozoic. If the answer was yes, the diversity increase would not necessarily have been artifactual, but it would put Valentine's hypothesis of exponential Phanerozoic diversity growth in serious jeopardy.

Among the time-dependent and time-independent biasing factors Raup identified, he focused his analysis on one source of time-dependent bias: the area, thickness, and volume of exposed rock (Raup 1976b, 292). When the area and volume of exposed sedimentary rock are tabulated and graphed according to the geological time scale, a pattern is produced that is "strikingly similar" to Raup's analysis of species diversity: values for both volume, area, and diversity all show a "net rise" in the Cenozoic from a low in the Cambrian, and all three graphs reflect spikes in the Devonian followed by pronounced drop-off in the Carboniferous. As Raup observed, the "simplest and most direct interpretation . . . is that species diversity is tracking the quantity of rock as a sampling phenomenon: more rocks produce more fossil specimens and given more specimens, an increased number of discovered species is virtually inevitable" (Raup 1976b, 292).

Despite the apparent strength of this correlation, Raup acknowledged a scenario in which changes in amount of exposed rock might represent environmental changes favorable to a genuine increase in global diversity. If changes in exposed sedimentary rock reflect increases in the area of continental shelves and epicontinental seas, the total habitable area for marine organisms may have increased during the Phanerozoic, which could account for a genuine and proportionate increase in species diversity. This was the explanation Valentine proposed in a 1973 paper in *Science*, and it was also the mechanism described in the 1974 Schopf and Simberloff papers on sea-floor spreading discussed in the last chapter (Valentine 1973b; Schopf 1974; Simberloff 1974). The MacArthur-Wilson species-area effect predicts that as habitable area increases, the number of species increases proportionately; as Raup noted, "there is mounting evidence that this relationship may be applicable to paleontological situations," meaning that "as outcrop area increases, so does the biological area sampled, at least in a statistical sense" (Raup 1976b, 294). The question, though, is whether rock volume is really a good measure of the habitable area. Raup asserted that "the burden of proof lies with those who would argue that diversity was indeed higher in the Cretaceous and Tertiary," and concluded that "there is no evidence for the existence of a long-term, worldwide trend towards increased diversity during the Phanerozoic" (Raup 1976b, 295–96.). Where Valentine argued for an exponential increase in species diversity, Raup contended that "diversity among marine invertebrates has been at a saturation or equilibrium level throughout most of the Phanerozoic." In other words, even the apparent 2.4× increase obtained by tabulating the species data from the *Zoological Record* was likely an artifact.

Raup's statistical analysis of preservation bias was not the only evidence that favored the equilibrium interpretation of Phanerozoic diversity patterns. In 1977, Richard Bambach weighed into the debate with a study of species richness in marine benthic habitats. As Arnold Miller has observed, an impasse had been reached by the late 1970s: "While there was growing agreement that the fossil record preserved evidence of a significant diversity increase through the Mesozoic and Cenozoic, the extent of the species-level increase remained contentious and, more importantly, the biological significance of the increase was being openly questioned because of the rock-volume problem" (Miller 2009, 371). There was no doubt that some change in diversity had taken place during the Phanerozoic; the debate, however, revolved around the extent of that increase, and the correct evolutionary/ecological interpretation of the change. Bambach's contribution was important because it drew on evidence from his detailed studies of individual paleocommunities which compose the global biotic environment. In other words, while Valentine's and Raup's studies analyzed "gamma diversity," or diversity of the entire global marine biota, Bambach focused on "alpha diversity," or the diversity within individual marine habitats. Bambach reasoned that because the global biosphere is a nested hierarchy (as Valentine had asserted in *Evolutionary Paleoecology of the Marine Biosphere*), the diversity of the global biosphere can be studied by extrapolating patterns from study of its individual components (Bambach 1977, 153).

The advantage Bambach cited for his own approach was that given the notoriously poor resolution of the fossil record, evolution of species diversity would be much more accurately obtained by examining particular smaller, local habitats in which species are well-preserved. Bambach considered data from 368 fossil communities, measuring changes in median species richness in individual communities from one period to the next rather than the aggregate number of species present in an interval. Bambach then grouped the community data into five large segments of time, which he analyzed for evidence of broader diversity patterns. In doing so, he identified an important set of trends. While the number of species in "high-stress" near-shore environments remained low throughout the Phanerozoic (reflecting the instability of highly competitive, environmentally variable habitats), species richness in more stable open marine environments rose dramatically in the Lower to Middle Paleozoic and again in the Mesozoic and Cenozoic (Bambach 1977, 158).

In extrapolating this pattern to global Phanerozoic diversity, Bambach reached two conclusions. In the first case, he found that within-habitat increases indicate a real increase in global marine diversity, and argued that Raup's assessment that no change took place "can not be accepted." On the other hand, he also concluded that Valentine's order-of-magnitude increase was also probably wrong. Bambach also found that the average number of species had likely doubled between the Middle Paleozoic through the Cenozoic, which was closer to the 2.4x diversity increase observed in Raup's initial tabulation. Adjusting his figure to account for estimates of a doubling of provinciality during this period (e.g., an increase in the number of communities) as proposed in paleoecological literature, Bambach's final estimate for global species diversity increase over the Phanerozoic was fourfold—more than Raup's, but considerably less than Valentine's (Bambach 1977, 163). Bambach also supported an interpretation that drew directly on the MacArthur-Wilson equilibrium model, arguing that "the very long intervals of stability in average numbers of species imply some sort of equilibrium in species richness, possibly like species-area equilibria seen in the study of island biogeography . . . with taxonomic turnover (evolution and extinction) balanced through time for particular community types" (Bambach 1977, 161).

When it was put in this perspective, the relationship between taxic analysis of diversity and survivorship analysis became clearer: if change in taxonomic diversity could be explained by an application of the MacArthur-Wilson insular model of arrival and extinction, then a study of the dynamics of survivorship was essential in understanding the mechanics of diversification. Raup brought these two themes together in a 1978 study entitled "Cohort Analysis of Generic Survivorship," in which he computed curves not for individual taxa, but rather for "cohorts" (a term borrowed from actuarial practice), which consisted of groups of taxa that originated in the same unit of geologic time. The survivorship curves for these cohorts, then, would depict the fates of "populations" of taxa over time, similar to the way in which an insurance actuary traces the longevities of all people born in a particular year. Because the cohort approach correlated survivorship with the absolute geological timescale, disturbances or "inflections" in the survivorship curves of individual cohorts could be interpreted either as changes produced by the normal "mortality pattern" of the cohort, or as potential "external events" such as sudden environmental changes. By constructing and comparing a series of sequential cohorts (e.g., representing successive geological periods) such inflections could be correlated to deduce their cause. In this way, Raup argued, cohort analysis "provides opportunities not otherwise possible for monitoring the tempo of evolution through geological time" (Raup 1978b, 2).

Raup applied cohort analysis to some 17,000 genera and subgenera identified in the *Treatise on Invertebrate Paleontology*, which he again transferred to computer tape. Seven survivorship curves were constructed, for cohorts originating from the Cambrian through the Triassic, and extant taxa were subtracted from the results. Raup argued that the slope of survivorship curves for higher taxa should be concave if species survivorship rates were assumed to be linear; in other words, if species survivorship is "stochastically constant," then a higher taxon will have a statistically longer duration than any of its constituent species, thus exhibiting an exponentially decaying survivorship curve. This assumption, he explained, was consistent with a "markovian branching model" in which "speciations and extinctions are treated as probabilistic events," akin to the MacArthur-Wilson insular model (or, by implication, the MBL model) (Raup 1978b, 5).

Raup's analysis showed that the seven survivorship curves did exhibit a general pattern of concavity, but two noticeable "inflections" appeared: the five curves that crossed the Permian-Triassic boundary showed a pronounced "turndown" (i.e., a steeper slope) during this interval, as did the four curves that crossed the Cretaceous-Tertiary boundary. Raup concluded that such periods of sudden, pronounced decrease in *all* of the cohorts present at a particular "moment" could be taken as evidence of some extrinsic, environmental mechanism producing mass extinctions (Raup 1978b, 7–8). Of course, the P-Tr and K-T boundaries had long been suspected as periods of mass extinction based on fossil evidence;

Raup's statistical analysis of cohort survivorship both confirmed this intuition and also introduced a new tool for identifying unusual diversity patterns against the background of expected survivorship rates.

In 1977, when Raup's paper was written, taxic paleobiology was entering a transitional phase. A general problem had been defined: What do broad patterns of diversification during the Phanerozoic reveal about the mechanisms of evolution and extinction? A variety of potential approaches for solving this question had been articulated, including the idealized models of the MBL simulations, statistical generalizations based on empirical data represented in the diversity tabulations of Valentine, Raup, and Bambach, and the survivorship analyses of Van Valen and Raup. Despite their differing conclusions, a common assumption in each approach was that ecology—and in particular the MacArthur-Wilson insular model—offered an important heuristic for interpreting these broad patterns in the history of life. But this problem also revealed something of an impasse: the MBL approach had shown the limitations of a model abstracted from real fossil data, but existing compilations of fossil data were too suspect to rely on. In his 1978 paper Raup expressed confidence "that the job of exploring the fossil record is far enough along to justify statistical analysis and synthesis," but emphasized that statistical analysis of the fossil record "cannot be applied validly unless data sets are large" (Raup 1978b, 14).

Jack Sepkoski: Reconstructing the Fossil Record

In order to fully explore the significance of the heuristic, models-oriented approach for analyzing the fossil record, taxic paleobiology required a massive database of originations and extinctions consisting of many thousands of taxa spanning hundreds of millions of years. While resources like the *Treatise on Invertebrate Paleontology* and *The Fossil Record* provided a rough initial survey, no such database existed. This is hardly surprising: in the first place, prior to the 1970s and the advent of digital computers with tape-storage drives, it is hard to imagine how such a database could even have been collected and used. Secondly, the task of assembling such a database would require an incredible amount of difficult, tedious work assembling, collating, and standardizing massive amounts of information from the diverse worldwide collection of journals, monographs, and museum collections where fossil data had accumulated. Nonetheless, by the early 1970s paleobiologists had begun to recognize the need for a complete overhaul of their databases. Gould was among the first to promote such a project, although he lacked the time and ambition to carry out the task himself. However, fortune had placed the ideal candidate right under his nose: J. John "Jack" Sepkoski, Jr., a Harvard graduate student who was interested in evolutionary problems, adept with mathematics and computers, appropriately ambitious, and, perhaps most important, increasingly disenchanted with his own very "idiographic" thesis on Cambrian stratigraphy.

One of the most vital tasks of any revolutionary movement is to recruit and indoctrinate new members of the community who will promote and advance the aims of the discipline. It was vital to the 1970s paleobiological movement to recruit younger members, and it is evident from their correspondence that Gould, Schopf, and others were consciously aware of this requirement. Writing to Gould in 1977, for example, Schopf commented on the importance of "local contact of intersecting circles" (such as the MBL collaboration) which, "if they happen by chance to control a press, or other influence . . . have an influence far beyond their number." Schopf expressed concern, however, that "the character of paleobiology groups seems to have changed markedly in the past 5 years," becoming "crystallized" and running the risk of losing "the element of chance and change" that had characterized the excitement earlier in the decade. Schopf related this concern directly to problems of attracting and assisting new students, speculating that "this is owing I surely believe to the fact that so few workers of any imagination are entering the field at present, and those who are have an incredible time getting even a mediocre job" (Schopf to Gould, October 19, 1977: PS-*P* 5, 14).

While Gould was less pessimistic, he conceded that "we may remain intellectually young, but if we have no successors the field will surely languish." Gould then meditated on the problem of recruitment and training using a metaphor drawn from Schopf's beloved stochastic paleontology:

> Would you regard it flip if I proposed a stochastic explanation[?]. Paleobiology is a tiny field (there may be a few thousand paleontologists about, but most were trained as stratigraphic geologists and only a half dozen or so schools turn out biologically trained paleontologists—and few of these have decent Biology departments to back the training). There aren't but a few dozen explicitly trained at any one time. . . . We are subject to sampling error, but you are operating on a scale where the individual molecule (i.e. paleobiol-

ogist) counts. With so few molecules, the only way to guarantee (on a stochastic basis) quality is to make sure that the pool of potential molecules ranges from silver to gold. This is what we now lack. The brightest evolutionary biologists do not choose paleontology, and ones that do (forgive my arrogance if I cite both of us here) do so from personal prior commitment. We (at least at Harvard) must choose our paleo students from people who generally lack biological training—and we must guess at what they can do and, even if bright, they come generally without that biological intuition sine qua non—and how do you teach it (Gould to Schopf, 25 November 1977: Schopf pap. 5, 14).

With such a depressing numbers game, it is clear that Gould keenly felt the importance of every potential recruit. This context is vital in recognizing the significance of his interaction with Jack Sepkoski—whom Gould described in the same letter as one of "only two good students in the past 10 years"—during the first half of the 1970s. Sepkoski's recruitment to the paleobiological agenda is a valuable case study of the programmatic efforts by Gould and others to advance their theoretical and institutional goals through pedagogy and mentoring.

Sepkoski came to the Harvard Geology Department in the fall of 1970 with—by Gould's standards—an unusual background and set of interests. As a geology major at Notre Dame, Sepkoski had developed fairly traditional interests in "petrography, paleoecology, and environmental and tectonic frameworks" of Paleozoic epicontinental or shelf environments, and he had conceived a plan "to pursue graduate study in the fields of stratigraphy, sedimentology, and paleontology" (Sepkoski, "NSF essay (final copy, no. 3)": Sepkoski pap.). Alongside these traditional interests, however, Sepkoski had pursued a mostly self-directed study of computer science and statistics, which included statistical studies of brachiopod fossils he had collected from local deposits. These interests culminated in a short paper on Q-mode cluster analysis of a Paleozoic reef environment published in the *Notre Dame Science Quarterly* in 1968, which he used as the basis for a cluster analysis program written in FORTRAN (Sepkoski, "Report on the Q-Mode Cluster Analysis Program for the Classification of Qualitative and Semi-Quantitative Data": Sepkoski pap. 20–21). This shows that, before he came to Harvard, Sepkoski's exposure to analytic methods was already fairly sophisticated (Sepkoski 2005).

Nonetheless, at this early stage Sepkoski's interests could hardly be described as typically paleobiological. In fact, he chose Harvard because he wanted to study with Bernhard Kummel, who despite having been Newell's first student at the University of Wisconsin was a fairly conventional stratigraphical paleontologist. According to Sepkoski's Notre Dame transcript, he took many courses in geology, chemistry, and physics, but none in biology (official Notre Dame transcript: Sepkoski pap.). Indeed, while Gould often later referred to Sepkoski as 'his student,' Sepkoski's advisor at Harvard was Kummel, and his eventual dissertation, "Dresbachian (Upper Cambrian) Stratigraphy in Montana, Wyoming, and South Dakota" was thoroughly traditional. Clearly, if Sepkoski was destined to be one of the rare paleobiological "molecules" Gould described in his letter to Schopf, Gould would have his work cut out for him.

To make matters worse, Sepkoski's initial experience at Harvard was quite unhappy, and after his first year he seriously considered transferring to UC Santa Cruz to study with Léo Laporte. In a letter to the graduate director at Santa Cruz, Sepkoski described both his disillusionment with the "pressure cooker" environment at Harvard, as well as his concern that "my particular interests do not coincide with any of the faculty here." He apparently felt that Laporte, an expert on Paleozoic reef communities, would make a better mentor, and that the "relaxed, personable atmosphere" at Santa Cruz "where professionalism is not emphasized to the exclusion of all else" was a better fit for his personality (Sepkoski to Robert Garrison, 18 April 1971: Sepkoski pap.). Part of the problem was that during Sepkoski's initial year at Harvard (1970–71), Gould was on sabbatical at Oxford and was not able to have much influence. The decisive factor in Sepkoski's decision to stay at Harvard, however, was likely a remarkable letter from Gould, who not only attempted to assuage Sepkoski's concerns about Harvard but also laid out a blueprint for the programmatic revolution in paleobiology Gould was in the process of fomenting:

> How can I advice [sic] you since your decision includes (apart from its truly intellectual proportions) so many emotional factors that I can neither assess or weigh. There is not a better man than Leo in that particular little area of Paleozoic paleoenvironments. Neither can I deny that there is probably more joy in California, both in the sun (literally and metaphorically) and in Leo's vitality and group approach vs., for example, my own kind of pedantry and reverence for an antiquated type of individualized scholarship. But Harvard does have considerable advantages. With a combination of people (including [Raymond] Siever, for example), you can surely gain advisers equal

in ability to what Leo does by himself. . . . But the main reason for Harvard is not this; it's rather the potential, if you seek it, for the most important ingredient in scientific innovation: stimulation from intelligent men in related fields. If you're just surrounded by geologists with geological training, you will do little more than an elegant piece of work along lines already explored. But there's a revolution going on in ecology and biogeography. It's related to an approach via deductive models (that you can comprehend, and many others of our grad students cannot) and much of it is centered at Harvard ([E. O.] Wilson, [William] Bossert; and it will be firmly lodged here if, as rumor (in the usual sense) has it, [Robert] McArthur [*sic*] comes here). The next great innovator in paleoecology will be the man who successfully learns to understand this revolution and transfer its insights into paleontology; it will not be the man who pursues geological study with geologists, however excellent (Gould to Sepkoski, 28 April 1971: Sepkoski pap. 10–13).

It is interesting that Gould focused his "pitch" on ecology and biogeography, since his own current project was the decidedly non-ecological punctuated equilibria paper he was (at the exact time of his correspondence with Sepkoski) finishing with Eldredge. Nonetheless, this letter shows that even at this early stage Gould recognized where the potential "revolution" for paleontology lay. Gould's appeal to Sepkoski also focused on Sepkoski's unique interest in "deductive models," which would certainly have been prominent in Gould's mind given his recent involvement with Schopf's Models project. Together, these factors were enough to change Sepkoski's mind; he wrote immediately to Léo Laporte to withdraw his application, reporting, "Since I last communicated with you on Santa Cruz, a number of considerations have risen which I was not regarding earlier" (Sepkoski to Laporte, undated draft [circa May 1971]: Sepkoski pap.). Gould's appeal to Sepkoski was not simply prescient: deductive models drawn from ecology and biogeography had a transformative, even revolutionary effect on paleontology largely because Gould helped to construct and promote a vision of paleobiology in which they played a central role. In using his prediction of a transforming field to recruit Sepkoski—who, as we will now see, played a major role in that transformation—Gould was able to engineer a kind of self-fulfilling prophecy.

While Sepkoski continued to work with Kummel on his traditional stratigraphic dissertation, upon returning from sabbatical Gould immediately began to draw Sepkoski under his wing. Sepkoski later wrote that "as a result of work with Steve and reading seminal publications like Valentine (1969, 1973)" his interests quickly began to turn from Cambrian paleoecology to the study of diversity patterns (Sepkoski 1994, 134). His education also took a much more biological direction than it had previously followed: in addition to standard courses in the geology department curriculum, Sepkoski took classes on evolution and behavior (with E. O. Wilson), ecology (with William Bossert), biogeography (also with Wilson), species diversity, and principles of evolutionary biology (in a team-taught course whose faculty included Ernst Mayr) (Sepkoski, course notes: Sepkoski pap. 25, 2 and 25, 4). He also participated in Gould's seminar "Quantitative Methods in Paleontology." This latter course introduced students to topics such as analysis of variance, matrix algebra, factor analysis, multiple regression and discriminant function analysis, multivariate numerical taxonomy, multivariate ontogeny and allometry, and multivariate study of phylogeny (Gould, syllabus, "Seminar: Quantitative Methods in Paleontology," spring 1969: courtesy of Roger Thomas). This course also reflected Gould's appreciation of the potential for quantitative analysis to enhance the status of paleontology: students were assigned Thomas Kuhn's article "The Function of Measurement in Modern Physical Science," which argued that the mathematization and quantification of a scientific discipline is an essential step towards making that discipline nomothetic (Kuhn 1961).

Perhaps more importantly, Sepkoski also began an ambitious program of critical self-study of literature in biometry, quantitative paleontology, and multivariate analysis. A series of notebooks he kept during this period (roughly 1971–74) show that Sepkoski rapidly acquired a knowledge of mathematical techniques that quickly surpassed Gould's own (Sepkoski, notebooks "Notes A–K" and "L–Z": Sepkoski pap. 24, 5). Gould drew on Sepkoski's growing expertise both in his teaching and research; fellow student Andrew Knoll recalls that when he himself took Gould's quantitative methods course in 1974, Sepkoski, although officially only the teaching assistant, actually taught most of the class (Andrew Knoll, personal communication, 14 July 2003). Likewise, Russ Lande (another Harvard product) remembers that while "some paleontologists may have learned basic statistics from Steve Gould . . . the gurus of the group for quantitative analysis and computer simulation of evolutionary processes were always Jack Sepkoski and David Raup" (Russell Lande, personal communication, 15 July 2003). In addition to relying on Sepkoski for assistance with his class, Gould enlisted him in a series of assignments supporting the early MBL work. For example, Sepkoski wrote the COLINK program used to compare phenetic and cladistic phylogenies in the original 1973 MBL paper, which, as I discussed in the last chapter, eventually brought Sepkoski into the MBL project as a full contributor.

Gould's most significant assignment for Sepkoski, however, came in the fall of 1973, when he hired him to begin to compile data on orders, families, and genera from existing compendia of fossils such as the *Treatise* and *The Fossil Record* in order to facilitate more accurate comparisons with MBL simulation runs. The initial project was to use the 67 stratigraphic intervals recognized in *The Fossil Record* to compile range data for as many marine taxa as possible. At the time, Sepkoski was beginning work on his dissertation on Cambrian paleoecology, and he recognized that by consulting additional data sources he could both improve stratigraphic resolution (by adding intervals of shorter duration) and add taxonomic information not present in the standard databases. As he recalled years later, "This gave me the first inkling that I could go beyond standard sources and improve taxonomic data." Ultimately, Sepkoski's project would balloon to become an entirely new database of marine taxa, which would take nearly a decade to complete and would provide the basis for his own and others' continuing diversity studies.[2] In this respect, although Gould could hardly have known where the project would ultimately lead, a simple work-study assignment was indirectly responsible for producing what would become one of the most significant empirical resources in modern paleontology.

Island Biogeography and the Fossil Record: "Crunching the Fossils"

One effect of Gould's intervention in Sepkoski's graduate career was that it diverted much of Sepkoski's attention and enthusiasm away from his original dissertation topic, with fairly negative results for Sepkoski's progress as a student. His original plan was to write a dissertation on community evolution in marine benthic communities during the first 100 million years of Phanerozoic evolution. However, after a frustrating summer of fieldwork in 1973, Sepkoski determined that he would be unable to identify gener-

---

2. For Jack Sepkoski's own reflections on this history, see Sepkoski 1994, 134ff. See also Sepkoski 1982 and Sepkoski 1993.

alized communities in the Cambrian strata he was investigating (Sepkoski 1994, 136). Placed in the awkward spot of being unable to pursue the paleoecological questions that had originally motivated the study, he essentially abandoned his dissertation for two years and threw himself into Gould's sideline projects. When, in 1974, he was offered a position at the University of Rochester, Sepkoski scrambled to finish his thesis, but he applied to Kummel to scale back the topic of the dissertation. Citing "strictly pragmatic" reasons, Sepkoski requested approval to "limit my dissertation to a discussion of only the stratigraphy" of the formations he had been studying, effectively eliminating all consideration of paleoecology (Sepkoski to Kummel, 23 April 1975: Sepkoski pap.). Kummel's curt reply made no secret of his displeasure: "Your thesis title of the revised dissertation has been approved. I hope Steve has emphasized in no uncertain terms that he and I are disappointed. On your next visit to Cambridge we need to have a long talk" (Kummel to Sepkoski, 6 May 1975: Sepkoski pap.).

Although the decision to settle "on writing a sedimentological dissertation" must have been disappointing, Sepkoski found himself increasingly drawn to equilibrium biogeographical studies, and it was out of this interest that his own contributions to taxic paleobiology emerged. Sepkoski had learned the principles of island biogeography and demography directly from Wilson, and his first published paper grew out of an application of the "stepping stones" model of island immigration MacArthur and Wilson had developed in the 1960s. That paper, cowritten in 1973 with Harvard ecology graduate student Michael Rex, investigated distribution patterns of living freshwater mussel species along coastal rivers to determine "species-area relationships, distance effects, and environmental factors" (Sepkoski and Rex 1974, 167). Although the paper did not deal with fossil taxa, it is important because it extended the insular model through analogy to scenarios that did not involve literal islands, a strategy which was being pursued at the time by members of the MBL group. The Sepkoski and Rex paper also shared with the MBL approach an interest in simulation models: in order to investigate the strength of correlations between species numbers and areas in the actual freshwater mussel data, Sepkoski and Rex "construct[ed] an independent stochastic model based on the processes of immigration along stepping stones" which allowed abstraction "from the real biogeographic system several factors known or thought to be important in influencing numbers of species." Like the MBL model of phylogeny, properties of the modeled system were idealized: "All species were considered to have identical biogeographic properties (i.e., immigration and extinction rates) and all islands to be of equal size and distance apart" (Sepkoski and Rex 1974, 175). After running a series of Monte Carlo simulations on a computer, Sepkoski and Rex found that the model "compare[s] favorably with actual data," and they concluded, "Our model does meet many of the requirements of MacArthur and Wilson's (1967) theory of island biogeography" (Sepkoski and Rex 1974, 177–81).

This study of neontologic biogeography is an early indication of the direction Sepkoski's thinking was taking regarding the application of biogeographic theory to questions about taxonomic diversity and distribution. It was a short step to move from conceiving of coastal rivers as "islands in a sea of land," to constructing a much broader analogy with the fossil record of the history of life (Ruse 1999a, 215). Sepkoski's work over the next decade would attempt to apply the logistic, equilibrium model to Phanerozoic diversity patterns, and it culminated in three landmark papers in *Paleobiology* that presented a "kinetic model" of Phanerozoic taxonomic diversity. Cumulatively, this work has come to be known simply as the Sepkoski model.

This project grew both from the task that Gould had set for Sepkoski, as well as from Sepkoski's aborted dissertation study of Cambrian paleoecology. Sepkoski's paleoecological ambitions had initially faltered when he attempted to extend Peter Bretsky's analysis of marine benthic paleocommunities to the early Paleozoic. As Sepkoski recalled years later, Bretsky "had shown that there was considerable environmental stability in the general composition and environmental distribution" of these communities from the Ordovician through the end of the Paleozoic, and Sepkoski wanted to examine whether these communities existed 50 million years earlier in the Upper Cambrian. However,

> a summer of fieldwork in 1973 showed me that Bretsky's generalized 'communities' did not exist in the Cambrian: the faunas were entirely different and the communities could not be traced up into the Ordovician—250 million years of stability, but no roots 50 million years earlier! I combed the literature to learn about this great faunal change and found little. I became convinced that the route to understanding would be from the top down, beginning with an analysis of global diversity patterns (Sepkoski 1994, 136).

Having identified the question, Sepkoski still needed to address the problem of data. This is where Gould's assignment to compile a new database influenced the story. Gould's original instructions had been to correlate range information for fossil taxa based on the intervals used in *The Fossil Record* with fossil data from the *Treatise*. Sepkoski's initial finding was that newer literature could be used to improve the resolution and precision of those resources, and he set to work cross-referencing and updating the data set. However, in 1974, he learned that a Czech paleontologist named Jarmila Kukalová-Peck had recently compiled a similar database, obviating the need for a duplicate study (Kukalova-Peck 1973). Gould lost interest in the project, but Sepkoski did not. Over the next five years, he continued annotating Kukalová-Peck's compilation with his own findings from his dissertation research and from a general review of taxonomic literature, and "this heavily marked book became the data base for my study of the diversity of marine orders" (Sepkoski 1994, 135). Eventually, after moving to Rochester in 1975, Sepkoski began two notebooks on family-level marine diversity: the first notebook reproduced existing data on early Paleozoic families from standard sources, which Sepkoski updated on the basis of new information that was being published on Cambrian and Ediacaran fauna recently uncovered in localities like the famous Burgess Shale of western Canada. The second notebook listed range data for all Phanerozoic marine fauna, and was completed some time around the spring of 1976. As Sepkoski recalls, "By the spring of 1977, I must have been sufficiently confident in the data, especially for the Cambrian, to generate Paleozoic diversity curves" (Sepkoski 1994, 136–37).

The "Kinetic Model" of Phanerozoic Taxonomic Diversification

The first publication to result from this compilation and analysis was "A Kinetic Model of Phanerozoic Taxonomic Diversity I: Analysis of Marine Orders," published in *Paleobiology* in the fall of 1978. The paper investigated whether "a simple mathematical model for the growth and maintenance of taxonomic diversity through geologic time" could be identified which would describe "interrelationships among a small number of variables, specifically origination rate, extinction rate, and number (or "diversity") of taxa, and show how these should vary with respect to one another and time." Sepkoski concluded that "the diversification of taxa within any large, relatively discrete ecological system should be fundamentally a logistic process"—or, in other words, that the best description of Phanerozoic taxonomic diversification is a logistic curve (Sepkoski 1978, 223). In essence, Sepkoski argued that diversification over evolutionary time is essentially an equilibrium process, and he explicitly situated his model as "an outgrowth of the seminal work of MacArthur and Wilson." He also placed his analysis of "evolutionary equilibria" in a tradition of paleontological investigation extending from recent work by Bambach, Raup, Simberloff, and Schopf back through pioneering studies by Simpson. The specific contribution Sepkoski's paper offered was "a more generalized and precise formulation of the concept of evolutionary equilibria" than previous studies, and it predicted "specific quantitative patterns in taxonomic diversity . . . that can be tested statistically with paleontological data" using Sepkoski's expanded Phanerozoic database. Because the model described "the behavior or relative 'motion' of diversity with respect to time," he christened it the "kinetic" model (Sepkoski 1978, 223–24).

In a number of ways, this paper was emblematic of the new paleobiology. In the first place, it was an exercise in analytical model building, considered by Schopf and Gould to be the essence of the "nomothetic" approach. Sepkoski's model was a simple mathematical description of diversity over time, expressed as the difference equation $\Delta D/\Delta t = r_d D$ (where D is diversity per interval t, and r_d the diversification rate, is the difference between the rates of origination and extinction), that was abstracted and generalized from fossil data (Sepkoski 1978, 224–25). Secondly, the basic model was imported from theoretical ecology and population biology, and took the heuristic approach of treating groups as if they were individuals. Sepkoski likened his modeling approach to Eldredge and Gould's punctuated equilibria, Stanley's 1975 paper on species selection, Van Valen's Red Queen's hypothesis, and the Raup et al. MBL work. Third, Sepkoski's analysis was empirically grounded; it attempted to reread the fossil record by exhaustively surveying taxonomic literature and compiling a database of fossil marine orders which could be tested against the predictions of a simple mathematical model. His strategy fell somewhat between nomothetic and idiographic approaches, and is thus characteristic of the taxic view more generally. By the late 1970s, the paleobiological community had come to recognize—despite notable holdouts like Schopf—that a purely nomothetic methodology was impracticable, and studies such as Sepkoski's pointed the way towards a fruitful compromise. Indeed, Gould wrote in his 1980 essay "The Promise of Paleobiology as a Nomothetic, Evolutionary Discipline" that

Sepkoski's model was "an interesting and fruitful interaction of nomothetics and idiographics," where "the form of the model remains nomothetic," while "idiographic factors determine the parameters and these
then enter as boundary conditions into a nomothetic model" (Gould
1980c, 115).

The specific contention of Sepkoski's 1978 paper was that the mathematical model for diversification could be applied to scenarios such as
the Cambrian "explosion" and evolutionary radiation, wherein "major
taxonomic groups radiate into new adaptive zones or previously 'unoccupied' ecospace" (Sepkoski 1978, 225). To test the model, Sepkoski analyzed his data on fossil metazoans from the Upper Precambrian to the
Lower Cambrian, and found that the data best fit the exponential diversification equation expressed in logarithmic form. Sepkoski reasoned
that during this interval, as a relatively few initial groups colonized previously uninhabited ecospace, exponential diversification came as a simple
consequence of the intrinsic rate of diversification r_d. Importantly, Sepkoski found that the per-taxon rate of diversification remained constant;
exponential growth in diversity was produced "simply from the multipli-
cative effect of the continuous addition of taxa." In other words, no environmental or extrinsic mechanism was required to trigger an increase
in diversity during the Cambrian explosion (such as changes in ocean
chemistry, atmospheric oxygen, climate, predation, etc.) as had previously been assumed. Rather, "Under the present hypothesis, the 'explosion' of taxa in the Cambrian becomes simply a rapid growth phase in a
continuous process of exponential diversification" that was initiated in
the Late Precambrian (Sepkoski 1978, 228).

Sepkoski's model thus predicted that diversification beginning from
a small initial population of taxa with plenty of unoccupied ecospace
should be exponential (a prediction that has been subsequently confirmed by Paleozoic fossil data). But what about the rest of the Phanerozoic? If the exponential rate of diversification had continued unabated,
Sepkoski noted that the oceans would now contain some half-billion
orders—which is clearly not the case. Following the Early Cambrian explosion, therefore, the rate of diversification evidently subsided; the question was, could Sepkoski's model account for this? Sepkoski argued that
the diversification rate is density-dependent—that is, it declines as the
number of taxa increase because of several factors, including increased
competition, reduction in local population size ("species packing"), and
reduction of niche sizes. These factors effectively reduce the origination

rate (r_s), which is one of the variables (along with extinction rate, or r_e) that determine the rate of diversification. A diversity-dependent model can be constructed by combining the graphs of the slopes of origination and extinction; when superimposed, the two functions meet at a point D, which is the "equilibrium diversity" for the system. Sepkoski explained that "at D, rate of origination by definition equals rate of extinction, so that their difference, the rate of diversification, is zero." As Sepkoski noted, this model is directly analogous to the MacArthur-Wilson model of island biogeography for describing total rates of origination and extinction; when per-taxon diversity is plotted $(r_d D)$, the curve produced "should be logistic with an initial sigmoidal phase of diversification followed by an indefinite period of steady-state diversity" (Sepkoski 1978, 229–32; fig. 8.5). Ultimately, this means that "the familiar logistic growth equations commonly used by ecologists to describe the temporal growth of single-species populations" can be extended to describe the diversification of multiple taxa over geologic time (Sepkoski 1978, 233).

Sepkoski recognized that his model was an idealization, since "the deterministic solution is, of course, overconstrained and can never be expected to appear in a natural system." In practice, the smoothly curving sigmoid function would have jagged edges as diversification rates fluctuate around the equilibrium diversity D. However, Sepkoski explained that this effect could be simulated to take "slight random components," such as environmental perturbations, into account by allowing origination and extinction rates to vary slightly; in the resulting curve "diversity increases with a perceptibly sigmoidal pattern and then fluctuates about the equilibrium in an irregular though limited fashion" (Sepkoski 1978, 233). This is precisely what the plot of the actual data for marine metazoan orders shows: beginning at a low point in the earliest Paleozoic, the curve climbs exponentially through the Vendian and Cambrian, reaching a peak in the Ordovician before settling down to a plateau that fluctuates around an equilibrium through the end of the Tertiary. Sepkoski concluded that "the entire Phanerozoic history of marine ordinal diversity is eminently comparable to this [simulated] stochastic curve," a similarity which is "so striking that most specific historical events, represented by knicks [sic] and bumps in the curve, appear no more important than simple stochastic fluctuations" (Sepkoski 1978, 233).

Nonetheless, Sepkoski acknowledged that a number of questions were left unanswered. Most significantly, his fossil data were for orders, and previous work by Valentine, Raup, and others had shown that it was

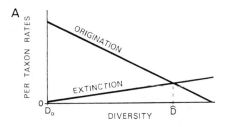

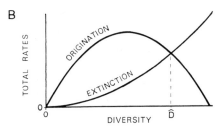

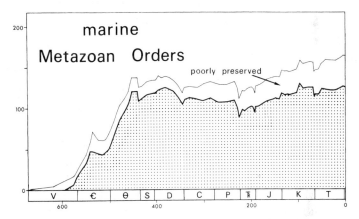

FIGURE 8.5. *Top*, Sepkoski's model of origination and extinction for both per taxon and total rates. The rates intersect at point $\hat{D}$, which is the equilibrium diversity point. *Bottom*, Sepkoski's initial "kinetic" model of Phanerozoic marine ordinal diversity. Note the logistic shape of the curve. J. John Sepkoski Jr., "A Kinetic Model of Phanerozoic Taxonomic Diversity I. Analysis of Marine Orders," *Paleobiology* 4 (1978): 232, 234. Courtesy of the Paleontological Society.

unclear whether diversity patterns in higher taxa were analogous to diversification at the level of species. Sepkoski noted that an ideal test of his model would be to apply it to species-level data, but conceded that "as is well known, this is rarely possible in paleontology." The relationship between diversity at higher and lower taxonomic levels would therefore need to be inferred in some way—a prospect that Sepkoski regarded as problematic, though not impossible. Even though higher taxa are often "large and arbitrary units subject to the idiosyncrasies of evolution and the whims of taxonomists. . . . these properties may not necessarily prevent higher taxa from paralleling the patterns and kinetics of species' diversification, even when information loss resulting from incomplete sampling of the fossil record masks these patterns in the species themselves" (Sepkoski 1978, 237). Sepkoski's approach to this problem involved a combination of idealized and generalized methods of data interpretation familiar from earlier studies—in the first place by constructing a computer model based on a variant of the MBL program to simulate the diversification of higher taxa and component species, and in the second by subjecting the results of the simulation to statistical correction for biases in the manner of Raup's rarefaction analysis.

For the initial simulation, Sepkoski modified the original MBL model in two ways. First, he substituted Raup's a priori "damping function" with equations for diversity-dependent origination and extinction rates. Second, he introduced a new routine for grouping completed phylogenies into monophyletic clades that are more variable, and hence more like "real" clades. Overall, this approach produced a more realistic simulation model than the MBL studies could achieve, which Sepkoski termed a "'biased random' taxonomy" (Sepkoski 1978, 238). He was then able to model and compare lineages, clades, and superclades, whose relationships could be considered analogous to species within higher taxa. The most significant element of this model was that, although the logistic pattern of lineage diversification was determined by constraints in the program (the diversity-dependent equations), diversification of clades and superclades was "only indirectly controlled through formation from lineages." Nonetheless, Sepkoski found that "the diversification of these 'higher taxa' is remarkably logistic, closely paralleling the diversification of lineages," and that the pattern of diversification of the simulated higher taxa closely matched the actual pattern of Phanerozoic Ordinal diversification (Sepkoski 1978, 239). In other words, if one assumes that species diversity is controlled by logistic MacArthur-Wilson population dynamics, Sepkoski's simulation showed that the expected result at higher taxonomic levels should *also* be logistic. This result helped to clarify questions raised by Raup and others about the relationship between species and higher taxonomic diversity: as Sepkoski concluded, "These initial results suggest that higher taxa may very well reflect the standing diversity of species" (Sepkoski 1978, 241).

But how could Sepkoski's prediction that species-level diversification should follow a logistic pattern be reconciled with previous estimates by Valentine, Raup, and Bambach, which had shown diversity to increase continuously from the Paleozoic to the Recent? Here Sepkoski was influenced by Raup's arguments about the potential effects of sampling bias: since fossil species are so much less likely to be preserved than higher taxa, Sepkoski applied a "partial sampling" function to the simulation, in which lineages were recovered only from some time units, with the probability of "preservation" decaying backwards exponentially in time. The results showed that "sampling most severely affects lineage diversity," distorting the original equilibrium pattern and generating the appearance of a linear rise to the end of the time series. However, diversity patterns at the higher levels were less affected, and retained their sigmoidal, logistic shape. Sepkoski concluded therefore that while diversity estimates from species-level data may be inherently skewed, patterns in higher taxa are more reliable and may even be enhanced by partial sampling. Therefore, "patterns of diversification observed for Phanerozoic orders may indeed reflect underlying patterns for species" (Sepkoski 1978, 244).

Sepkoski's initial model would prove to be as controversial as it was innovative, and as we will see Sepkoski himself modified it significantly over the next several years. Critics have raised objections to everything from his underlying assumptions about the effects of bias and statistical tests of significance in the model, to the quality and resolution of the actual data used to construct the empirical pattern. However, leaving those objections aside, I want to briefly call attention to the significance of this paper in the terms I have defined as central for the paleobiological movement of the 1970s. The initial Sepkoski model attempted to combine all the elements for explaining Phanerozoic marine diversification within a simple model derived from standard assumptions in ecology and population biology—it is thus admirably nomothetic. The basic assumption that phylogenies behave like populations was drawn directly from suggestions in the ecological and paleobiological literature of the early 1970s. The tabulation of marine metazoan orders was classic taxic paleobiology—Sepkoski assumed that a more accurate count of ordinal originations and extinctions would produce a better estimate of evolutionary patterns. Sepkoski's use of simulation also extended the project of the "Models in Paleobiology" symposium and the MBL group, but it modified the idealized model to reduce oversimplification and make its simulation more realistic. And manipulating the results of the simulation to reflect potential biases in the fossil data exemplified the statistical or 'generalized' approach to rereading presented in earlier studies by Van Valen, Raup, and others. Sepkoski's paper was thus one of the first to reflect the hybrid nomothetic/idiographic approach that would characterize much of paleobiology in its later, post-revolutionary phase. As Sepkoski explained, while his model and analysis were "admittedly oversimplified,"

> what I have been seeking is not a complete causative account of the history of life but rather a description of the fundamental patterns in the temporal behavior of taxonomic diversity. The kinetic model is a first approximation of this behavior but one that can be modified and expanded to describe secondary patterns of diversification and also to test certain causal hypotheses (Sepkoski 1978, 245).

While Sepkoski's paper contained strongly nomothetic elements, it did not seek to completely resolve the history of life into a set of causal 'gas laws.' Nonetheless, it embraced the heuristic value of models and simulations, provided they were shown to be adequately sensitive to and testable against the actual fossil record.

From the Kinetic Model to Faunal Succession

The 1978 kinetic model paper was Sepkoski's breakthrough publication, and helped establish him as one of the up-and-coming young paleontologists. In his review of the manuscript for *Paleobiology*, Dan Simberloff rated the paper "better than 'excellent'—exciting!" and commented "I felt I was reading an instant classic as I read it" (Simberloff, "Review of Sepkoski: A Kinetic Model. . . .": PS-*P*). Publication of this paper also coincided with Sepkoski's move from the University of Rochester to the University of Chicago, where he was hired in 1978 to fill the vacancy in invertebrate paleontology left after Ralph Johnson's death. Schopf had been angling to bring another "nomothetic" paleobiologist to Chicago for several years; in 1973 he had been behind an unsuccessful attempt to lure Gould from Harvard, and in 1974 he spearheaded an effort to appoint Raup as a university professor which, even after two favorable votes at the departmental level, was frustrated by a negative decision from the provost's office (Schopf to Charles E. Oxnard, 22 January 1973: Schopf pap. 5, 14). Sepkoski's hiring, therefore, represented a significant victory for Schopf in advancing his institutional goals in his own department.

Having finally completed his dissertation the year before—"freeing me of an albatross"—Sepkoski turned his attention full-time to compiling and analyzing his database, resulting in a string of important publications that ultimately led to the resolution of the disagreement between Raup, Valentine, and Bambach over the character of Phanerozoic marine diversity (Sepkoski 1994, 138). Sepkoski's initial kinetic model paper seemed to support an interpretation that was most in line with Raup's; as Sepkoski wrote to Raup prior to publishing the first paper, "I think 150 m.y. of equilibrium in the Phanerozoic is a nearly undeniable conclusion." He agreed that "the linear increase following the Triassic to double the Paleozoic plateau may be caused by your 'pull of the Recent.'" However, Sepkoski also noted that plots of his data against Raup's and Bambach's showed better correlation with Bambach's community data, which he found "disturbing" (Sepkoski to Raup, 1 June 1977: Sepkoski pap.). Even as the first kinetic model paper was being prepared for publication, Sepkoski therefore was entertaining doubts about whether a simple logistic model of ordinal diversity offered an accurate depiction of global marine diversity over the Phanerozoic.

At this same time, Sepkoski began updating his family-level compilation of taxa, which he ultimately transferred to computer punch-cards that he used to input the data for a series of factor analyses. As he later recalled, this was performed during the spring of 1977, partially in an effort to drain his University of Rochester computing account (Sepkoski 1994, 138). Over the next year, Sepkoski used the familial data to revise his conclusion that a single equilibrium characterized Phanerozoic diversity. In letters to colleagues, he admitted these doubts frankly: writing to Jeffrey Levinton, he revealed the "disturbing fact" that while his original model predicted that "survivorship curves should become more concave upon attainment of equilibrium," further refinement of his data revealed that at the family level, Phanerozoic marine invertebrates "become consistently *less* concave in time," which "isn't simply a 'pull of the recent effect'" (Sepkoski to Jeffrey Levinton, 28 April 1978: Sepkoski pap.). Several weeks later, Sepkoski wrote to Gould to report that he was at work on a "second kinetic model paper" in which he was describing "multiple equilibria (which I now know how to model, or at least describe, mathematically)" (Sepkoski to Gould, 20 June 1978: Sepkoski pap.). And in September of 1979, just before the second paper was published, he admitted to Dale Russell that he now realized that "I overstated my case in arguing a constant equilibrium in marine diversity throughout the Phanerozoic and will lay bets that lower taxonomic diversity indeed has increased since the Paleozoic," although "considerably less than Valentine's order of magnitude" (Sepkoski to Dale Russell, 18 September 1979: Sepkoski pap.).

The major argument of the second "kinetic model" paper, published in *Paleobiology* in the fall of 1978, was that analysis of data at the family level showed that not one, but *two* equilibria characterized the Phanerozoic marine record (Sepkoski 1979). At the outset, Sepkoski acknowledged that his earlier study of ordinal diversity "breaks down" when "greater detail" was applied to the analysis of diversity. That detail, he reported, came from an analysis of 1,600-plus metazoan families across the Paleozoic, which because of their "greater numbers and shorter average durations . . . are capable of revealing subtle, short-term patterns and trends in diversity that are not clearly visible with higher taxa" (Sepkoski 1979, 225–26). The central feature of the evolution of diversity at the familial level that Sepkoski's analysis revealed was a second, post-Cambrian radiation that "greatly altered both diversity and faunal composition in the world ocean." This second radiation also followed a logistic pattern of initial exponential diversification succeeded by equilibrium, but it had a higher initial rate of diversification and peaked at a plateau some three times the maximum familial diversity of the Cambrian. This distinctive "Paleozoic shelly fauna" followed a separate evolutionary trajectory from the earlier Cambrian fauna; as the Paleozoic fauna expanded, its Cambrian counterpart began a slow decline from the Ordovician to virtual extinction in the Permian (Sepkoski 1979, 235–36; fig. 8.6).

From an interpretive standpoint, Sepkoski argued that the "two-phase" kinetic model suggested that Paleozoic diversification was more "heterogeneous" than previously suspected. However, he noted that the

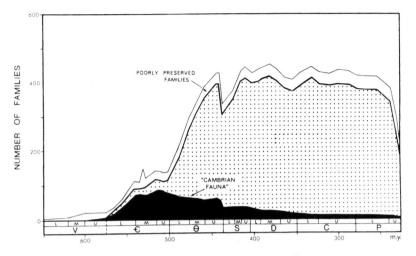

FIGURE 8.6. Sepkoski's two-curve kinetic model for marine families. J. John Sepkoski Jr., "A Kinetic Model of Phanerozoic Taxonomic Diversity II. Early Phanerozoic Families and Multiple Equilibria," *Paleobiology* 5 (1979): 235. Courtesy of the Paleontological Society.

two-phase model corresponds well with the "multiple equilibria" model of community evolution presented by Bambach in 1977, which had proposed that "some sort of equilibrium in species richness, possibly like species-area equilibria seen in the study of island biogeography" had governed the evolution of diversity in successive communities (Bambach 1977, 161; Sepkoski 1979, 238). Having demonstrated the mathematical validity of the two-phase logistic model, Sepkoski admitted that a deeper "evolutionary and ecological" question remained unanswered: "Why should so many phylogenetically unrelated taxa be segregated into two distinct groups or 'phases' with distinct evolutionary characteristics and histories?" Although a "rigorous" answer could not yet be offered, he pointed to "anecdotal" evidence that, in ecological-adaptive terms, the Cambrian fauna was more generalized than the later Paleozoic fauna, rendering its members well adapted for radiation into the relatively empty ecospace of the Early Cambrian, but less so for survival in the crowded environment of the later Paleozoic seas. Sepkoski demonstrated that the mathematics of such a model was analogous to the two-species ecological dynamics modeled in the 1930s by G. F. Gause, which predicted the relative survival of two species sharing the same environment as a function of competitive replacement (Sepkoski 1979, 244; Gause 1934).

If the Paleozoic radiation could be decomposed into two separate logistic diversification curves, an implicit question remained: with more
complete data, would additional faunas, with their own characteristic diversity patterns, appear during the Phanerozoic? At the end of his
1979 paper, Sepkoski openly acknowledged the possibility that one or
more additional equilibria might yet be uncovered, although he was also
aware that acquiring data of sufficient resolution would be difficult. Two
years later, he offered a solution to this problem by subjecting some 2,800
Phanerozoic marine families to factor analysis to attempt to determine
whether statistical tests could sort out remaining questions in the history
of marine diversity (Sepkoski 1981). This approach exemplified Sepkoski's methodology for reading the fossil record: he began the paper by
describing the fossil record as "an extremely complex, multi-component
system" whose complexity "often seems overwhelming, and almost infinite, making rigorous generalizations nearly impossible to construct."
However, he argued that paleobiologists, including Valentine, Eldredge,
and Gould, had pointed to a way in which thinking about complex systems (such as paleoenvironments or phylogenetic trees) as nested hierarchies permits "valid generalizations about processes and properties at
any particular level" to be made "without requiring complete knowledge
of analogous processes and properties at other levels." In this way, "The
history of the earth's biota also may be amenable to hierarchical study.
It may be possible in various situations to collect historical entities together into subsystems and then trace and even explain the general behavior of these subsystems through time without necessarily understanding the behavior of the individual entities" (Sepkoski 1981, 36–37).

Such a hierarchical model recalls some of the nomothetic character
of Schopf's "gas laws" approach. Unlike the idealized approach taken
in the MBL work, however, Sepkoski's generalizations were statistically
derived from the fossil record itself, and not set as initial conditions in
an abstract model. As Sepkoski explained in a letter to Valentine a year
before publishing this paper, there was "a peculiar kind of reductionism going on in paleontology. . . . Rather than going after the fundamental building blocks of the science, such as the molecular biologists are
doing, paleobiologists seem to be first trying to build the most general
and all encompassing (e.g. holistic) models possible, and then modifying
these in small steps to make them more and more realistic, but still general." He described his own view as one of "trying to get some grasp on
the boundary conditions of historical evolution in the seas" prior to at-

tempting to explain the actual mechanisms that produce the evolution of diversity at specific moments. One problem with the MBL attempt to "describe phylogenetic patterns in their simplest and most general form: as stochastic branching processes," was that the complexity of the actual fossil record "cannot be reproduced in a very simple branching process." Ultimately, then, "to produce a more realistic pattern," constraints needed to be added to the system to bring the model closer in line with 'reality' (Sepkoski to Valentine, 24 February 1980: Sepkoski pap.). This is what Sepkoski's series of kinetic model series of papers attempted to do.

Sepkoski's next installment in the series, in 1981, essentially repeated the approach of Flessa and Imbrie's 1973 analysis of "evolutionary pulsations" which had grouped the marine record into 10 "diversity associations" using Q-mode factor analysis. However, Sepkoski's study drew on his improved family database and modified some of the statistical treatments used by Flessa and Imbrie. In the first step, Sepkoski organized the 2,800 families, representing 91 classes across 82 stratigraphic stages from the Venidan to the Pleistocene, into clade or "spindle" diagrams that offered a visual depiction of diversification of classes over time (fig. 8.7). This visual representation was reminiscent of the approach taken by the MBL papers, and Sepkoski noted that "simple statistical generalizations" were possible from such evidence (Sepkoski 1981, 39). The most obvious characteristic of this figure was the unevenness of the distribution of diversity: many classes contain very few families, while the vast majority of marine families in the known fossil record are distributed among a few, very large classes. Times of first appearance of classes are also unevenly distributed, with some two-thirds appearing in the first quarter of the Phanerozoic, and only 4% emerging in the last half. Finally, a majority of the clades that appeared in the early Paleozoic never achieved significant diversity and died out by the Permian, making the Paleozoic "appear as the 'age of small clades' and the Mesozoic and Cenozoic Eras as the age of grand-old, established clades" (Sepkoski 1981, 37).

However, beyond these broad generalizations, Sepkoski acknowledged that it was difficult to determine much about the dynamics of diversification as an actual historical process from qualitative visual summary of clade shape. To probe more deeply, he turned to multivariate factor analysis to measure the relative strength of correlations between diversification in heterogeneous groups and time. "Positive" factor scores indicated that a particular taxon contributed significantly to diversity in

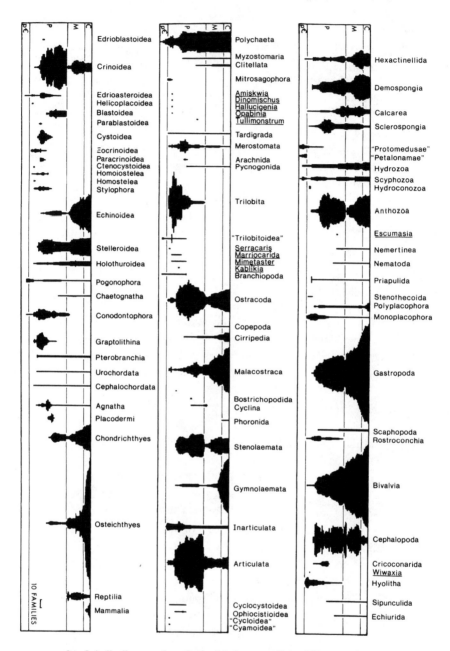

FIGURE 8.7. Spindle diagrams from Sepkoski's factor analysis of Phanerozoic marine taxonomic diversification. J. John Sepkoski Jr., "A Factor Analytic Description of the Phanerozoic Marine Fossil Record," *Paleobiology* 7 (1981): 38. Courtesy of the Paleontological Society.

a particular sample, and taxa with high positive scores were grouped into "diversity associations" following the same procedure used by Flessa and Imbrie (Sepkoski 1981, 40). This analysis revealed three factors with the highest "factor loadings" (or positive scores), accounting for roughly 91% of the taxa analyzed; according to Sepkoski, these factors represented "the 'three great evolutionary faunas' of the Phanerozoic marine fossil record" (Sepkoski 1981, 43; fig. 8.8). The first fauna, corresponding to the initial Paleozoic radiation, encompassed the characteristic Cambrian taxa such as trilobites and annelid worms that dominated before rapidly declining in the Ordovician. This fauna was then succeeded by the second, more taxonomically diverse Paleozoic "shelly fauna," which experienced a much steeper exponential growth in the Ordovician and plateaued until the end of the Paleozoic. However, the great extinction event at the end of the Permian marked the decline of the second fauna, ushering in a third major diversity association, the Mesozoic-Cenozoic or "modern" fauna, which exhibited the greatest diversity of the three and began an exponential rise in the Triassic that continued mostly unabated through the end of the Phanerozoic (Sepkoski 1981, 45).

Sepkoski claimed that the strength of these associations indicated "a fundamental simplicity to all the faunal change we see in the marine fossil record" that belies the appearance of "almost chaotic variation" in the raw fossil data. The repeating pattern of logistic growth followed by equilibrial stability is also an apparent departure from what could be predicted using a simple stochastic branching simulation, indicating that some historical, deterministic processes must have shaped the evolution of Phanerozoic marine diversity (Sepkoski 1981, 44). Sepkoski argued that since the timings of the three evolutionary fauna "appear to be intimately related to patterns of total diversity in the Phanerozoic oceans," each fauna may indicate changes in the ecological-evolutionary characteristics of the Phanerozoic marine environment. In other words, each fauna can be "characterized by a different 'style' of diversification," indicating that the history of marine diversity has been shaped by specific, contingent historical events, such as the great Permian extinction, that selected the characteristics of the organisms that dominated during particular periods of life's history (Sepkoski 1981, 49–50). A central implication of this conclusion was that while heuristic models (such as the logistic equation) could offer insight into the general dynamics of evolutionary diversification, the history of life could not be explained by a stochastic model or a set of paleontological "gas laws."

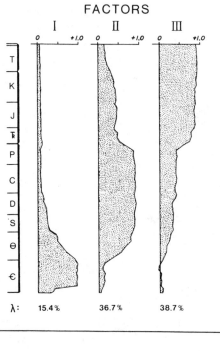

FACTORS

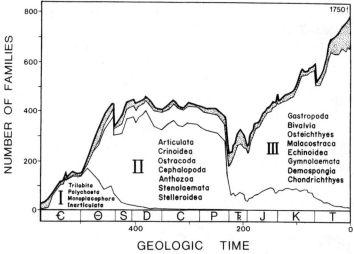

FIGURE 8.8. *Top*, Sepkoski's factor loadings for the three stages of Phanerozoic marine diversity. *Bottom*, the three "evolutionary faunas" represented as overlapping diversity curves. J. John Sepkoski Jr., "A Factor Analytic Description of the Phanerozoic Marine Fossil Record," *Paleobiology* 7 (1981): 43, 49. Courtesy of the Paleontological Society.

The graph Sepkoski produced, which represents the three evolution-ary faunas as a succession of logistic diversity curves, would acquire iconic status in paleobiology, and would come to epitomize the move-ment in taxic paleobiology that dominated the field during the 1980s and 1990s. While Sepkoski's analysis has been debated and reexamined (by Sepkoski and others) many times, it also provided a blueprint for a new kind of analytic paleobiology that has characterized the field ever since. Continued development and study of his growing database also con-vinced Sepkoski that major extinction events played a decisive, deter-ministic role in the evolution of marine diversity. He explicitly acknowl-edged this fact in a letter to Norman Newell, in which he praised Newell for showing "the American audience how mass extinctions could be rec-ognized and analyzed," to which "my contribution really does noth-ing more than add a few bits of new information" (Sepkoski to Newell, 12 August 1982: Sepkoski pap.).

In 1984, Sepkoski completed the kinetic model series with a paper that examined the three-fauna model explicitly in the context of mass extinctions, which as the next chapter will discuss had become an in-creasing focus of paleobiological research in the early 1980s. In this last paper, Sepkoski did not consider the broader significance of his model, arguing that its "precise and parsimonious description of familial diver-sification" showed how large-scale patterns in the fossil record "may be simple consequences of a small number of evolutionary parameters that appear to remain nearly constant through time" (Sepkoski 1984, 247). However, he argued for the important role of mass extinctions as "per-turbations" of the underlying diversity dynamics "produced by forces external to and basically independent of the long-term growth and decay of the evolutionary faunas," which acted to clear ecospace for successive episodes of diversification (Sepkoski 1984, 257). Sepkoski explained how causally independent extinction events could perturb the logistic pattern in a predictable way by temporarily altering per-taxon extinction and origination rates, interrupting periods of evolutionary equilibrium and introducing new episodes of exponential expansion that led to new equi-librium conditions (Sepkoski 1984, 258; fig. 8.9). As Sepkoski put it in the conclusion:

The fossil record does appear highly structured, at least at the scale of changes in familial diversity within suites of higher taxa. Given that evolu-tionary faunas behave, to a first approximation, like discrete entities, much

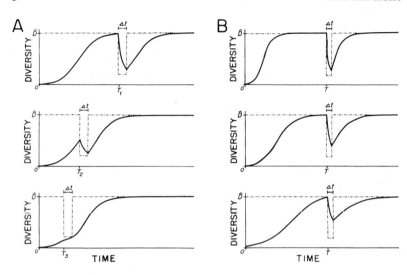

FIGURE 8.9. Sepkoski's "simple model" of mass extinction as a perturbation of a logisti-cally diversifying system. This model predicts (A) that the closer a system is to equilib-rium, the greater the effects of the perturbation will be, and (B) that systems with more rapid rates of diversification will experience greater declines in diversity following a per-turbation. J. John Sepkoski Jr., "A Kinetic Model of Phanerozoic Taxonomic Diversity. III. Post-Paleozoic Families and Mass Extinctions," *Paleobiology* 10 (1984): 260. Courtesy of the Paleontological Society.

of the evolutionary change in diversity and faunal composition that we see in the Phanerozoic oceans seems to be a consequence of a very small num-ber of parameters and their functional interrelationships. . . . This holds true even for the changes effected by mass extinctions; although the timing and in-tensity of these events appear to be controlled by forces external to the fau-nal system, the responses of the faunas during decline and rebound seem to be properties of their previously set initial diversification rates and equilib-ria. Therefore, much of what appears complex, or even random, in the his-tory of familial diversity and faunal change may not originate from any in-herent complexity but rather from the nonlinear functional interrelationships among the set parameters (Sepkoski 1984, 262).

In other words, Sepkoski was describing a system in which a basic ecolog-ical model—the logistic equation—functioned as an internal constraint on the evolution of diversity, while a series of external constraints—mass extinctions—operated contingently to perturb the system with partly predictable results. Sepkoski's overall approach to reading the fossil record was therefore, as Gould noted in 1980, "nomothetic" in the sense that it was generalized and predictive, but also "idiographic" in the sense that it responded to specific historical circumstances which are not deductive consequences of that model.

Aftermath: "Consensus" and Consensus?

Sepkoski's model also helped engineer a tentative solution to the disagreement between Raup, Valentine, and Bambach about the nature of Phanerozoic marine diversity, particularly in answering the original question "Is the appearance of diversity increase over the Phanerozoic an artifact of bias or a real biological signal?" In late 1981, Sepkoski, Bambach, Raup, and Valentine jointly published a paper titled "Phanerozoic Marine Diversity and the Fossil Record" in the journal *Nature*, which has been nicknamed the "consensus" or "kiss and make up" paper (Sepkoski et al. 1981).[3] As early as 1977, Sepkoski had become aware that his new database might shed light on the debate, and he had sent a letter to Bambach outlining the close correspondence between his estimate of global marine familial diversity and Bambach's own tabulation of within-species community diversity. In both Sepkoski's histogram (an early version of the "3-faunas" graph he would publish in 1981) and Bambach's tabulation, diversity did indeed appear to increase over the Phanerozoic, which would seem to settle the issue. Nonetheless, Sepkoski was hesitant to jump to conclusions about his data, since, as he explained to Bambach, "I am not yet convinced that the post-Paleozoic rise evident in all our data is real. . . . We must get around to doing some rarefaction analyses of Phanerozoic community data!" (Sepkoski to Bambach, 1 June 1977, quoted in Miller 2009, 374). Miller suspects that Sepkoski's insecurity about his data, as well as his allegiance to Raup's arguments about rock-volume bias (reflected in Sepkoski's own 1976 study of species-area effects), likely prevented him from publishing the analysis at that time. However, in 1980 Sepkoski and Bambach had occasion to revisit the issue during a conference in Germany where they discussed the problem with eminent German paleontologist Adolph Seilacher, who had conducted his own study of diversity based on trace fossils (Miller 2009, 374).

3. See Miller 2000 and Miller 2009 for a discussion of the history of this collaboration and the source of the nicknames.

As Miller has reconstructed, the following year Sepkoski drafted the first version of what was to be a jointly authored paper with Bambach, the "centerpiece" of which was a five-way comparison of Seilacher's, Valentine's, Raup's, Bambach's, and Sepkoski's estimates of Phanerozoic marine diversity. Eventually, however, invitations to coauthor were extended to Valentine and Raup (and to Seilacher, who declined); the final paper was published as a four-way collaboration. The paper's central conclusion was that, based on the comparison of histograms representing the five different diversity estimates, "all five estimates of Phanerozoic taxonomic diversity are measuring a single underlying pattern that is reflected at both local and global scales and at various taxonomic levels in the marine fossil record" (Scpkoski ct al. 1981, 436; fig. 8.10). That pattern consists of low diversity during the Cambrian, followed by higher increasing diversity in the post-Cambrian Paleozoic, interrupted by Mesozoic diversity trough following the Permian-Triassic event, and concluding with steadily increasing diversity through the Cenozoic to a Tertiary maximum. The authors noted that this pattern was supported not only by their own analyses, but also by earlier studies by Newell and even by Phillips' 1860 diversity curves (Sepkoski et al. 1981, 437).

Beyond settling the debate between Raup and Valentine (Valentine tacitly accepted that Raup's three-fold diversity increase was closer to the truth than his own order of magnitude rise), this "consensus paper" was an important vindication of taxic paleobiology. One of the central problems associated with "rereading" the fossil record was determining whether statistical generalizations offered reliable insight into actual biological phenomena. Sepkoski et al. argued that the "remarkably consistent pattern" in their "empirical estimates" of diversity suggested "that a strong evolutionary signal underlies the preservational and taxonomic noise of the known fossil record" (Sepkoski et al. 1981, 435). Miller contends that this result led to "an explosion of research emphasizing the macroevolutionary and paleoecological processes responsible for the major features of Phanerozoic diversification and extinction," all of which drew on the implicit assumption "that the patterns that researchers were trying so hard to explain were, after all, *real*" (Miller 2009, 380). While debates have persisted about the *correct* interpretation of these patterns, it is fair to say that a central operating assumption in evolutionary paleobiology ever since has been that statistical generalizations based on the empirical fossil record are capable of detecting genuine, biological evolutionary signals, and that some patterns in the fossil record are real.

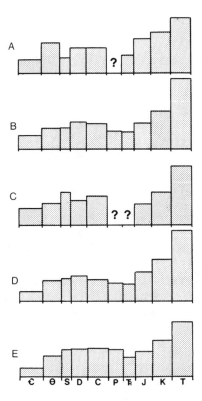

FIGURE 8.10. The five datasets compared in the "Consensus Paper." The histograms represent (A) Seilacher's data on trace fossils, (B) Raup's raw data on number of species per geological system, (C) Bambach's data on community species richness, (D) Raup's modified data on generic diversity per system, and (E) Sepkoski's weighted familial diversity data. J. John Sepkoski Jr. et al., "Phanerozoic Marine Diversity and the Fossil Record," *Nature* 293, no. 5832 (1981): 436. Reprinted by permission from Macmillan Publishers Ltd.

In many ways, this "victory" for the taxic, generalized approach to reading the fossil record represented a validation of the earlier model of analytic paleobiology championed by Newell, Imbrie, and Simpson in the 1950s and 60s. This tradition—rich in both careful empirical documentation *and* complex mathematical analysis—was what Sepkoski sometimes, half-jokingly, described as "the natural history of data." One of Sepkoski's later students, Michael Foote, has commented on the contrasting themes in this approach, which was "marked by meticulous collection of data on a monumental scale and by an interplay between mathematical modeling and rigorous, insightful data analysis," and above all exhibited a determination to investigate "how the ecological and evolutionary properties of different groups, the quirks of their fossil record, and the habits of their monographers, combined to yield synoptic, global signals in the data" (Foote 1999, 236). In an important sense, this tradition— and Sepkoski's particular contribution to it—helped reconcile some of those opposite tensions that had characterized paleobiology in the early 1970s: it brought together random *and* deterministic processes, models *and* empirics, extrinsic (abiotic) *and* intrinsic constraints, stochasticity *and* selectivity, idealization *and* historicity, or as Gould put it in 1980, "nomothetics *and* idiographics." In resolving some of this tension, the taxic approach gave paleobiology a way forward that built on the excitement of new methods and technologies, without totally abandoning that most priceless and traditional resource of paleontology—the fossil record itself.

Writing to Sepkoski to comment on the consensus paper, paleontologist Phil Signor expressed approval and "firm agreement" with the conclusion "that there is a strong signal coming from the fossil record," but also gleefully commented that he "welcome[d] this final spike through the collective hearts of the four horsemen of the MBL" (Phil Signor to Sepkoski, 7 June 1981: Sepkoski pap.). As evidenced by Gould's supportive comments about the "nomothetic-idiographic" approach in Sepkoski's work, as well as by Raup's tacit approval and collaboration in the consensus paper itself, most of those original "horsemen" had in fact all but abandoned the dream of an idealized, stochastic paleobiology in which historical events were subordinate to "gas laws" governing individual paleontological "molecules." Schopf, however, was unable to make this compromise, and spent the final years of his life increasingly marginalized from the growing consensus in paleobiology. The sad irony, of course, is that he ended up alienated from a movement that he, more than any other individual, did most to build. But as will be discussed in the final two chapters of this book, by the early 1980s paleobiology had entered a phase of its development in which consolidating its gains and cementing its institutional position—both within the discipline of paleontology and within the larger community of evolutionary biology— became paramount. The "revolutionary phase" was effectively over, and the future of paleobiology would increasingly belong to those who were able to articulate a vision that established paleobiology at the "high table" of evolutionary theory.

CHAPTER NINE

The Dynamics of Mass Extinctions

During the 1970s, the study of extinction became a more and more prominent part of paleobiologists' agenda for rereading the fossil record. However, many of the most active proponents of paleobiology during this time expressed frustration at the lack of a coherent theory that explained the role of mass extinctions in macroevolution. For example, it was clear that extinctions influenced the patterns of diversity that paleontologists like Valentine, Raup, and Sepkoski were investigating, and even the MBL simulation project recognized extinction as an important variable in evolution. Norman Newell's pioneering work in the 1960s established a new legitimacy for analysis of mass extinctions, and helped relieve some of the stigma traditionally associated with "catastrophism." But important questions about extinction—and especially mass extinction—persisted: How frequent were major episodes of extinction? What were their causes and mechanisms? How geographically widespread were the largest extinctions? How could major extinctions be reliably recognized—separated from background noise—in large fossil data sets? And, most broadly, what role did extinction have in determining the patterns of life's history?

These questions were central to the project of rereading the fossil record. They also went to the heart of the revision of the traditional synthetic view of evolution that many paleobiologists were trying to establish. Darwin himself downplayed the evolutionary significance of extinction in *Origin of Species*: extinction, he believed, was essentially just the flip side of natural selection—the elimination of the "unfit"—and therefore required no independent causal explanation or analysis. In particular, Darwin rejected most evidence of mass extinctions as examples of the imperfection of paleontological data; episodes where large numbers of taxa appeared to disappear suddenly from the fossil record could be attributed, he thought, to the extreme imperfection of the geological record. There are a variety of good historical reasons for his position. In the first place, mass extinctions were, in the 19th century, still closely connected with the taint of catastrophism that many scientists, particularly in Britain, associated with natural theology. Secondly, the known fossil record in Darwin's day *was* poor, and Darwin was rightly skeptical of broad claims (such as John Phillips's model of faunal succession) made on the basis of such an imperfect resource. The upshot, however, was that Darwin's view of extinction became the received wisdom in evolutionary biology, and even in the late 1970s and early 1980s there was resistance to treating extinction as a significant independent evolutionary phenomenon. One of the signature accomplishments of the paleobiological movement was to change that state of affairs.

Mass Extinction, the Fossil Record, and Evolutionary Biology

"Bad Genes or Bad Luck": Raup on Extinction

As he did in so many other areas of the paleobiological movement, Raup played a vital role in the reconsideration of the function and dynamics of mass extinctions. His serious interest in extinction began as an offshoot of his work in both the MBL project and the study of diversity patterns. The basic philosophy of the MBL simulations (and of stochastic paleontology generally) was that extinction could be modeled as an essentially random process (that is, simply as a statistical matter, extinction could be modeled as random input). A key element in this view was the work, inspired by Van Valen's survivorship studies, that showed that in many lineages the probability of extinction did not change over time. The question, then, was to figure out what that meant, and to find a way to accurately model such a phenomenon. In fact, Raup would change his mind more than once about the causality and significance of mass extinctions—much to the chagrin of Schopf—but the "stochastic view" was the starting place for his investigation the subject.

In 1977, Raup used the occasion of his election as Paleontological

Society president to lecture on the problem of extinction.[1] In recognition of the changing attitudes in the profession his election represented, he began almost defensively, remarking on being "the first president of The Society who has never defined a species," but he assured his audience "the work that I, and other so-called paleobiologists, have been doing for several years differs only in emphasis from what most paleontologists have been doing for the past couple of centuries" (Raup 1978a, 517). Theoretical paleontology, he maintained, was not the enemy of descriptive practice, and he stressed that "a lot of the theoreticians would be out of business if traditional systematics were to stop." The topic of his address, "the extinction of higher taxa as a general evolutionary process," was an example of this relationship: "It is not traditional in its approach but is completely dependent upon traditional approaches for its implementation" (Raup 1978a, 517).

Raup drew attention to two assumptions he felt were common to "most discussions" about extinction in the paleontological literature. The first was the tendency to view extinction of a higher taxon as "a special or unusual event, something that needs to be explained." The second was the assumption that higher taxa become extinct as the result of "a common failure or deficiency" among their constituent species. The result of these assumptions was that when a higher taxon becomes extinct, "we are saying that the critical element in the extinction is to be found among the set of morphologic or ecologic characteristics common to the whole group. Why did the echinoids survive the Permo-Triassic event and the blastoids not? We assume that 'echinoidness' was somehow better than 'blastoidness'" (Raup 1978a, 517–18). Raup acknowledged that some extinctions could be explained this way, but insisted that such rea-

---

1. One sign of the success of paleobiology was the increasing recognition of its proponents by their peers. For example, between 1974 and 1984 the Charles Schuchert award, given by the Paleontological Society annually to an outstanding paleontologist under 40 years of age, was presented to Raup, Gould, Schopf, Stanley, Eldredge, and Sepkoski in close succession. In 1979, Newell was awarded the Paleontological Society Medal, the highest honor of the society, for career accomplishments (Valentine, Raup, Bambach, Gould, Stanley, and Eldredge would each eventually be awarded the PS Medal as well). And, in 1977, Raup was elected President of the society, an honor for which he was preceded by Jim Valentine and would be succeeded in later years by Gould, Stanley, and Sepkoski (A list of past awardees of Paleontological Society honors is available at http://www.paleosoc.org/awards.html).

soning often led to a fruitless "search for common denominators of failure" and the proposal of "torturous" explanations. Rather, he proposed
to show, using statistical arguments, that "extinction is the expected fate
of a higher category, and that the extinction of a taxonomic group of species does not require a common cause" (Raup 1978a, 518).

Raup used a logical argument to demonstrate that extinction is a statistical inevitability in any phylogenetic branching process. As he explained, even the largest, most "successful" groups will, given enough
time, become extinct, because while extinction guarantees immunity
from success (putting the group "out of the game"), success provides no
corresponding immunity from eventual extinction. As an illustration,
Raup presented an MBL-style hypothetical branching simulation to
show that most simulated phylogenies, which begin small, do not survive
for very long, while a very few persist for a much longer time and branch
repeatedly to contain many subordinate members. While this pattern
gives the appearance of winners and losers, Raup pointed out that, in
the absence of any selective factors (i.e., inherent qualities that predis-
pose for survival), statistically one would predict "a complete gradation
between the successful and the unsuccessful"—and that this is exactly
what extensive runs of the simulation showed. In other words, a completely random simulation produced patterns very similar to those observed in the actual fossil record without the necessity of invoking adaptive characteristics to explain it (Raup 1978a, 520).

Turning to the fossil record, Raup pointed out that there were limits to an analogy between real and hypothetical lineages; empirical studies by Simpson and others had shown that turnover rates—the tempo
of evolution—vary significantly between major groups of animals (e.g.,
clams versus mammals). "Clamness" and "mammalness," in other words,
appear to matter. The most likely scenario, Raup argued, is that among
some groups, or at some levels of taxonomic resolution, extinction is effectively random, while at others it is not. The task of theoretical paleobiology was to develop models and tests to distinguish between the
two (Raup 1978a, 521). Beginning with the null hypothesis that taxonomic extinction is a random process, Raup then presented the specific
case of the echinoderms. At the beginning of the Carboniferous, there
were nine classes or subclasses of echinoderms; by the Triassic, four of
these classes had died out, including the blastoids. The question was, is
there something about those groups which predisposed them to extinction? Using rarefaction, Raup analyzed the statistical likelihood that one

group would survive rather than another. He found that, contrary to traditional paleontological expectation but quite within the scope of the null hypothesis, all nine original echinoderm classes had a more or less equal expectation of survival: "The fact that the blastoids did not survive, or that the echinoids did survive, is not a statistically significant departure from expectation" (Raup 1978a, 522).

Does that mean, therefore, that extinctions are genuinely random? Raup was not prepared to go that far. The exercise he presented proved only that "there is nothing in the numbers to enable us to reject the hypothesis that extinction was just bad luck," or to assert "that echinoids were any more fit that blastoids." Importantly, Raup stressed that "only good functional morphology" could answer whether there was some "deterministic" or selective reason for the comparative success of the two groups—in other words, traditional descriptive paleontology still had an important role to play. However, Raup did acknowledge the striking and perhaps radical implication of the analysis: "that if the Carboniferous and Permian were to be reinacted [sic], the blastoids might have survived instead," which "suggests that just plain luck could have played a larger role than is normally believed" (Raup 1978a, 522).

In 1977, then, Raup appears to have been perfectly willing to accept, with some qualifications, that extinction was often—and perhaps even usually—a matter of bad luck. However, in just a few years his attitude would drastically change. As the result of his participation in a conference in Barcelona on the subject of extinctions, Raup published a short paper in 1981 titled "Extinction: Bad Genes or Bad Luck?" in the relatively obscure journal *Acta Geologica Hispanica*. This paper proposed to investigate essentially the same problem he had covered in his 1977 address: to "play the devil's advocate" and investigate the validity of the "conventional wisdom (attributable to Darwin and Lyell)" that extinction is a "positive" evolutionary process ("in the sense of representing the replacement of less well adapted types by better adapted types"). The alternative explanation Raup proposed to investigate was "that extinctions are randomly distributed with respect to overall fitness (or adaptiveness of the organism) and that extinction of a given species or higher group is more bad luck than bad genes" (Raup 1981, 25–26).

Raup began the paper by reviewing the dominant view in biology and ecology on extinction, focusing on the birth-death model of replacement on which MacArthur and Wilson had drawn so heavily. According to this model, under normal conditions local extinction for groups be-

low a certain population threshold "becomes probable as a purely stochastic phenomenon." This process can be extrapolated to the so-called
time homogeneous model (such as the MBL model) where evolution is a
stochastic process and the probabilities of branching and extinction are
constant (cf. Van Valen's Red Queen's hypothesis). Under this assumption, Raup noted, "it is conceivable that groups of organisms can, over
geological time, drift to extinction just because of an accidental excess of
extinctions over speciations" among their members. If this were true—as
some of Raup's own earlier papers had suggested—then "a search for
causes of extinction in the conventional sense would be meaningless"
(Raup 1981, 27). Raup's test case for this paper was the famous example
of the trilobites, which, during the Cambrian, accounted for some 75%
of all fossil species described, but by the end of the Permian had become
entirely extinct. Was this a matter of bad genes or bad luck?

Raup calculated the problem as follows: assuming a value of .09 as the
mean extinction rate (μ) for trilobites (which Raup derived from his own
earlier study of extinction rates among marine invertebrates during the
Phanerozoic), he used the formula

$$P_o = \left[\frac{\mu t}{1 + \mu t} \right]^2,$$

to estimate the probability of total extinction (P_o), where μ is the number
of constituent species at the start of the Cambrian and t is the time span
being considered. Assuming a value of about 6,000 species for μ, Raup
calculated a probability of 4×10^{-82}, a value "so near zero that we can
conclude with confidence that the time homogeneous model will *not* explain the trilobite extinction" (Raup 1981, 28). Raup tested this result using a computer simulation in which a pre-determined standing diversity
of "species" was distributed unevenly across 10 "higher taxa" and subjected to random "extinctions" over a large number of time units. No
matter how he adjusted the program's settings, he could not reproduce
a pattern similar to the actual Phanerozoic record, where a very large
group—like the trilobites—became completely extinct while other much
smaller groups survived. His conclusion was that there were simply too
many trilobites for a string of bad luck to have done them in. Therefore,
Raup concluded that "we can entertain more seriously the possibility
that trilobites were in fact selected against compared with other marine
invertebrates," and more broadly that "the extinction of once successful groups such as the trilobites is thus most reasonably explained on the

basis of bad genes rather than bad luck, at least in the present state of knowledge" (Raup 1981, 28–32).

This is essentially the same conclusion that Raup had reached in his 1979 letter to Schopf I discussed at the end of chapter 7, where he contended that "the chances are nil that the trilobites could have drifted to extinction and one must conclude that . . . [the] Paleozoic extinctions [were] actually selective . . . meaning in turn that trilobites were inferior beings and deserved to die" (Raup to Schopf, 28 January 1979: Schopf pap. 3, 30). Although this conclusion infuriated Schopf, Raup stuck to his guns, and the Barcelona paper records the start of Raup's public reconsideration of the problem. As he explained to Schopf in a letter written the same year as the Barcelona meeting, "I have been through a major change of mind on randomness of extinction. Through much of the 70's, I had a strong hunch that most extinction was taxonomically random. And I used all the tricks (innuendo, etc.) in my papers to support that view. Now, however, I have done much more rigorous analysis of extinction data and can only conclude that the major extinctions were highly selective—and statistically significantly so. There is no alternative but to discard my earlier conclusion" (Raup to Schopf, 29 October 1981: Schopf pap. 3, 30). While this conclusion deepened a growing rift between Raup and Schopf, it also signaled the beginning of one of the most exciting and controversial episodes in the establishment of the paleobiological movement.

"Extinctions Are In"

There are two specific factors that can help contextualize Raup's reconsideration of his position on extinction: the refinement and availability of Sepkoski's massive new database of the Phanerozoic marine fossil record, and new evidence that suggested an impact mechanism for that most famous of all mass extinctions, the dinosaurs. Paleontologists had long recognized periods in the history of life where the fossil record seemed to indicate great loss of standing diversity: two of the most famous were at the boundary of the Cretaceous and Tertiary (K-T) periods, when the dinosaurs disappeared, and the earlier—and apparently larger—extinction event at the end of the Permian (P-Tr) that saw the demise of the trilobites and many other groups. Prima facie, these episodes appeared to be evidence of massive, catastrophic mass extinctions. However, firm conclusions about the characteristics of such events were hampered by several factors: first, there were difficulties in dating extinction events with enough precision to determine whether they had been truly instantaneous in ecological time, as opposed to merely rapid with respect to the geological timescale. A second, and related, problem was preservation bias, which might produce the appearance of that a group had become suddenly extinct when in fact it may have persisted, unrecorded or detected, for millions or tens of millions of years after the "event." Third, it was not clear what kind of mechanism could produce rapid and wholesale extinction of diverse groups on a truly global scale; candidates included massive volcanism, changes in sea level, cosmic radiation, and extraterrestrial impacts, but little hard evidence existed to prefer one over another. Fourth, most extinction analysis was based on higher taxa—families, orders, and classes—but species extinction was extremely difficult to reliably estimate, thus making estimates of genuine loss of standing diversity imprecise. And finally, there was the lingering cultural bias against catastrophism that Newell had campaigned against in the 1960s but which still subtly discouraged hypotheses that invoked cataclysmic events.

Quite suddenly, all of this changed. In 1978, while working in Gubbio, Italy, geologist Walter Alvarez and his father, the Nobel-prize-winning physicist Luis Alvarez, detected abnormally high levels of the element iridium in a rock formation was located exactly at the K-T boundary, precisely when the dinosaurs dropped off of the fossil record. This was significant because iridium is an element found only in trace amounts in terrestrial rocks. The immediate and obvious conclusion was that the iridium had been deposited by the impact of an extraterrestrial body, and that this impact had triggered the extinction of the dinosaurs. This result was published in *Science* in 1980, and it touched off immediate excitement and debate in the geological and astronomical communities (Alvarez et al. 1980). At the same time, paleobiologists were using the databases and techniques developed over the previous decade to significantly improve their analyses of the dating and magnitude of mass extinctions. For example, in 1978 Raup published a paper in *Science* in which he used data from Sepkoski's nascent compendium of the marine fossil record to examine the mass extinction at the P-Tr boundary. Using rarefaction analysis to estimate the actual number of species that were killed (Sepkoski's data was for higher taxa), Raup concluded that a staggering 88% of genera and 96% of marine species died out, "suggesting a mass extinction of truly dramatic proportions, possibly approaching (though of course not reaching) complete extinction of marine life" (Raup 1978c, 218).

Amidst the mounting evidence that mass extinctions were genuine biological phenomena that might, in some cases, have been caused by extraterrestrial bolide impacts, a number of interdisciplinary meetings were convened in the early 1980s to debate evidence, discuss hypotheses, and weigh evolutionary implications. One of the earliest and most important of these was a four-day meeting, in 1981, convened jointly by the National Academy of Sciences and the Lunar and Planetary Institute at the Snowbird ski resort in Utah, at which some 55 papers were presented on various aspects of the problem of large-body impacts in the history of life. Writing afterwards for the Current Happenings section of *Paleobiology*, Raup reported on the "awesome array of disciplines" represented at the conference, from physics to geology to oceanography to paleontology. The star of the show was the Alvarez iridium data, which Raup claimed led even those "who are not inclined toward catastrophic extinction" to agree that "there probably was a large body impact at the end of the Cretaceous" (Raup 1982, 1). Other topics were discussed as well, including the frequency, dating, physical evidence, and effects of large impacts, as well as the correspondence between these events and episodes of apparent mass extinction in the fossil record. Raup concluded that the evidence was exciting, though equivocal. "When all the returns are in," he wrote, "we may have little more than an important new tool for stratigraphic correlation . . . *or* we may have totally new paradigms for geological and paleobiological interpretation of the Phanerozoic" (Raup 1982, 3). Privately, to Schopf, Raup expressed slightly less reservation: "To me, this is one really exciting time to be alive in geology and evolutionary biology. We may be witnessing a major revolution—or perhaps a large red herring" (Raup to Schopf, 25 October 1981: Schopf pap. 3, 30).

However, this new and mounting evidence was actively resisted by some, including Schopf, who clearly saw it as a threat to his cherished dream of a stochastic paleontology. Immediately after the Snowbird meeting (which he had attended), Schopf wrote to Raup to complain about the "charlatan-like quality" of the "assertive and dogmatic way" the impact hypothesis was being promoted by "non-paleontologists" (here he was specifically targeting the Alvarez group, which consisted of geologists, physicists, and astronomers, but had no paleontologists). Schopf was convinced that "the extinction of the dinosaurs (or any individual species) is a trivial problem in paleontology," and he was unrepentant about the vocal opposition he raised at the meeting, stating "if this makes me an unpleasant person in that group, or in any group, then so be it." And he closed the letter with an angry handwritten postscript: "It's a *fad*. It will lead, in due course, to a myth . . . a just-so story about how the dinosaurs were killed off" (Schopf to Raup, 23 October 1981: Schopf pap. 3, 30). Apparently, Raup and Schopf had angry words after the meeting, because Raup's reply offered "to amplify some of the nasty remarks about your character that I made on the phone." Aside from objecting to Schopf's behavior at the meeting and the "condescending" tone of the paper Schopf presented (which Raup described as "something one might expect in a personal letter from Simpson"), Raup was troubled by his friend's apparent inability to countenance dissenting viewpoints, observing that Schopf gave the impression he believed "one should make up one's mind about a scientific question and stick to it, come hell or high water." Raup closed with a warning: "If you don't change your stance somewhat I fear that your scientific papers will lose their credibility" (Raup to Schopf, 25 October 1981: Schopf pap. 3, 30).

Schopf's reply was uncharacteristically contrite: he admitted he "was embarrassed by the implications," and acknowledged, "I may be wrong on the Ir/Extinction argument" (Schopf to Raup, 30 October 1981: Schopf pap. 3, 30). However, just a few months later relations between the two had become strained again. This time, the trigger was Schopf's response to Raup's Barcelona paper, which Schopf contended invoked the old idea of "racial senescence" to explain the demise of the trilobites. Pointedly, Schopf inquired whether Raup's "simulation of trilobite extinction—i.e. bad genes—is yet another go at 'racism'—i.e. biological determinism, i.e., innate, in-born propensity for 'inferiority.'" It is clear that Schopf's criticism was closely tied to defending his own "stochastic view" of life, but also to his own personal and political beliefs about the inherent worth of individuals. Schopf went on, "Philosophically, Dave, one either has a stochastic view of life, or, one searches for a silver bullet which grants immortality to a 'good set' of genes. I personally believe that *all* species are out there trying as hard as they can—just like all individuals within a species are" (Schopf to Raup, 14 December 1981: Schopf pap. 8, 31). This sentiment is very similar to the one Schopf communicated to Gould at about the same time while discussing his objections to punctuated equilibria (where he advocated a "luck of the draw" view of life), and in other papers Schopf—who constantly felt the need to justify his own accomplishments to his friends, department colleagues, and the profession at large—often described his own life and career in similar terms (Schopf to Gould, 4 December 1981: Schopf pap. 8, 31).

In any event, Raup insisted that his Barcelona paper had nothing to do with senescence—much less "racism"—and suggested that the basis of Schopf's concern was his interpretation that Raup's "bad genes" explanation was "a serious departure from your world view which says that all clades have indistinguishable extinction and speciation rates." But Raup went on to point out an irony in Schopf's criticism of the dinosaur impact extinction hypothesis, noting that Schopf's preferred mechanism—sea level changes—implied just the kind of deterministic explanation he campaigned against in other contexts.

> You interpret the Maestrichtian extinctions as being caused by "ecologic constriction precipitated by changing of habitat and climate." In other words, the environment changed and the dinosaurs could not cope—so they died. This was bad luck in the sense that they were " . . . in the wrong place at the wrong time . . .". But it can also be seen as bad genes in the sense that had the dinosaurs of the Western Interior had different capabilities they could have resisted extinction. . . . With this as a base, and in the framework of my Barcelona paper, let me submit that your interpretation of dinosaurian extinction involves bad genes and not bad luck (Raup to Schopf, 17 March 1982: Schopf pap. 8, 32).

And one suspects that Raup may have been deliberately baiting Schopf when he closed the letter by "applaud[ing]" Schopf's "clear departure from a strict species-as-particles model."

What this episode shows, though, is the power of Schopf's idée fixe to lead him into self-contradiction, which further marginalized him from the direction paleobiology was headed in the early 1980s. In previous chapters we saw how this same tendency played out in his debate with Gould over punctuated equilibria. Schopf loathed what he saw as "determinism" in the work of others, but was blind to it in his own explanations. But paleobiology was moving beyond the all-or-nothing stance of its early activist phase towards a more pluralistic stance. This was a view expressed by Gould in his 1980 essay where he applauded the combination of "nomothetic" and "idiographic" approaches in the recent studies of Phanerozoic diversity; it was evident in Raup's comment that "some situations are appropriate for a species-as-particles approach (such as my phenetic clock idea) and some are not (such as trilobites and dinosaurs)" (Raup to Schopf, 17 March 1982: Schopf pap. 8, 32). And it would be prominent in further discussions over the magnitude, regularity, and evolutionary effect of mass extinctions that unfolded over the next several years.

Periodicity in Extinction

One of the most striking institutional developments in the growth of paleobiology during the 1970s and 1980s was the meteoric rise to prominence of the paleontology group at the University of Chicago. From the start, Schopf pushed hard to expand the role and visibility of paleontologists within the Geophysical Sciences department at Chicago. Following Ralph Johnson's death in 1976, Schopf strongly advocated hiring Jack Sepkoski for this replacement, arguing that Sepkoski had "the greatest potential for significant contributions to paleontology of any of the younger group whom I know or have heard of" (Schopf to J. V. Smith, 10 January 1977: Schopf pap. 4, 60). A few years later, Schopf's long-frustrated goal of bringing Raup to the department was achieved when, in 1983, Raup was lured away from the Field Museum of Natural History to take over as chair of geophysical sciences at Chicago. It was, therefore, not without justification that Schopf claimed, in a 1983 letter to his dean, to be "chiefly responsible for having built the paleontology group at Chicago into the best in the world" (Schopf to Stuart A. Rice, 2 September 1983: Schopf pap. 5, 11). The final piece of the puzzle came, in the fall of 1985, when David Jablonski and Susan Kidwell were hired away from the University of Arizona. Chicago now had the greatest concentration of quantitatively oriented analytical paleobiologists in the world. This group—the "Chicago School" or "Chicago Mafia"—would set the tone for paleobiological studies of the patterns of evolution and extinction over the next two decades and beyond, and would become the most important center for graduate training in analytical paleobiology in the world.

One of the consequences of Raup's move to Chicago was the resumption of an extremely close working relationship with Sepkoski. This relationship would produce one of paleobiology's signature ideas of the 1980s: the proposal that mass extinctions, through geologic time, exhibited a regular, recurring "periodicity." This idea became closely associated with the Alvarez impact hypothesis, eventually being invoked to support the proposal that a companion star to the earth's sun, nicknamed "Nemesis," orbited the solar system every 26 million years, causing disturbances in the Oort comet cloud that triggered catastrophic impacts like the one that allegedly killed off the dinosaurs. However, the real roots of periodicity lay in the completion of the fossil database Sepkoski had begun as a graduate student under Gould, which he finally refined and published in 1982. Sepkoski and Raup's statistical analysis of this database was one of the most visible and well publicized examples of an approach to rereading the fossil record that came to dominate paleobiology through the 1980s and 1990s.

As I have discussed in the previous chapter, the initial fruit of Sepkoski's huge database project was a list of about 3,500 fossil marine families, published as *A Compendium of Fossil Marine Families* (Sepkoski 1982). Up to this point, Sepkoski had not been particularly interested in mass extinctions, except as they might influence the patterns of diversification he had been studying. It was with this in mind that he attended the 1981 Snowbird meeting on impacts and extinction "as an uninvited skeptic" where, in his own words, "I had my paleontologic worldview changed." From the presentations and discussions at Snowbird, Sepkoski made a series of important realizations: "I learned (or, more aptly, realized) that large-body impacts were frequent over geologic time and therefore, in a sense, were uniformitarian; I learned that the mortal effects of impact could be specified in some detail, something other hypotheses of mass extinction could not do; and I learned that this data set I had been compiling had taught me a lot about mass extinction" (Sepkoski 1994, 141). Prompted by a question from the floor about the frequency and magnitude of mass extinction events following one presentation, Sepkoski stood up and gave an impromptu lecture on a "three-category ranking" of extinctions in the Phanerozoic record. This was Sepkoski's first inkling that his Phanerozoic family data might be useful for something besides diversity studies.

At the same time, Raup had received a pre-publication copy of Sepkoski's *Compendium*, which he immediately began analyzing. This database was basically an alphabetical list of fossil marine families, with information about the first and last stages in which they appeared in the record. It was just now possible to easily transfer portions of the database onto a computer using a simple word processor, and Raup began sorting the database to examine which groups went extinct during which intervals. The result of his initial experimentation was a "remarkable graph" that showed extinction rates plotted against time for the entire Phanerozoic; as Sepkoski later put it, "with fresh vision, Dave had produced a stage-level graph that clearly showed the five big mass extinctions of the Phanerozoic as well as a secular decline in background extinction in between" (Sepkoski 1994, 141). In other words, Raup's analysis showed quantitatively that extinctions during the Phanerozoic were clustered around significant peaks, and were not evenly distributed over time. Sepkoski joined Raup in further refining this analysis, and began to notice that additional, smaller extinction peaks could be resolved from the data in locations that other paleontologists had described as "extinction events" using more traditional, stratigraphic methods. This was exciting, since "the magnitudes of those peaks were in what I had previously assumed to be the noise level of the data" (Sepkoski 1994, 142).

Sepkoski quickly prepared an abstract for a symposium on macroevolution at the upcoming North American Paleontological Convention with the title "Macro Extinction." One of the major conclusions of this presentation was that mass extinctions were more common than had been previously assumed, and could be separated into three classes: major extinctions (such as the P-T event), intermediate extinctions (which accounted for the remaining four of the 'big five' extinctions), and smaller peaks. A common feature in all of these events, though, was that they "were geologically short termed and most appear to have occurred within a single stratigraphic stage" (Sepkoski and Raup 1982, 25–26). Sepkoski further noted that each of these extinction events was characterized by a precipitous drop in standing diversity, followed by a rebound; he concluded that "extinction events seem to represent perturbations of a stable equilibrial system rather than collapse of an unstable, chaotic system" (Sepkoski and Raup 1982, 26). This last point was important, as it seemed to suggest that mass extinction was a regular occurrence in the normal fluctuation of the global biota—in other words, as Sepkoski would put it later, extinctions, "in a sense, were uniformitarian."

Sepkoski and Raup announced their initial findings in a 1982 paper in *Science*, "Mass Extinctions in the Marine Fossil Record." This paper drew explicit attention to the evolutionary implications of mass extinctions, arguing that a number of them "have 'reset' major parts of the evolutionary system during the Phanerozoic," although it also discussed the problem of "rigorously" distinguishing genuine mass extinction events from background statistical noise (Raup and Sepkoski 1982, 1501). Raup and Sepkoski also demonstrated that the graph of major extinction peaks could be superimposed onto Sepkoski's faunal diversity curve, where extinctions appeared as sharp drops in standing diversity, followed by equally sharp rebounds. The paper did note, however, that the analysis did not reveal any insight into "what stresses caused the mass extinctions," nor could their data achieve enough resolution to determine whether the events were "short-lived in human or ecological time" (Raup and Sepkoski 1982, 1502).

This paper was the first to attract wider attention to the Raup-Sepkoski extinction analysis, but it was also the beginning of controversy and debate over the statistical validity of that analysis itself. In a response, published in *Science* the next year, James Quinn questioned Raup and Sepkoski's use of linear regression to identify extinction peaks, and made a number of criticisms of their statistical methodology. In their response, Raup and Sepkoski enlisted prominent University of Chicago statistician Stephen Stigler, and argued that while their "implicit linear model and untransformed data" may have been flawed, the statistical significance of their extinction peaks was still valid. However, they did acknowledge that "higher resolution data and more precise extinction models are needed before any such speculations can be tested definitively" (Quinn et al. 1983). Indeed, Sepkoski later admitted that his early analysis had been imprecise, especially because of the limited resolution of his familial data set; it was for this reason that he began compiling a second, even more massive database on fossil genera, which he would continue to update and refine for the rest of his career (Sepkoski 1994, 143; Sepkoski, Jablonski, and Foote 2002).

Meanwhile, Raup and Sepkoski began playing with graphical output from their analysis of extinction peaks. One of the questions they were investigating was implicitly raised by Quinn's critique, namely, whether extinctions were regularly or randomly distributed in time. Raup described this process as one of visual pattern recognition: "We were looking at the computer output mostly as a series of pictures—looking for a gestalt that could lead us in interesting directions" (Raup 1986b, 114). As Sepkoski recalled, Raup produced "a magnificent strip of CalComp paper that looked like a seismic reflection profile," in which 12 extinction peaks were not only immediately visible above the background noise, but appeared to be regularly spaced in time—one mass extinction roughly every 26 million years (Sepkoski 1994, 142–43). This was a surprising, even shocking result, because there was no known mechanism that could produce mass extinctions with such regularity. However, Raup and Sepkoski were immediately suspicious of the pattern, since there were a variety of reasons why it could have been produced as a statistical artifact. The two spent much of 1983 testing and retesting their data in an attempt to falsify this result; the basic question concerned the likelihood that the extinction points could have coincidentally formed the 26-ma pattern due to random fluctuations. But their tests of statistical significance resisted this attempt: repeatedly, the results came back with over 99% confidence (Raup 1986b, 116–22).

In early 1984, Raup and Sepkoski published a paper "Periodicity of Extinctions in the Geologic Past" in *Proceedings of the National Academy of Sciences* (Raup and Sepkoski 1984). The venue for publication was not accidental; at the time, *PNAS* did not require peer review if one of the paper's authors was a member, so the paper could be published without the usual delay experienced in most journals. Raup later defended this decision, pointing out the necessity for swift publication since their ideas had become widely known informally (Raup 1986b, 127–29). The paper itself was only five pages long, and described the tests to which Raup and Sepkoski had subjected their putative 26-ma pattern. In the first place, the authors defended the admittedly crude resolution of their familial database; here they drew on the evidence of the 1981 "consensus paper" that suggested patterns in familial diversification closely matched species diversity (Raup and Sepkoski 1984, 801). They also ruled out concerns that extinction rates varied with length of the geological stage, which could produce the appearance of greater extinction during longer stages. This conclusion reinforced the view that mass extinctions were genuinely episodic. Finally, they explained that they had eliminated families with low resolution ranges and questionable designations, as well as living families which might be subject to the preservation bias known as "the pull of the Recent," in order to reduce as much dubious data as possible from their analysis.

The normalized analysis of the raw extinction data "expressing the number of families becoming extinct as a percentage of the number present" yielded a now-famous figure depicting the fluctuation of extinction spikes distributed evenly across the last 250 million years (Raup and Sepkoski 1984, 802; fig. 9.1). Despite the striking appearance of periodicity in the graph, Raup and Sepkoski remained cautious: extinction points were plotted at the end of each time stage, which somewhat artificially lumped all extinctions at stage boundaries, possibly exaggerating the extent to which each peak represented a discrete event. On the other hand, they noted that since many stage boundaries were originally estab-

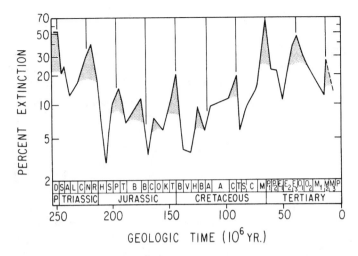

FIGURE 9.1. Raup and Sepkoski's graph of periodicity in mass extinctions. David M. Raup and J. John Sepkoski Jr., "Periodicity of Extinctions in the Geologic Past," *Proceedings of the National Academy of Sciences of the United States of America* 81 (1984): 802.

lished on the basis of faunal turnover in the fossil record, they defended this choice as "the best general inference in the present state of knowledge" (Raup and Sepkoski 1984, 802). The problem, as the authors put it, was that "because the data represent interval estimates of the times of extinction and because no two peaks of extinction can be recognized if they are separated by less than one stage, qualitative impressions may be misleading." The data therefore required a variety of statistical tests to assess the reliability of the periodicity in the time series, and Raup and Sepkoski applied Fourier analysis, nonparametric analysis, and Monte Carlo simulations. They found that the data appeared to exhibit a genuine signal at the 26-ma periodicity, and concluded from these tests that "it seems inescapable that the post-Late Permian extinction record contains a 26-ma periodicity" (Raup and Sepkoski 1984, 804).

The paper concluded with a short discussion of the implications of periodicity in extinction, which were described as "broad and fundamental." There was no known biotic cause that could be invoked to explain such regularly-spaced mass extinctions millions of years apart, but Raup and Sepkoski suggested that an extraterrestrial "forcing agent" might produce such a phenomenon. The authors were ambivalent about the exact mechanics of this process: they noted that the oscillation of our solar system through the galactic arms of the Milky Way had been estimated to follow a periodicity that roughly matched the 26-ma cycle, and cited astronomer Eugene Shoemaker's suggestion that this motion might produce unusual comet activity. They also invoked the Alvarez team's recent impact hypothesis, and pointed out that two of the Alvarez impact events—at the Late Cretaceous and the Late Eocene—matched closely with their extinction peaks (Raup and Sepkoski 1984, 805). While they acknowledged that much further investigation was required, Raup and Sepkoski described the potential implications for evolutionary biology as "profound": "The most obvious is that the evolutionary system is not 'alone' in the sense that it is partially dependent upon external influences more profound than the local and regional environmental changes normally considered." Paleontologists had recognized that major mass extinctions would produce evolutionary "bottlenecks" which would substantially alter the adaptive requirements for the global biota and have a profound impact on the course of subsequent diversity and evolution. The periodicity data suggested that these events were much more common—even regular—in the history of life. Raup and Sepkoski concluded, "Without these perturbations, the general course of macroevolution could have been very different" (Raup and Sepkoski 1984, 805).

Mass Extinctions, Paleobiology, and the Media

On its own, Raup and Sepkoski's analysis would almost certainly have drawn significant attention. However, a series of events transpired in the wake of their announcement that brought almost unprecedented scientific and media attention to paleobiology. This media coverage certainly fanned the flames of scientific controversy as much as it brought wider attention to the science itself. The precipitating event came, in January of 1984, when Marc Davis, Piet Hut, and Richard A. Muller published a paper in *Nature* that suggested an exciting but controversial agent for periodic mass extinctions: cyclical comet showers that struck the earth roughly every 26 million years, to coincide with the periodic cycle of extinctions proposed by Raup and Sepkoski. The most spectacular element of the paper was the proposed mechanism for producing periodic comet showers: a hypothetical "dark companion to the sun, traveling in a moderately eccentric orbit" which the group colorfully named "Nemesis" "after the Greek goddess who relentlessly punishes the excessively rich, proud, and powerful" (Davis, Hut, and Muller 1984, 715).

At about this time, the popular media suddenly discovered periodicity. The first example was an article in the *New York Times* by veteran science journalist John Noble Wilford, who published a piece on the 26-ma periodicity, which was published two months *before* the PNAS article actually appeared (Wilford 1983). Wilford had been in attendance at the Flagstaff meeting where Sepkoski presented his preliminary results, and his article enthusiastically trumpeted the research as a "potentially revolutionary" influence on paleontology. He also drew attention to the "catastrophic" implications of the theory, writing that "the idea of a cyclical pattern to mass extinction calls into question some assumptions about the slow, steady workings of nature and elevates the importance of rare, catastrophic events in the course of life." While he was careful to note that the idea "does not upset any basic theories of nature, including Darwin's," Wilford suggested that "periodic catastrophies have probably altered the pace of evolution, making it more episodic and even more subject to chance." A year later, Wilford published another positive report about the idea and the accompanying Nemesis hypothesis, which he called "a provocative and promising attempt" that "may lead to a new view of mass extinctions and their possibly decisive role in evolution" (Wilford 1984).

However, not all media coverage was this measured, and some was quite sensationalist. In an otherwise generally sound article on periodicity and Nemesis, the Toronto *Globe and Mail* called periodicity potentially "one of the most important non-molecular evolutionary findings of the century," and concluded dramatically, "If this scenario is correct, then our evolutionary descent from primitive structures to reasoning human beings was one long roller-coaster ride through a cosmic shooting gallery" (York 1985). *The Washington Post* also hyped the significance of mass extinction studies, lumping them together with other "current challenges to orthodox Darwinism" that included punctuated equilibria (Rensberger 1984). But an editorial by *Christian Science Monitor* science editor Robert Cowen derided the work as "half-baked theorizing," and urged "a great deal of skepticism" towards the claims, citing criticism by British paleontologist Anthony Hallam of the Raup and Sepkoski analysis (Cowen 1984). And, in a fairly remarkable series, the *Times* published three editorials—two of which were opinion pieces written anonymously by the editorial staff "debunking" Nemesis and periodicity. The first, in 1984, argued, "The new catastrophism may be headed for a crash landing," and called the Raup-Sepkoski analysis "far from rock solid" (Wade 1984). The next year, the editorial staff confidently opined that "the apparent periodicity of extinctions in the fossil record is probably an accidental artifact of the method used to count them," arguing that the pattern was most likely produced by random sorting (*New York Times* 1985b). And, in what Gould later called "one of the most disturbing and blatantly anti-intellectual statements I've read in years," a third editorial chided scientists investigating extraterrestrial causes for mass extinctions with the rebuke, "Astronomers should leave to astrologers the task of seeking the causes of earthly events in the stars" (*New York Times* 1985a).

The weekly newsmagazines also got in on the fun. *Newsweek* first reported on extinction theory in a piece focusing on evolution in demanding environments—"evolutionary crucibles"—and quoted both Sepkoski and Jablonski (Begley and Carey 2 January 1984). This was followed up a few months later by a short item that introduced the Nemesis hypothesis and asserted that "survival of a species would depend as much on luck as fitness" (*Newsweek* 1984). One of the longest popular pieces was a 1985 cover article in *Time* that went even further to suggest that impact theory "could shake the foundations of evolutionary biology and call into question the current concept of natural selection" (*Time* 1985).

Journalists were not alone, however, in feeding the hype surrounding mass extinctions—scientists and journal editors got in on the act as well. In a 1984 column in *Natural History*, Gould examined some of the philosophical consequences of periodicity in extinction, and somewhat facetiously suggested an alternative name for the solar companion alleged to cause the cometary showers. Gould entered a mock plea to Davis, Hut, and Muller: "If Thalia, the goddess of good cheer, smiles upon you and you find the sun's companion star, please do not name it (as you plan) for her colleague Nemesis" (Gould 1984a, 18). Nemesis, Gould pointed out, "is the personification of righteous anger," who punishes those who by their own actions incur the wrath of the gods; this conception of a mechanism for extinction "represents everything that our new view of mass extinction is struggling to replace—predictable, deterministic causes afflicting those who deserve it." Gould argued that Sepkoski and Raup's conclusions suggested that periodic mass extinctions were none of these; extinction randomly affected species regardless of their fitness in their current environment, and the only determinable cause of the evolution of complex life might be the very instability that mass extinctions produce. He went on to suggest an alternative to the name Nemesis: Siva, the Hindu god of destruction, who "does not attack specific targets for cause or for punishment," but whose "placid face records the absolute tranquility and serenity of a neutral process, directed toward no one but responsible for the maintenance and order of our world" (Gould 1984a, 19). Ultimately, Gould argued that the periodicity hypothesis might produce "the greatest revision of cosmology (at least for our little corner of the heavens) since Galileo" (Gould 1984a, 19).

In the mainstream scientific journals there was a pronounced and perhaps cultural split: the coverage in *Science* was overwhelmingly positive, and in *Nature* it was almost completely negative. *Science* news editor and columnist Roger Lewin proved to be a friend to paleobiologists, giving extensive coverage to periodicity and Nemesis in no fewer than seven essays between 1983 and 1988. While he accurately reported the details of the theories, he was unabashed in his enthusiasm: for example, reporting on a 1985 Berlin conference on mass extinctions, he claimed that the meeting signaled "an important turning point in the further development of evolutionary biology," citing a growing "rapprochement" among scientists that potentially "extends, but not replaces, conventional neo-Darwinism" (Lewin 1985b, 151–53). Across the Atlantic, however, *Nature* editor John Maddox took a much dimmer view of the new paleontological theories. In two sharply worded pieces, Maddox cast doubt on Nemesis and periodicity, in each case endorsing alternative explanations presented by scientists in accompanying opinion essays. His first piece, published in early 1984 (alongside Hallam's critique of periodicity), alleged that "the intellectual climate has changed in favour of catastrophism," while his second, a year later, endorsed Antoni Hoffman's critique of the statistical tests for periodicity and called the Raup-Sepkoski work "a little spurious" (Maddox 1985, 627; Maddox 1984, 685). In a fairly unusual step, Lewin directly rebutted this argument in *Science*, where he explicitly singled out both Maddox and Hoffman in a short note titled "Catastrophism Not Yet Dead," and defended the Raup-Sepkoski statistical tests of periodicity (Lewin 1985a, 640).

As far as popular coverage is concerned, the final item worth mentioning was a 1985 essay by Gould in *Discover* magazine titled "All the News That's Fit to Print and Some Opinions That Aren't," which examined and sharply criticized much of the journalistic hype surrounding periodicity and Nemesis. Gould reserved particular disgust for the

editorials in the *New York Times* ("my beloved daily paper from my
youth"), which he chastised for blindly accepting criticisms by Hoffman
and others without properly examining or presenting the actual arguments. "The editorial pages of newspapers," he remarked, "are just not
the places to resolve complex scientific issues." He concluded by asking,
"Are we now to decide intricate factual issues by vote, or by bald mar-
shalling of public opinion?" (Gould Stephen 1985, 90).

What did all this press coverage mean for periodicity, and paleobiology more generally? In his 1986 popular book on the extinction debate,
The Nemesis Affair, Raup reflected on that very question, wondering
whether the attention was a "good or bad influence." On the one hand,
Raup believed that journalists helped spread the scientific ideas more
quickly and to a wider audience than would otherwise have been available. On the other, "a lot of misconceptions" and "numerous mistakes"
were produced by uninformed editors more interested in a splashy story
than an accurate account. Overall, though, he felt that "the positive results probably outweigh the mistakes" (Raup 1986b, 178–79). In a 1986
essay review in *Paleobiology*, Karl Flessa commented that while the
"pizzazz of the impact hypothesis. . . . makes for good headlines, . . . all
this media attention has done more than simply make media stars out of
some of our colleagues. It has made biologists, geochemists, geophysi-
cists, and even astronomers take notice of paleobiology. . . . There is undeniably a wide and increased interest in what the fossil record of extinction has to say" (Flessa 1986, 329).

On the whole, the publicity was probably also good for paleobiology
as well. Although he contributed no scientific input to the theory, Gould
frequently championed the idea, and as periodicity coincided with the
period of Gould's own rise to genuine fame, the public was hearing a
lot more about paleontology. More to the point, periodicity and Nemesis helped raise the profile of paleontology. While the theory continued
to be hotly debated by paleontologists, biologists, and astronomers—
and ultimately failed to achieve lasting acceptance—for a short time it
drew the kind of public attention normally reserved for big discoveries
in physics or medicine. Suddenly, paleontology had something exciting,
even vital, to say to a broad audience. It is not difficult to guess why.
As Gould himself observed, "Something about catastrophism rubs people the wrong way, and rubs a lot harder than most other irreverences"
(Gould 1985a, 91). A theory that potentially revises our understanding

about the basic forces that shape the history of life—and perhaps the role of chance in that history—was exactly the kind of big, sexy idea Gould and other paleobiologists had been campaigning for.

The Meaning of Mass Extinction

From an intellectual standpoint, the most important aspect of extinction theory was that it contributed to a potential revision of evolutionary theory that depended on uniquely paleontological data and analysis. Gould, in particular, quickly folded this research into a broader campaign to argue for a macroevolutionary revision of evolutionary biology, which I will discuss in chapter 10. Right from the start, the central extinction theorists themselves—Raup, Sepkoski, Jablonski—were explicit in emphasizing that it was the *evolutionary* consequences of mass extinctions that were so important and exciting. This had an important disciplinary consequence as well: mass extinctions, more than any other paleobiological subject or theory, was actually taken seriously and accepted as a unique and important contribution to evolutionary theory by biologists. Indeed, when geneticist John Maynard Smith famously welcomed paleontologists back to the "high table" in a 1984 *Nature* essay, he was explicitly recognizing the importance of extinction studies. In describing "the greatest impact that palaeontology is having on the way we see the mechanisms of evolution," Maynard Smith acknowledged, "It is now apparent that massive extinctions, involving many different taxa, have been a repeated feature of evolution." This realization, he argued, asked a profound question of evolutionary biology: If "the impact of these extinctions is not random," did that make "extinctions, then, a necessary motive force of evolution?" If so, then basic biological assumptions might be called into question: "Ecologists tend to see nature as dominated by competition. They would therefore expect the extinction of one species, or group of species, to be caused by competition from another taxon. Most palaeontologists read the fossil record differently. The Dinosaurs, they believe, became extinct for reasons that had little to do with competition from the mammals. Only subsequently did the mammals, which had been around for as long as the Dinosaurs, radiate to fill the empty space. The same general pattern, they think, has held for other major taxonomic replacements." In recognition of this contribution, Maynard Smith closed his essay by concluding, "The palaeontologists have too long been missing from the high table. Welcome back" (Maynard Smith 1984, 402).

Gould wasted little time in turning this admission to his advantage. In early 1985 he published a paper in *Paleobiology*, entitled "The Paradox of the First Tier: An Agenda for Paleobiology," that articulated a revised vision of paleobiology with mass extinction as a central component. The general thrust of his argument was that evolutionary biology is the study of patterns of organic change that operate on three distinct, hierarchical levels, or "tiers." Different disciplinary approaches are more or less suited to investigation of these tiers: the first tier involves "evolutionary events of the ecological moment," while the second "encompasses the evolutionary trends within lineages and clades that occur during millions of years in 'normal' geological time." According to orthodox neo-Darwinism, "no causal differences separate these phenomena, only a quantitative increment in time and effect" (Gould 1985c, 2–3). While Gould acknowledged that much of classic Darwinism applied to both tiers, he maintained that paleobiology offers special insight into the second one, since "if trends occur primarily within a pattern of punctuated equilibrium, and if the differential success of species that must power such trends arises from truly emergent, species-level selection . . . then the Darwinian model of macroevolution as extrapolated selection among organisms must fail and trends have legitimate autonomy" (Gould 1985c, 3). Additionally, Gould announced, we now see that there is a third tier as well: "The most exciting subject in paleobiology today, and the source (I suspect) of its principal agenda for the 1980s, lies in our recent recognition that one of our best-recognized and most puzzling phenomena, mass extinction, is not merely more and quicker of the same, but a third distinct tier with rules and principles of its own" (Gould 1985c, 3). This realization had potentially radical consequences for the synthetic theory of evolution: "The dilemma of the modern synthesis for paleobiology," Gould wrote, "lay in its claim that all theory could be extrapolated from the first tier, thus converting macroevolution from a source of theory to a simple phenomenology—a body of information to document and to render consistent with a theoretical edifice derived elsewhere." This was, in effect, the traditional descriptive role assigned to paleontology. But, as Gould continued, "if the tiers of life create pattern by emergent rules not predictable from processes and activities at lower tiers, then paleobiology adds its special insights without contradicting principles of lower tiers." Mass extinction ensures the independence of the third tier because "it restructures the biosphere either randomly or by rules different from those operating in normal times"— and it belongs to paleobiology because it can *only* be studied using the data and timescale of the fossil record (Gould 1985c, 3). As Gould explained, "New views on mass extinction argue that, whatever happens at the second tier, mass extinctions are sufficiently frequent, intense, and different in impact to undo and reset any pattern that might accumulate during normal times" (Gould 1985c, 7). In other words, processes at the third tier have total causal independence from the first two tiers, irrevocably breaking any patterns that might accumulate as a result of extrapolation of macroevolutionary processes onto higher levels of taxonomic organization and into geologic time. Gould suggested that the frequency, magnitude, and importance of catastrophic mass extinctions may not only foreclose on our cultural hopes for evolutionary progress, but may also invalidate uniformitarianism itself (Gould 1985c, 8).

The "agenda for paleobiology" promised in Gould's title, then, centered on making the study of mass extinction part of the centerpiece of a revived paleontology. Gould emphasized that the study of mass extinctions was not only a source of evidentiary and methodological renewal for paleontology, but also an antidote to paleontologists' historical marginalization in relation to evolutionary biology. By recognizing mass extinction as a third tier of evolutionary process, Gould wrote, "we have reclaimed our proper role among the evolutionary sciences—as generators of testable and irreducible theories about pattern in the history of life, not passive recordkeepers of a phenomenology fully explained at the first tier" (Gould 1985c, 10). In particular, Gould identified a number of "testable predictions" that followed from the Raup-Sepkoski periodicity model, having to do with the ways that vacant adaptive space was re-populated following extinctions, and dealing with the interaction between extraterrestrial and terrestrial forcing agents during periods of catastrophe. "More generally," Gould continued, "the quantitative theory of periodic, catastrophic impact challenges paleobiology to construct a general theory for shifting higher-order taxonomic pattern in life's history." By the end of the essay, Gould's excitement had spilled over into unabashed triumphalism:

We have affirmed the theoretical independence of our discipline by recognizing hierarchies of structure and tiers of time—and we have set the basis

for fruitful junction with neontology at the first tier by studying interactions
between levels and by recognizing the value of analogy when extrapolation
doesn't work and causal unity when it does . . . If the solar companion exists,
paleobiology shall be the impetus for the greatest revision of cosmology, at
least for our corner of the heavens, since Galileo. And if it doesn't, *tant pis.*
Our own star is riding high enough (Gould 1985c, 11).

One sense in which Gould was surely correct was that paleobiology's fortunes did not depend on a particular theory of mass extinction—
Nemesis or periodicity—to remain in the evolutionary spotlight. Even as
the debate dragged on over the plausibility of a companion star and the
validity of Raup and Sepkoski's statistics, it became increasingly clear
that study and appreciation for the evolutionary importance of mass extinctions was going to become a thriving industry. From the very start,
paleobiologists had recognized that the study of mass extinction had the
potential to pose and possibly answer questions of broad evolutionary significance. For example, in 1983 Flessa and Jablonski published a report
of a recent conference in the "Current Happenings" section of *Paleobiology* entitled "Extinction is Here to Stay," in which they reviewed
some of these potential implications. One of the most important lessons
extinction teaches us, they argued, is that "we often lose sight of the fact
that there are always survivors." They pointed out that "paleontologists
have much to learn" not only about why groups become extinct, but also
about "the species and clades that make it through an extinction event,"
and "the prerequisites for survival" (Flessa and Jablonski 1983, 320).

Sepkoski explored some of these questions in a 1985 essay entitled
"Some Implications of Mass Extinctions for the Evolution of Complex
Life," in which he emphasized the dual function of mass extinctions,
both in terminating lineages and existing genes, and also in making
room for new specializations (Sepkoski Jr. 1985, 223). The study of extinction, he argued, is intimately related to the study of diversity, since
"each extinction event is manifested in a sharp drop in familial diversity . . . then followed immediately by rapid rebounds, representing radiations of new lineages into the 'ecospace' cleared by the mass extinctions" (Sepkoski Jr. 1985, 225). Here Sepkoski was framing extinction
as an essentially ecological problem, since mass extinctions presumably
have major environmental consequences that can be studied using the
kinds of ecological models he had used in his own diversity studies. Sepkoski also emphasized that extinctions are not necessarily "bad" for evo-

lution. Mass extinction seems, at least sometimes, to promote rapid cladogenesis (perhaps supporting the punctuational model of speciation); when extinctions are plotted against diversity patterns, it is apparent that sharp drops in diversity are almost immediately followed, in many cases, by even sharper rebounds. Sepkoski concluded, "The evolutionary resilience of the Earth's marine ecosystem suggests that it could have tolerated even more frequent and severe perturbations without collapsing entirely," and he suggested that "even more fundamentally, it may prove that total stability is actually detrimental to the evolution of complex life" (Sepkoski Jr. 1985, 229–30). In fact, from an ecological perspective, mass extinctions may even be a *requirement* for developing complex ecosystems like the earth's: Sepkoski's diversity analysis revealed that per-capita rates of "normal" background extinction among marine families during the Phanerozoic were in a lengthy asymptotic decline towards zero. "In the absence of mass extinction," Sepkoski argued, "this situation would mean that macroevolution would be confined to the slow process of anagenesis and evolutionary novelties would appear rarely at best." He concluded, "Only mass extinction would break this stagnation by clearing ecospace for the radiation of new lineages. Thus, over the long histories of evolutionary systems, perturbations of the biotic environment may not just be tolerated but may actually be essential to ensure the continuation of evolutionary experiment and the further development of complex life" (Sepkoski Jr. 1985, 230). The message was simple: From an evolutionary-ecological perspective, mass extinction is good.

Over the next several years—and indeed up through the present—the study of the evolutionary consequences of mass extinction became one of the most important foci of paleobiological research. Jablonski quickly became one of its most important proponents. Picking up on the work of his University of Chicago colleagues Raup and Sepkoski, he published an important paper in *Science* in 1986 that drew attention to the interaction—or alternation—of the distinct "macroevolutionary regimes" present during extinction under normal and catastrophic conditions. In emphasizing the point that "mass extinctions are not simply intensifications of processes operating during background times," but rather processes which are "qualitatively as well as quantitatively different in their effects," Jablonski argued that it is in fact the alternation of these distinct regimes that may be responsible for "shap[ing] large-scale evolutionary patterns in the history of life" (Jablonski 1986, 129).

Jablonski focused his analysis on shallow-water marine organisms such as bivalves and gastropods, whose evolutionary history is extremely well-represented in the fossil record. What he found was, in effect, that during the distinct regimes, different clade-level traits, such as geographic range, longevity, and species richness, were favored differently. For example, during background regimes, clades with species diversity and broad range for individual species were successful; however, under mass extinction regimes, survival favored groups with maximal geographic range at the clade level were favored, independently of whether those clades contained many species with broad individual species ranges (Jablonski 1986, 129–30). Another way of putting it is to say that while background regimes favor clades that are highly diverse and widely distributed, those same traits are not favored during mass extinctions. From the point of view of evolutionary process, Jablonski saw this as supporting a hierarchical view of macroevolution, since mass extinctions enforce "a selectivity manifested at a different hierarchical level from those characteristic of background times" (Jablonski 1986, 130).

What this meant for the study of macroevolution, Jablonski argued, is that mass extinctions imposed qualitatively different selective requirements on organisms, requiring a modified understanding of the role of selection in evolution. "Many traits of individuals and species that had enhanced the survival and proliferation of species and clades during background times," he maintained, "become ineffective during mass extinctions," while other kinds of traits "that were not closely correlated with survivorship differences become influential." The important consequence is that during mass extinction regimes, "evolution is channeled in directions that could not have been predicted on the basis of patterns that prevailed during background times" (Jablonski 1986, 132). Selectivity during mass extinctions was not completely random, but it would be fair to say that it is random—or "indifferent"—with respect to the selective conditions that applied during normal times. Here Jablonski also invoked Raup's concept of "nonconstructive selectivity," which referred to the fact that since "mass extinctions are the result of environmental stresses so rare as to be beyond the "experience" of the organisms, extinction may be just a matter of the chance susceptibility of the organisms to these rare stresses." As Raup put it, "this model provides for profound effects on evolving systems, but the effects are not constructive in the usual Darwinian sense" (Raup 1986a, 1532). Jablonski also stressed, however, that mass extinctions might have "a larger role than is generally appreciated" for enabling faunal change, by allowing groups who were "previously unimportant but with traits enhancing survivorship during mass extinctions" to radiate and diversify. In a sense, mass extinctions become a mechanism for explaining the kind of dramatic faunal turn-over described in Sepkoski's three-fauna model that is difficult to explain using traditional Darwinian selection.

Overall, Jablonski's model also supported the emerging paleobiological compromise between "randomness" and "determinism" that Gould, among others, had come to favor, and which Schopf had so strongly resisted. This returns us to Raup's question about whether "bad luck" or "bad genes" caused extinction. While genes obviously play a role in determining what traits an organism has, mass extinctions may cause extinction of otherwise successful, well-adapted taxa through "no fault of their own." One might say, to revise Raup's stance in the early 1980s, that for many groups it was "bad luck to have bad genes." This became Raup's own, modified view in the late 1980s and 1990s (Raup interview). But because traditional Darwinian microevolution cannot always predict the outcome under such circumstances, Jablonski regarded his model as an important endorsement of a macroevolutionary hierarchy, "in which selection, drift, and other evolutionary processes operate at a variety of focal levels, with consequences both upward and downward within a genealogical hierarchy from gene to clade" (Jablonski 1986, 132). Ultimately, Jablonski concluded, "the alternation of these macroevolutionary regimes disrupts any smooth extrapolation of microevolutionary or macroevolutionary processes across the sweep of geologic time," meaning that "a complete theory of evolution must incorporate the different sets of random processes that characterize the background and mass extinction regimes" (Jablonski 1986, 133). This "hierarchical view," as I will discuss in the next and final chapter, was at the heart of paleobiology's challenge to the received view of evolutionary theory.

CHAPTER TEN

Toward a New
Macroevolutionary Synthesis

Perhaps no event signaled paleobiology's entry into the mainstream of evolutionary biology more than John Maynard Smith's 1984 *Nature* essay "Palaeontology at the High Table." The occasion for Maynard Smith's piece was Stephen Jay Gould's presentation of the 1984 Tanner Lectures at Claire Hall, Cambridge, on the subject "Challenges to Neo-Darwinism and Their Meaning for a Revised View of Human Consciousness." In his lecture, Gould offered a summary of paleobiological challenges to "the hegemony of Neo-Darwinism," which included critiques of evolutionary determinism, reductionism, and adaptationism (Gould 1985b). Surprisingly, Maynard Smith—one of the staunchest defenders of the modern synthesis—responded quite positively. Beginning his essay by lamenting the relative lack of evolutionary contribution from paleontologists from the 1940s onward, he characterized the typical response of his colleagues in evolutionary biology archly and succinctly: "the attitude of population geneticists to any palaeontologist rash enough to offer a contribution to evolutionary theory has been to tell him to go away and find another fossil, and not to bother the grown-ups" (Maynard Smith 1984, 401). However, Maynard Smith reported that over the past ten years that attitude had changed, thanks in large part to the work being done by paleobiologists like Gould. He concluded the essay with a statement that has become legendary among paleobiologists: "The palaeontologists have too long been missing from the high table. Welcome back" (Maynard Smith 1984, 402).

If the aim of this book were to simplistically document the heroic and triumphant efforts of paleontologists to establish their discipline as

an equal partner in evolutionary biology, then we could end here, with
what appears to be a dramatic affirmation of this success by one of the
leading geneticists and evolutionary theorists of the later 20th century,
published in what is arguably the world's most important general scientific journal. But paleobiology's proponents were not "heroes," and
Maynard Smith's acknowledgement was not quite the unmitigated approval it seems. I will argue that, by the mid-1980s, the "paleobiological
revolution" was effectively completed, albeit in a qualified sense: If the
main work of the revolution—establishing a theoretical and institutional
platform for paleobiology—was coming to completion by the end of the
1970s, then the early to mid-1980s saw a campaign, led by Gould, to establish paleobiology's place at the "high table" of evolutionary theory as
part of a necessary reformulation of the modern evolutionary synthesis.
This "new synthesis" was primarily a theoretical/ideological campaign,
and it would create space for the study of hierarchical features of macroevolution, including species selection, punctuation, and (as we have
seen) mass extinction. It saw the completion of many of the programmatic goals of paleobiological reformers like Schopf, Gould, and Raup,
but it did not necessarily ensure the complete acceptance of paleobiology by the rest of the evolutionary biology community.

The Promise of Paleobiology

It was during this last phase of paleobiology's establishment that Gould
fully emerged as the central spokesman for the discipline and, not coincidentally, as a widely recognized public figure. Increasingly, he focused
his efforts on championing and popularizing the cause of paleobiology
in mainstream journals, popular essays, and books while leaving the details of the science itself to his longtime compatriots and younger colleagues. Both Gould's popular essays and his technical papers during
this period frequently and explicitly invoked the history of science—and
paleontology's place in that history—as part of a self-conscious strategy
to frame the paleobiology of the previous decade as a major revision of
existing evolutionary theory. Always an astute student of history, Gould
realized from the example of Ernst Mayr and other framers of the original modern synthesis that a revolution was never complete until the historical record was rewritten to acknowledge its success. Gould's revisionist campaign may have only been partially successful, but elements

of the program he endorsed—particularly macroevolution and extinction dynamics—were genuinely embraced by the wider biological community, and formed the core of paleobiology's ongoing orientation.

As I discussed briefly in chapter 6, in 1980, at Schopf's invitation, Gould published two remarkable essays in a special issue of *Paleobiology* celebrating past successes and looking ahead at the new decade. These papers—"The Promise of Paleobiology as a Nomothetic, Evolutionary Discipline" and "Is a New and General Theory of Evolution Emerging?"—were written at Tom Schopf's request that Gould "prepare for Paleobiology a review of progress in invertebrate paleontology during the recent past . . . and its prospects for the immediate future" (Gould to Schopf, 4 June 1979: Schopf pap. 5, 14). Apparently, Gould had too much to say on the subject for a single paper, and he described the essays as an "article (in two parts)" (Gould 1980c, 96, n. 1). In effect, he treated his task of assessing the state of the discipline as an opportunity to offer "a partisan statement" about his particular vision for paleobiology as part of a new synthetic view of evolution. These two papers were Gould's first serious public effort at revising the narrative of contemporary evolutionary biology in order to cast paleobiology in the central role he thought it deserved.

The first of the two papers, "The Promise of Paleobiology," was an unreserved criticism of the tentative role paleontologists had taken in asserting the status of their profession. Gould began by giving credit to G. G. Simpson for initiating a new approach to "infusing" paleontology with biological concepts, but lamented that while "Simpson played the prince in 1944 with the publication of *Tempo and Mode in Evolution*, his shoe did not fit comfortably for another 20 years or so." Gould argued that paleontology was limited not by the quality of its data (a now familiar rebuttal to Darwin's dilemma), but rather by the imagination of its practitioners: the employment of a "restrictive methodology . . . still controlling much of our work," and "an approach to biology that condemns us to imitation and exemplification, rather than encouraging novelty." Here Gould returned to a theme he first explored in his 1972 paper with Eldredge, the claim that an adherence to "inductivism" was the central limiting factor on the conceptual growth of paleontology (Gould 1980c, 96–98). According to Gould, "A general theory of paleontology can only emerge from its status as guardian of the record for vast times and effects" when it could document the broad patterns of macroevolutionary change "on a hierarchy of levels," rather than simply acquiring empirical data for its own sake. It would only be then that "paleontol-

ogy may become an equal partner among the evolutionary disciplines"
(Gould 1980c, 98–99).

One of the most interesting features of Gould's critique was the explicitly institutional context in which he framed it. In a conscious reflection on his own efforts to promote paleobiology over the previous decade, he remarked, "All good scholars know that the way a discipline
chooses to define its subfields both reflects its institutional attitudes and,
more importantly, channels thought and effort in particular directions."
Gould noted that in the case of paleobiology Simpson's work provided
the initial impetus for change, but he assigned the credit for achieving
that transformation to the collection *Models in Paleobiology*, as the pur-
poseful "exemplification" of "Simpson's style," and to the journal *Paleobiology* itself as "its conscious embodiment." Not content to rest on his
laurels, however, Gould asserted that paleobiology's roots were "still
shallow," since "much paleobiological work continues in the 'empirical
law' tradition," a path which "will furnish no new or expansive generalizations" (Gould 1980c, 100–101).

Gould then devoted the remainder of his essay to defining and defending the scope of paleontology's unique contribution to evolutionary theory. To begin with, he targeted both the traditionally "passive" stance of
paleontologists towards evolutionary theory, and also the lingering concern that paleontological data is an insufficient source for significant theoretical conclusions. Characterizing the typical approach to evolutionary questions by paleontologists as "fundamentally uncreative," Gould
attacked the "passive stance before evolutionary theory" as "derivative
and dull" (Gould 1980c, 102–4). The conundrum faced by contemporary
paleobiologists, he explained, was adherence to "the principle of extrapolation": the notion, revolutionary when first introduced by Simpson in
the 1940s and 1950s, that "the fossil record demanded no principles beyond what geneticists had discovered by studying modern populations."
While he acknowledged that the extrapolation principle was critical in
establishing a basis for paleontologists' participation in the modern synthesis, Gould argued that a consequence was the "distressing" result
that "Simpson's vision of continuity and extrapolation hardened and has
come, finally, to restrict paleontological theory rather than to enlarge it"
(Gould 1980c, 104–5). The antidote to this dilemma, Gould maintained,
was to accept the proposition that macroevolution has "some reasonable
claim to theoretical independence" from the mechanisms, processes, and
data supporting the synthetic theory of microevolution. Macroevolution,

which Gould defined as "differential success of species," should be "decoupled" from microevolution, and macroevolutionary "units" (species and higher taxa) should be treated as "irreducible entities" in a hierarchical evolutionary theory. This would have the major consequence, he argued, of opening new approaches to the tempo and mode of evolution to paleontological study, including identification and explanation of trends and the investigation of stochastic and nonadaptive mechanisms for evolutionary change that were visible only at the temporal resolution of the fossil record. Here Gould offered his own theory of punctuated equilibria and the principle of "species selection" as examples of such uniquely paleobiological contributions to macroevolutionary theory (Gould 1980c, 106–7).

The final few pages of the essay were an eclectic review of responses to a questionnaire Gould had sent to 20 colleagues in invertebrate paleontology, soliciting nominations for "the 5 subjects of invertebrate paleontology that, in their opinion, had been most fruitful and most disappointing since the Darwinian centennial of 1959" (Gould 1980c, 108). The survey results were a mixed bag: the group sampled was nearly universal in "cheering" the arrival of plate tectonics, the use of new imaging and preparation techniques, and the arrival of new journals, and in "booing" numerical taxonomy, "reductionist" treatment of fossils, and declining systematics practice. But the survey respondents were divided in their appreciation of the areas most representative of the new paleobiology—punctuated equilibria, theoretical morphology, and stochastic modeling—demonstrating the lingering ambivalence within the field towards the very revisions Gould's essay was trying to announce (Gould 1980c, 108–12).

Perhaps for this reason, Gould closed this first essay with a defense of the proposition that paleobiology is a "nomothetic" (law-producing) science, using the stochastic modeling work he had produced with Raup and Schopf as his primary example. There was nothing really new in this section—Gould was rehashing many of the arguments he had made in papers over the past several years, discussed in chapter 7—except for his more explicit acknowledgement of the *balance*, rather than opposition, between historical and stochastic approaches to the fossil record. Here Gould tied his growing interest in hierarchy to his formulation of what a nomothetic paleobiology would look like. While admitting that it would be "insane" to maintain that the "uniqueness" of individual fossil groups did not contribute to explanations of their differential success and failure in local settings, Gould stressed that "the question, rather, is one of scale": "Just as the determinism of a single cast [of dice] yields to the stochastic run at a higher level, so too might the population of species—despite its heterogeneity of taxon and time—be treated with profit as a collection of particles possessing no individual uniqueness" (Gould 1980c, 114). Here Gould was clearly groping towards his later stance on evolutionary contingency, where he would make the case that life's history is controlled both by chance and determinism. It is also, Gould stressed, the proper way to understand the methodological relationship between the historical and nomothetic approaches to the fossil record, where "idiographic factors determine the parameters and then enter as boundary conditions into a nomothetic model" (Gould 1980c, 115). In the end, Gould rather triumphantly proclaimed, the history of life "is the story of millions of species all governed by an overarching body of theory," and he concluded that "evolutionary theory must be the center of a nomothetic paleontology," while "*paleobiology* must be the locus of its construction" (Gould 1980c, 116).

Having defined his personal vision for the future direction of paleobiology, the task of Gould's second essay, "Is a New and General Theory of Evolution Emerging?," was to articulate its place in the wider constellation of evolutionary biology. Here Gould's revisionist historicizing was in full effect, and his explicit context was the modern synthesis, which he set up as the foil against which to oppose his image of a new, more expansive and inclusive formulation of evolutionary theory. Critics quickly noted that Gould's argument required him to define the synthesis rather narrowly, and even unfairly. This is a legitimate point; Gould's characterization of the synthetic theory in this essay was in many ways a straw man. However, Gould's real point was to stress the ways in which paleobiology offered perspectives that were not always recognized by evolutionary biologists outside of paleontology, and he may have consciously exaggerated the rigidity of the synthesis in order to emphasize those contributions. In the end, many biologists—including Maynard Smith—would concede his basic point, even while they disapproved of his argumentative style.

The essay began by defining the version of the synthesis Gould hoped to correct, which he argued rested on "two major premises": that random micromutations are the source of all variability, and that natural selection guides all evolutionary change. In other words, as he put it, "Gradualism, continuity and evolutionary change by the transformation of populations," guided by "selection leading to adaptation" (Gould 1980b, 120–21). Gould acknowledged that exceptions were sometimes allowed—such as random drift—but stressed that nonadaptive mechanisms were nearly always considered "epiphenomena of adaptive genetic changes." This, he argued, did little more than provide biologists with the superficial appearance of pluralism: "Thus, a synthesist could always deny a charge of rigidity by invoking these official exceptions, even though their circumscription, both in frequency and effect, actually guaranteed the hegemony of the two cardinal principles" (Gould 1980b, 120). Gould went on to explain that the *original* version of the synthesis may have indeed been genuinely pluralistic—he cited the first edition of Dobzhansky's *Genetics and the Origin of Species* and Simpson's *Tempo and Mode in Evolution* as examples—but that over succeeding decades the synthesis had "hardened," and even "later editions of the same classics are more rigid in their insistence upon micromutation, gradual transformation and adaptation guided by selection." In support of this view, Gould quoted Mayr's 1963 proclamation that "the proponents of the synthetic theory maintain that all evolution is due to the accumulation of small genetic changes, guided by natural selection, and that transspecific evolution is nothing but an extrapolation and magnification of the events that take place within populations and species" (Mayr 1963, 586, quoted in Gould 1980b, 120).

In invoking a kind of "synthetic originalism," Gould sought to position his corrective not strictly as a rejection of Darwinian evolutionary theory, but rather as a kind of return to its framers' intent. This was a rhetorical strategy Gould maintained up through the end of his career, in which he essentially argued he was attacking the theory in order to save it. It is in this context that Gould penned a line that would generate more controversy than anything he had yet written about punctuation, adaptationism, or macroevolution, and which, fairly or unfairly, would color many critics' views of his radicalism. It is worth reproducing in its full context:

> I well remember how the synthetic theory beguiled me with its unifying power when I was a graduate student in the mid-1960's. Since then I have been watching it slowly unravel as a universal description of evolution. The molecular assault came first, followed quickly by renewed attention to unorthodox theories of speciation and by challenges at the level of macroevolution itself. I have been reluctant to admit it—since beguiling is often forever—

but if Mayr's characterization of the synthetic theory is accurate, then that
theory, as a general proposition, is effectively dead, despite its persistence as
textbook orthodoxy (Gould 1980b, 120).

Gould himself was later quick to point out the qualifying language in
this statement: the words "if," "Mayr's characterization," "general proposition," "effectively dead," and "textbook orthodoxy" all signal his re-
luctance to allege that the majority of his professional colleagues were
professing an empty theory. But in this case, Gould may have been (for
once) too subtle, and for the rest of his career he would be dogged by al-
legations that he had hearalded the demise of Darwinism.

The rest of the essay laid out what, exactly, Gould thought was wrong
with the existing theory, and explained how he proposed to modify
it. He argued that underlying the synthesis was a misplaced "faith in
reductionism"—that is, the idea that natural selection could be extrapolated from the level of the gene up to the higher taxa as the source of
all evolutionary change. "The general alternative," he argued, "is a concept of hierarchy—a world constructed not as a smooth and seamless
continuum," but rather "as a series of ascending levels, each bound to
the one below it in some ways and independent in others." According
to Gould, such reductionism fails because it cannot account for observable patterns, available primarily at the resolution of higher taxa and
over millions of years, of speciation and macroevolution. These patterns,
he maintained, are controlled by forces—some of which are perhaps
stochastic—that cannot be produced merely by mutation and selection.
In the case of speciation, Gould argued that Mayr's proposal of allopatry (plus founder effects and drift) did not truly challenge classic neo-
Darwinism, since (1) all changes were still understood to be adaptive,
and (2) selection is still cumulative and sequential—"it is, if you will,
Darwinism a little faster." Instead, Gould proposed that "many, if not
most" species originate stochastically through a combination of adaptive
and nonadaptive processes, and that a fundamental "discontinuity"—
which he provocatively labeled the "Goldschmidt break"—characterized
the appearance of new species (Gould 1980b, 121–23).

Gould applied the same logic to the problem of macroevolution, for
which he invoked an analogous "Wright break" that explained how "just
as mutation is random with respect to a population, so too might speciation be random with respect to the direction of a macroevolutionary

trend" (Gould 1980b, 126). Here Gould emphasized the idea that macro-evolutionary trends could be produced by differential success among species, rather than by adaptation by individuals, which he now identified as a direct consequence of punctuated equilibria. The result was the "higher form of selection"—species selection or sorting—proposed by Steven Stanley that effectively "decoupled" macroevolution from microevolution. Likewise, the infamous absence of major transitions in the fossil record might indicate "a potential saltational origin for the essential features of key adaptations," produced perhaps by mutation in key regulatory genes that guided ontogenetic development (Gould 1980b, 127).

Overall, Gould argued that his proposals pointed to a "new theory" that would "be rooted in a hierarchical view of nature." He emphasized that this theory did not entail rejecting Darwinian processes, but rather that it would "recognize that they work in characteristically different ways upon the material at different levels." It would also, he argued, "restore to biology a concept of organism," since it recognized that "organisms are not billiard balls, struck in deterministic fashion by the cue of natural selection," but rather that "they influence their own destiny in interesting, complex, and comprehensible ways" (Gould 1980b, 129). Here Gould invoked the explanatory power of evolutionary constraint: he rejected a "strict selectionism" that would imply that adaptation is available to the organism in "all directions," favoring instead a view that "constraints exerted by the developmental integration of organisms themselves" provide "a more powerful influence upon evolutionary directions than the external push [natural selection] itself." And true to his vision of a genuinely hierarchical theory, Gould imagined a primary role for different kinds of constraints operating at different levels of hierarchy, including "punctuational change at all levels . . . essential non-adaption, even in major parts of the phenotype (change in an integrated organism often have effects that reverberate throughout the system); channeling of direction by constraints of history and developmental architecture" (Gould 1980b, 129).

Gould's critique drew a nearly immediate—and critical—response from several prominent evolutionary biologists. The most noteworthy of these rejoinders was published in *Science* the following year, co-authored by one of the major contributors to the modern synthesis, G. Ledyard Stebbins, and one of Dobzhansky's most prominent students, Francisco Ayala. Titled "Is a New Evolutionary Synthesis Necessary?," this paper rejected Gould's characterization of the synthesis as a "straw man," and claimed that existing neo-Darwinism was a far more pluralistic and subtle enterprise than Gould had acknowledged. However, the paper provokes two important observations. The first is that Stebbins and Ayala's reply appeared in a mainstream journal with international circulation, indicating that its authors considered Gould's critique worth the attention of the broader scientific community. This had the consequence of bringing Gould's ideas to a much wider audience than his original essays, and eventually gave him a platform in the same journal for his own response, "Darwinism and the Expansion of Evolutionary Theory," which essentially condensed and recapitulated the arguments he had made in *Paleobiology* (Gould 1982).

Second, and perhaps more revealing, is the fact that despite their critical tone, Stebbins and Ayala conceded that Gould's "appeal to the pluralistic structure of evolutionary theory, to the hierarchical nature of the evolutionary process, and to the distinctive contributions made by the study of macroevolutionary phenomena deserve attention" (Stebbins and Ayala 1981, 967). The authors explicitly challenged Gould's assertion that the modern synthesis downplayed nonselective factors such as drift. They were also highly critical of punctuated equilibria, which they argued conflated morphological change with genetic change (thus assuming that morphological "stasis" indicated genetic stasis) and also mischaracterized the essential compatibility of "instantaneous" change at the geological time scale with "gradual" change at the ecological time scale (Stebbins and Ayala 1981, 968). But they acknowledged that "whether macroevolution occurs according to the punctualist or the gradual model is something to be decided empirically," and conceded that the issue was not whether punctuational change occured, but rather its "relative importance." Stebbins and Ayala were even more receptive to Gould's assertion of the independence of macroevolution from microevolution: while they maintained that no fundamentally new mechanisms should be invoked to "decouple" the two processes, they admitted the existence of "one sense (which epistemologically is most important) in which macroevolution and microevolution are decoupled, namely, in the sense that macroevolution is an autonomous field of study that must develop and test its own theories. In other words, macroevolutionary theories are not reducible (at least in the present state of knowledge and probably in principle) to microevolutionary theories" (Stebbins and Ayala 1981, 970).

Bringing Macroevolution to the Fore:
The 1980 Chicago Meeting

Stebbins and Ayala were not the only biologists beginning to acknowledge the potential independence of macroevolution as a field of study. The efforts of Gould, Eldredge, Stanley, and other paleontologists bore fruit when, in 1980, a large group of prominent paleontologists and biologists met at the Field Museum of Natural History in Chicago for a "Conference on Macroevolution." The legacy of this meeting is somewhat cloudy; among paleobiologists of the right generation, it is widely remembered as an important event. However, it is difficult to pinpoint any particular consensus or agreement that emerged from the conference, which did not produce a volume of published proceedings, and personal recollections of the meeting tend to be selective and contradictory. The most lasting impression seems to be that it was an occasion where paleobiological theories—especially punctuated equilibria and species selection—first began to penetrate the zeitgeist of evolutionary biology and to engender significant controversy and debate. This effect was almost certainly magnified by the fact that several science journalists attended the conference and published independent accounts of the proceedings that served to ramp up the controversial nature of the issues discussed by the participants.

The conference itself was the brainchild of Eldredge and SUNY Stony Brook paleontologist Jeffrey Levinton, who began initial planning in early 1979 for a meeting entitled "Evolution in Geologic Time." Eldredge and Levinton originally proposed the meeting as a Geological Society of America Penrose Conference, citing the "recent attempts to restructure evolutionary theory" made by paleontologists (Eldredge to Frye, 23 January 1979: AMNH Newell 5). The Penrose Conferences were an annual series established by the GSA in 1969 to "provide the opportunity for exchange of current information and exciting ideas pertaining to the science of geology and related fields," and Eldredge proposed the meeting as "a confluence of paleontologists, systematists and geneticists presenting their disparate views on the various theoretical issues currently occupying the time of active evolutionary theorists" (Eldredge to Frye, 23 January 1979: AMNH Newell 5).[1] Norman Newell,

1. Information on the Penrose Conferences was obtained at http://www.geosociety.org/penrose/submitProposal.htm.

Eldredge's former advisor and colleague at the AMNH, was added to the list of organizers in the formal proposal to the GSA, and the proposal argued that given the "large number of paleontologists and neontologists" currently "engaged in a variety of controversies" over "major evolutionary patterns in the fossil record . . . . The time is now ripe for a synthesis of many points of view" (Norman Newell, Jeffrey S. Levinton, and Niles Eldredge, "Proposed Penrose Meeting on Macroevolution": AMNH Newell 5). Newell eventually dropped out during the later stages of organization because of a scheduling conflict that prevented his attendance at the conference itself. The organizers listed five major questions to be addressed at the meeting: (1) whether climate is related to evolutionary patterns and biotic change; (2) whether patterns in the fossil record are tied to specific events or produced stochastically; (3) whether population-level variation is significant to explain macroevolution; (4) whether evolution is gradual or punctuational; and (5) whether evolution is guided by biological constraints, such as development and gene regulation, or solely determined by environmental conditions. The proposal emphasized the interdisciplinarity of the meeting, citing the "remarkable set of interactions that have linked paleobiologists, stratigraphers, population geneticists, evolutionary biologists, systematics theoreticians and biogeographers," and expressed the hope of establishing "a proper perspective for all of these workers" and a "consensus on important issues." It concluded with a list of roughly 30 planned invitees, including paleobiologists such as Gould, Simpson, Stanley, Sepkoski, Raup, Schopf, Valentine, and Van Valen, and biologists including Mayr, Maynard Smith, Michael Rosenzweig, and Walter Fitch (Newell, Levinton, and Eldredge, "Proposed Penrose Meeting": AMNH Newell 5).

No record of the Penrose Committee's decision exists, but the proposal appears to have been rejected by the GSA. Plans for the meeting continued, however, and in early 1980 the organizing group had grown by three members: Joel Cracraft, Russ Lande, and Raup. An important consideration seems to have been the location of the conference, and all three of the new organizers represented Chicago institutions (Cracraft at the University of Illinois at Chicago, Lande at the University of Chicago, and Raup at the Field Museum). In February a letter was mailed out to some 75 potential attendees, and a revised version of the Penrose proposal was submitted to the National Science Foundation. In their solicitation letter the organizers described "a largely informal meeting devoted to the subject of Macroevolution" where scientists from different disciplines could "articulate our objections and disagreements" (Jeffrey S. Levinton to Norman Newell, "Sample Invitation Letter," 18 February 1980: AMNH Newell 5). Rather than inviting attendants to speak on preselected topics, the invitation offered six broad themes and asked invitees to nominate their area of preferred participation; the organizers noted that unfortunately not all attendees would be able to speak due to space and time limitations.

At least one invitee found fault with the organizers' proposal. In a blistering reply, Simpson attacked everything from the format of the meeting, to the selection of the speakers, to the list of the participants and the planned use of NSF funds. He even doubted the very premise of the meeting itself, noting that the program "covers many aspects of evolution that cannot by any definition be considered as macroevolution." He wondered, "How on earth do you expect to cover all these subjects in useful summaries, let alone to educe novel or important ideas on any considerable number of those subjects or perhaps any of them?" Simpson also questioned whether the tight scheduling of the meeting would leave time for genuinely "free and open discussion," and expressed indignation at being asked to submit a proposal for a paper rather than being invited outright to speak. He also noted what he considered an imbalance in the institutional representation of attendees, pointing out that it was likely that "from 40 to 50 percent" would be from the AMNH, Stony Brook, and the University of Chicago, and wondering whether those institutions were "really that predominant in the field." Not surprisingly, Simpson concluded his letter by declining to attend, with typical archness: "I have been to many conferences. Some were great. A few were not and I regretted going. I declined invitations to some others, and did not regret doing so. I suppose that all I am accomplishing here is explaining to myself and to you, all highly valued colleagues, why I am inclined to think that I would not regret declining this one" (Simpson to Cracraft, Eldredge, Lande, Levinton, Newell, and Raup, 5 March 1980: AMNH Newell 5).

While some of Simpson's complaints can be put down to crankiness, his allegation about institutional selectivity is worth considering. Among the list of participants, the AMNH/Columbia, Stony Brook, and University of Chicago did indeed lead the field, with nine, five, and five invitations respectively. However, several institutions were not far behind—UC Berkeley, Harvard, and Michigan each received four—and the total list of 76 invitees represented nearly 40 different institutions, both in the US and abroad. Moreover, the final program did not appear to show a significant bias towards the institutions Simpson mentioned. The format of the meeting called for 12 sessions, each consisting of one 30 minute paper followed by two 10-minute responses and a 25-minute period of open discussion. Of the 12 formal papers, scientists from Berkeley, Harvard, and UC Davis presented two each, and out of the 24 scheduled responses, Chicago and the AMNH represented only five collectively, with zero from Stony Brook ("Program: Conference on Macroevolution, October 16–19, 1980").[2] Furthermore, program itself shows very little bias towards any particular theoretical predisposition, disciplinary orientation, or generational status. As table 10.1 shows, the organizers seem to have been careful to balance the program between older, established figures (such as Stebbins, Olson, Bock, and Wright) and "young Turks" (Stanley, Sepkoski, and Vrba). Nor did the meeting appear overly biased towards paleobiology, despite the participation of Raup and Eldredge on the organizing committee.

The meeting took place at the Field Museum from 16 to 19 October. What actually happened there is the subject of some debate. Although the organizers initially planned to have the proceedings published in a volume afterwards—apparently a representative from Columbia University Press attended the conference—the book never materialized ("Research Conference Proposal to National Science Foundation," Jeffrey S. Levinton, PI [undated spring 1980]: AMNH Newell 5). Personal accounts of the meeting are somewhat in conflict. In a letter to Raup shortly afterwards, Schopf expressed relief that the meeting was over, and confessed doubts about whether any such "meeting of biologists and paleontologists" could be "worthwhile scientifically." Yet he went on to say, "Such a meeting had to occur; if for no other reason, to make the effort for conventional explanation. I liken it to the first day we spent at Woods Hole with Steve and Dan in 1972, trying by induction to settle the Permo-Triassic extinctions. It didn't work" (Schopf to Raup, 3 November 1980: Schopf pap. 8, 30). Schopf's main gripe seems to have been the lack of discussion of stochastic explanations of macroevolution, and he complained about the "believers in the literal reading of the fossil record [who] can play games by inserting special explanations at will," whom he evidently felt dominated the conference. However, he also

2. My copy of the program was provided by Roy Plotnick, who attended the meeting as a graduate student at the University of Chicago.

TABLE 10.1. **A tabulation of the program from the 1980 conference on macroevolution at the Field Museum of Natural History in Chicago.**

Session	Paper	Discussants
Session I. Hypotheses in Macroevolutionary Analysis	Richard Lewontin (Harvard), "Phenotypic, Genotypic, and Adaptive Spaces in Macroevolutionary Differentiation"	Norman Platnick (AMNH) John Maynard Smith (Sussex)
	G. Ledyard Stebbins (UC Davis), "Macroevolution: A Holistic Approach to Evolutionary Theory"	Stanley Salthe (Brooklyn College) Steven M. Stanley (Johns Hopkins)
Session II. The Data of Macroevolution	James S. Farris (SUNY), "When is a pattern not a pattern?"	Walter J. Bock (Columbia) James Hopson (Chicago)
	E. C. Olson (UCLA), "Macroevolution: Patterns and Data from the Fossil Record"	Geerat Vermeij (Maryland) J. John Sepkoski Jr. (Chicago)
Session III. Speciation and Macroevolution	Edward O. Wiley (Kansas), "Process and Prediction in Speciation"	John Endler (Utah) Keith S. Thomson (Yale)
	Francisco J. Ayala (UC Davis), "The Geometry of the Speciation Process"	Edwin Bryant (Houston) Hampton L. Carson (Hawaii)
Session IV. Rates of Speciation and Extinction	David Jablonski (UC Berkeley), "Extinction and Speciation Rates in Late Cretaceous Mollusks: Governing Factors and Macroevolutionary Implications"	Karl J. Niklas (Cornell) Elizabeth S. Vrba (Transvaal Museum)
	Michael Rosenzweig (Arizona), "Filling Niches Quickly and Carefully: Speciation and Extinction Rates in an Ordovician Assemblage"	Thomas J. M. Schopf (Chicago) John A. Wiens (New Mexico)
Session V. Processes of Phenotypic Change	Pedro Alberch (Harvard), "An Epigenetic Approach to Macroevolution"	Stuart Kauffman (Penn) Søren Løvtrup (Umea)
	Russell Lande (Chicago), "The Genetic Basis of Phenotypic Change: From Mutation to Macroevolution"	Stephen Jay Gould (Harvard) William Atchley (Wisconsin)
Plenary Lecture	Sewall Wright (Wisconsin), "Character Change, Speciation and the Higher Taxa"	
Session VI. The Origination of Functional Design	George V. Lauder (Harvard), "Morphology and the Evolution of Functional Design: Alternative Approaches to the Analysis of Form"	Daniel Fisher (Michigan) George McGhee (Rutgers)
	George Oster (UC Berkeley), "Mechanics, Morphogenesis, and Evolution	Bobb Schaeffer (AMNH) Arnold Kluge (Michigan)

expressed hope: "This meeting sort of clears the air, and allows for a direct assault on a full-fledged stochastic accounting of macroevolutionary
phenomena."

Recollections by other participants indicate that little agreement or
consensus was reached between paleontologists and biologists. Valentine recalls that "I didn't think that it was very important . . . it didn't impress me one way or the other very much" (Valentine interview). Others recalled that punctuated equilibria dominated discussions, but again
that little in the way of productive debate took place. Gould recounts
that the meeting was "inspired in good part (though by no means entirely, or even mainly) by the developing debate over punctuated equilibrium" (Gould 2002, 981). Eldredge remembers the conference as his first
meeting with Elizabeth Vrba, with whom he and Gould would later collaborate on hierarchical solutions to macroevolution. He also recalls that
"to a certain extent, it really was a meeting hinged around punc[tuated]
eq[uilibria]," though "it was not set up that way," which made him "ter-
ribly nervous the whole time"—and that as a result, he "had a hard time
enjoying it at all" (Eldredge interview). Stanley also remembers punctuated equilibria being the main subject of debate, but in his interpretation "it was a setup—it was trying to bash into punctuation" (Stanley interview). This allegation may be somewhat surprising, considering that
one of the authors of punctuated equilibria was on the organizing committee, but Stanley felt that Levinton, who was an outspoken critic of the
theory, had orchestrated the unfavorable reception. Whatever the organizers' true intent, most discussion of punctuated equilibria must have
taken place during the response periods and free discussion, since none
of the formal papers explicitly focused on the theory, nor were any of the
main papers presented by obvious supporters of the idea.

It is very likely that whatever discussion of punctuated equilibria actually took place at the meeting was retrospectively magnified by the rather
remarkable press coverage the conference received. It is not unusual for
science journalists to attend large or important scientific gatherings, and
journals often publish short summaries of meetings as "current happenings" items. But the coverage of the 1980 macroevolution meeting seems
somewhat out of proportion with the significance of the event. Looking
back, Gould connects this to broader cultural factors at the time: "The
Chicago meeting escalated to become something of a cultural *cause cé-
lèbre* because, and quite coincidentally, the symposium occurred at
the height of renewed political influence for the creationist movement

in America" (Gould 2002, 981). Indeed, shortly before the conference took place, an article in *Discover* magazine connected the conference with "growing dissent from the prevailing view of Darwinism," and suggested that the meeting would be an opportunity for evolutionary theory itself to be questioned (Gorman 1980, quoted in Gould 2002, 982). The impression that a major reconsideration of evolution was underway was strengthened when, shortly after the meeting, *Newsweek* published an article titled "Is Man a Subtle Accident?," which alleged that "evidence from fossils now points overwhelmingly away from the classical Darwinism which most Americans learned in high school." The article went on to claim that the Chicago meeting demonstrated that "the majority of 160 of the world's top paleontologists, anatomists, evolutionary geneticists and developmental biologists supported some version of this theory of 'punctuated equilibria,'" and characterized the debate as hinging on "the significance of hopeful monsters" that "flout the law of natural selection." While the piece noted that "little common ground" was achieved at the meeting (rather strangely, given the supposed consensus over punctuated equilibria that it proclaimed), it grossly exaggerated the stakes in the conflict by characterizing opponents of punctuated equilibria as "fighting a rearguard action on behalf of Darwinism" (Adler and Carey 3 November 1980). A more reasonable account of the conference appeared at the same time in the *New York Times*, but even that article suggested that "biology's understanding of how evolution works" was "undergoing its broadest and deepest revolution in nearly 50 years," and it connected this "revolution" mainly to the ideas of Eldredge and Gould, whom it quoted extensively (Rensberger 1980).

These popular accounts undoubtedly aggravated many who attended the conference; even Gould acknowledges that "this kind of reporting kindled the understandable wrath of orthodox Darwinians and champions of the Modern Synthesis," although he also maintained that the journalistic assertions made on behalf of punctuated equilibria were "outrageous" and false (Gould 2002, 983). However, what seems to have tipped the scales was an essay that appeared several weeks later in *Science* under the title "Evolutionary Theory under Fire." The article's author was Roger Lewin, the news editor for *Science* and a widely respected science journalist with a PhD in anthropology, whose treatment of the meeting was significantly lengthier and more subtle than those in the popular press. Nonetheless, Lewin's assessment agreed with the *Newsweek* and *Times* articles in two crucial regards. First, he considered the meeting to have been "one of the most important conferences on evolutionary biology for more than 30 years," and second, he characterized "the central question of the Chicago conference" as "whether the mechanisms underlying microevolution can be extrapolated to explain the phenomena of macroevolution," whose "answer can be given as a clear, No" (Lewin 1980, 883).

Lewin's coverage undeniably represented a significant achievement for paleobiologists, whose ideas were now presented in one of the world's most important science journals as possibly producing "a turning point in the history of evolutionary theory" (Lewin 1980, 883). However, while this piece may coincide with the beginning of the arrival of punctuated equilibria—and paleobiology, more generally—as a broader cultural phenomenon, it also marks the renewal and escalation of hostilities between biologists and paleontologists. In many ways, paleobiology had flown under the radar of mainstream biology during the 1960s and 1970s; in achieving the notice that many paleobiologists had been campaigning for, the discipline would now also be subject to much more intense scrutiny and criticism from both outside and within.

Lewin probably did not do paleobiologists many favors by the way he quoted the paleontologists and biologists he interviewed, and this opened him to charges of bias and misinterpretation. For example, he characterized Olson's statement "I take a dim view of the fossil record as a source of data" as an "ancient lament intoned by some," and opposed these "defeatist views" with much lengthier positive quotations from Sepkoski, Stanley, and others (Lewin 1980, 883–84). This kind of editorializing appears throughout the essay: Stebbins's claim that most participants had agreed that macroevolution did not require additional mechanisms was termed "surely . . . a polarized view of what actually transpired"; George Oster's quip to Maynard Smith, "You may have had the wheel, John, but you didn't invent it," was labeled approvingly as "a telling metaphor"; Ayala's apparent concession that "stasis is a real phenomenon" was described as "a generous admission"; and so on (Lewin 1980, 884–87). More troubling to some participants was the impression that Lewin had distorted what actually transpired at the meeting by misquoting or otherwise misrepresenting the views of many in attendance. For instance, Lewin quoted Ayala as admitting, "We would not have predicted stasis from population genetics, but I am now convinced from what the paleontologists say that small changes do not accumulate'" (Lewin 1980, 884). Ayala remains indignantly insistent that his words were misquoted: "How could I say that?" he asked.

"From the purely logical sense, if small changes keep happening, they obviously will accumulate. . . . I don't know where he got that" (Ayala interview).

Ayala's objections were not isolated. Lewin's piece attracted a number of sharp replies, several of which were printed in *Science*. One letter, signed by Douglas Futuyma, Richard Lewontin, and several other participants, decried the "bias in Lewin's account" which "is especially evident in his choice of quotations and in the interpretations he puts on those quotations." It chastised Lewin for presenting "a simplistic caricature of the modern synthesis, render[ing] condescending judgments on its defenders, and repeatedly giv[ing] the last, longer, and stronger word to the advocates of saltationism." While Futuyma and the letter's other authors characterized "the current debate on macroevolution [as] useful and healthy," they rebuked Lewin for "taking it on himself to arbitrate a scientific debate" and for "encourage[ing] widespread misunderstanding of a particular set of ideas and, more generally, of the way science actually works" (Futuyma et al. 1981, 770). Alan Templeton and L. Val Giddings echoed these remarks and quipped, "The macroevolution meeting at Chicago was not so much an historic challenge to evolutionary theory as it was a challenge to the history of evolutionary theory" (Templeton and Giddings 1981, 773). And, invoking a degree of vituperation reminiscent of T. H. Morgan and others, biologist Hampton Carson dismissively sneered, "Forty years ago, the modern followers of Darwin (Fisher, Haldane, Wright, Dobzhansky, and Mayr) stole the evolutionary spotlight from the paleontologists. This conference saw an attempt by a few fossil zealots who are able to charm reporters to regain attention. Most unfortunately, the ideas they used have neither data base nor innovation" (Carson 1981, 773). Nor were biologists the only participants to object to Lewin's sunny portrayal of the meeting. Paleontologist Søren Løvtrup published an editorial in *Systematic Zoology* wherein he mocked the notion that "we are facing a paradigm shift in evolutionary biology" as a result of punctuated equilibria, and quipped, "If the canonization of this theory was the outcome of the Chicago meeting, then the neo-Darwinians have nothing to fear in the nearest future" (Lovtrup 1981, 498–500).[3]

3. For the record, Roger Lewin continues to defend his piece. He explained to me, "My view of the role of a serious journalist is that he/she is not just a mirror of events, reporting precisely what is going on, without insight. For one thing, that would be rather boring.

Species Selection, Sorting, and the Hierarchical Theory of Macroevolution

While the outcome of the macroevolution conference itself is unclear, it is nonetheless unquestionable that it and the media coverage it received attracted attention to paleobiology and raised the profile of paleobiological studies of macroevolution. Recall from chapter 1 that the term "macroevolution" itself was from the very beginning closely linked with paleontology. However, the undeniable watershed for paleobiology came in the 1970s, when punctuated equilibria emerged as the first serious theory to challenge the synthetic view that gradual accumulation of microevolutionary change was sufficient to account for change at higher taxonomic levels. While this idea was certainly controversial, it opened the floodgates for a new wave of macroevolutionary investigation, ultimately leading to the "hierarchical view of life." This view had two essential premises: first, that evolution is understood to operate at different levels of a nested hierarchy (e.g., the gene, the organism, the population, the species, the genus) through different kinds of processes. Second, and more controversially, it held that these processes are often causally independent of one another, leading to what Stanley famously called the "decoupling" of macroevolution from microevolution. It was this hierarchical view that Gould promoted in his two *Paleobiology* essays of 1980, and which would form the basis of his agenda for paleobiology during its emergent phase in the 1980s.

Punctuated equilibria may have opened the door to hierarchical macroevolution, but the first formulation of the theory in 1972 was, as even its authors acknowledge, essentially extrapolationist—that is, it followed the dominant logic of synthetic biology which held that macroevolution requires no mechanisms other than those of mutation, drift, and natural selection (although later articulations of the theory would revise this stance). A crucial modification of this theory came in 1975, when Stanley published a short paper entitled "A Theory of Evolution above the Species Level," which proposed the concept of "species selection" as

---

More important, a serious journalist should have an ear (or eye or nose, whatever metaphor you prefer) for the direction in which a particular intellectual issue is moving. If the journalist is unable to do that, they should be in a different line of business, in my view." Roger Lewin, personal communication, 14 May 2010.

an explicit consequence of punctuation. As Stanley put it, "If most evolutionary change occurs during speciation events and if speciation events are largely random, natural selection, long viewed as the process guiding evolutionary change, cannot play a significant role in determining the overall course of evolution" (Stanley 1975, 648). Eldredge has justifiably noted that this conclusion was implicit in the 1972 paper, but that Stanley made the point more explicit, drawing out the broader implication that "macroevolution is decoupled from microevolution, and we must envision this process governing its course as being analogous to natural selection but operating at a higher level of biological organization. In this higher level process species become analogous to individuals, and speciation replaces reproduction. The random aspects of speciation take the place of mutation." Stanley labeled this process "species selection," and argued that "whereas natural selection operates on individuals within populations . . . *species selection* operates upon species within higher taxa, determining statistical trends" (Stanley 1975, 648).

Stanley's paper, merely five pages long, only hinted at the mechanisms and dynamics of this process. But four years later, in 1979, he expanded his insight to a book-length treatment of the subject, *Macroevolution: Pattern and Process*. Here Stanley set the theory of species selection in the broader context of the development of evolutionary biology, and attempted not only to defend a punctuational view of speciation in which species were discrete, stable units, but also to argue that this position had been endorsed by central synthetic authors such as Wright, Simpson, and Mayr (Stanley 1979, 22–26). Stanley's book went on to defend the principle of species selection at great length, and to reiterate his call for the "decoupling" of macroevolutionary processes from microevolutionary ones using a variety of empirical examples. Nonetheless, Stanley did not *reject* microevolutionary processes like mutation and natural selection; rather, he argued that "it is the fate of adaptations, once established, that is determined by species selection" (Stanley 1979, 191).

While Stanley's proposals were not universally accepted, his book drew considerable attention to the emerging hierarchical view, and should be considered an important contribution to the controversy over punctuated equilibria that surfaced in the 1980 Chicago meeting. The book was widely reviewed, and it drew strong if mixed responses from both paleontologists and biologists. In his review for *Paleobiology*, Russ Lande called it "the most comprehensive text written by a paleontologist

since Simpson's *The Major Features of Evolution*," although he also criticized many of Stanley's empirical claims (Lande 1980, 233). In his review
for *Science*, the biologist David Woodruff called *Macroevolution* "a welcome addition to the literature of evolutionary biology," and lauded paleontology's "exciting rejuvenation," wherein "Stanley and his fellow paleobiologists . . . have introduced some scientific rigor into a traditionally
descriptive field" (Woodruff 1980, 716). Woodruff concluded by noting,
"The new paleobiological view of evolution, based on the application of
theoretical ecology to the fossil record by Gould, Raup, Schopf, Sepkoski, Stanley, Valentine, Van Valen, and others, is intuitively appealing.
. . . Evolutionary biologists can no longer ignore the fossil record on the
ground that it is imperfect. As Stanley shows, it is highly relevant to the
elucidation of Darwin's mystery of mysteries—the origin of species and
the diversification of life" (Woodruff 1980, 717).

Even other, less flattering reviews acknowledged the importance of
the challenge Stanley and others were mounting. For example, Michael
Novacek reported that "Stanley and others have effectively exposed the
problems that arise with uncritical acceptance" of "neodarwinism." Novacek lauded the attempt to integrate macroevolutionary and microevolutionary theories, although he was "left wondering where and how that
bridge would be constructed" (Novacek 1980, 224). Likewise, despite
reservations, Levinton called the book "important . . . for the challenge
it poses to the modern synthesis," and predicted it would set "the stage
for a controversy that will be with us for many years" (Levinton 1980,
433). In fact, one of the only wholly negative reviews came from Van Valen, one of the paleontologists cited in Woodruff's favorable review, who
judged that "no reasonably important argument in the book is persua-
sive," and remarked that "the ploy of piling up one bad argument on another leads not to an accumulation of evidence but to a larger pile of rub-
bish." Still, even Van Valen grudgingly acknowledged that "the thesis of
the book may, with luck, be largely correct," although "that will be for
other work to show" (Van Valen 1980, 620).

At the same time, during the late 1970s, Gould was pursuing his own
exploration of alternative mechanisms for macroevolutionary patterns.
From his early work on Bermudan land snails, Gould became interested
in the role and timing of development in producing evolutionary change.
For example, his study of the genus *Poecilozonites* focused in part on the
evolution of paedomorphs, species whose juvenile characteristics were
retained into adulthood, and this data was used in the original 1972 pa-

per on punctuated equilibria. In 1977, Gould published his first book, *Ontogeny and Phylogeny*, which he described as one long argument for the evolutionary importance of heterochrony, or "changes in the relative time of appearance and rate of development for characters already present in ancestors . . . *changes in developmental timing* that produce *parallels* between the stages of ontogeny and phylogeny" (Gould 1977, 2). Here Gould focused especially on two aspects of paedomorphosis as potential mechanisms for producing rapid speciation: neoteny, the retention of juvenile ancestral characteristics by retardation of somatic development, and progenesis, precocious sexual development through accelerated somatic development. Both of these phenomena, he maintained, could function as causal mechanisms for explaining the rapid morphological changes he claimed were documented in the fossil record. In particular, he argued, "Progenesis is a perfectly orthodox (though unfamiliar) mechanism that permits rapid transition for very little initial genetic input, and that frees morphology to experiment not only by releasing selective control altogether (and abandoning Darwinism), but by directing it elsewhere." Ultimately, he believed, such developmental phenomena "might even save Darwinism from an embarrassing situation usually swept under the rug of orthodoxy—the difficulty of explaining transitions between major groups if the transitions must be gradual and under the continual control of selection upon morphology" (Gould 1977, 340–41).

The basic thrust of Gould's argument was that "orthodox" Darwinism assumes that every adaptation is selected from random variations in response to environmental conditions. Those adaptations that prove successful survive. Gould modified this notion by suggesting that internal constraints, such as developmental "pathways" encoded by genes or guided by behaviors, can prevent certain kinds of modifications in individual organisms—in other words, organisms can resist some of the force of selection. The next phase of this argument was his proposal, coauthored with Richard Lewontin, that not all features of organisms are the result of adaptations. Using an architectural analogy to a kind of leftover space created in the construction of domed arches in medieval cathedrals, Gould and Lewontin argued that certain evolutionary features are "spandrels" (an architectural term), or accidental by-products of some other combination of adaptations. In a controversial turn of phrase, Gould and Lewontin cheekily labeled the traditional "adaptationist" program the "Panglossian paradigm," after the character in Voltaire's *Candide* who resolutely maintains (despite mounting evidence to the contrary) that we live in the best of all possible worlds. The world we live in, Gould stressed repeatedly, is one shaped not by inevitability but by unpredictable contingency (Gould and Lewontin 1979).

When, in 1979, Gould and Lewontin published their now infamous paper "The Spandrels of San Marco and the Panglossian Paradigm: A Critique of the Adaptationist Programme," Gould had therefore already established a consistent interest in examining alternatives to selection as the sole directing force for phyletic evolution. His use of the term "spandrels" merely articulated this commitment in a rather attention-grabbing way, which perhaps explains why critics have usually overlooked its antecedents in his work. The paper begins by pointing out the tendency for evolutionary biologists to ignore constraints on development and to look only for direct adaptations. This was not a new point for Gould; he had been interested in morphological constraints ever since he wrote his allometry review as a graduate student, and his insight was honed by his work on random morphology with Raup and Schopf. While Gould and Lewontin castigated Panglossian adaptationism, they did not deny the importance of adaptation or selection. In fact, their proposal was fairly moderate: they granted the traditional efficacy of both adaptation and selection, but added the caveat that initial modifications often arise not as a direct response to environment but instead as "a secondary utilization of parts present for reasons of architecture, development or history" (Gould and Lewontin 1979, 593). In other words, many features of organisms first appear as nonadaptive consequences of "architectural" constraints on size and body plan, and only later develop adaptive significance. For this reason, Gould and Lewontin urged biologists to eschew "just so stories" that attempt to provide causal explanations for the existence of every morphological feature of an organism. It is worth noting that in offering their theory as a "pluralistic" compromise, Gould and Lewontin repeatedly invoked Darwin's own appreciation for pluralism (they presented their alternative to neo-Darwinian adaptationism "in Darwin's spirit"). The point, as they concluded, was not to establish an anti-Darwinian dogma, but rather to mitigate a tendency in evolutionary biology to focus only on the level of the gene: "A pluralistic view could put organisms, with all their recalcitrant, yet intelligible, complexity, back into evolutionary theory" (Gould and Lewontin 1979, 597).

Gould's next major foray into this territory was a paper he coauthored with Elizabeth Vrba that put a name on the "secondary utilization of parts" he had described with Lewontin: "exaptations," which he and Vrba defined as "such characters, evolved for other usages (or no function at all), and later 'co-opted' for their current role" (Gould and Vrba 1982, 6). The importance of exaptation for Gould's developing view of macroevolution was that the concept further explored the importance of processes which are, at least at some level of the evolutionary hierarchy, random with respect to selection. As Gould and Vrba explained, many features of organisms did not evolve because of any selective value to those organisms, and selection often acts only *after* such features are already present. "At the level of the phenotype," they argued, there exists a "nonaptive pool [which] is an analog of mutation—a source of raw material for further selection." These features "can be regarded as randomly produced with respect to any potential co-option by further regimes of selection," and "*originate* randomly with respect to their effects" (Gould and Vrba 1982, 12). Here Gould and Vrba explicitly cited Gould's study of random simulations of morphology with Raup, and argued for a view of evolution in which much of the "direction" of morphology is produced not by deterministic selection, but by constraints imposed by features whose origins are essentially "random" with respect to fitness at any given time: "In short, the codification of exaptation not only identifies a common flaw in much evolutionary reasoning—the inference of historical genesis from current utility. It also focuses attention upon the neglected but paramount role of nonaptive features in both constraining and facilitating the path of evolution. The argument is not anti-selectionist, and we view this paper as a contribution to Darwinism, not as a skirmish in a nihilistic vendetta" (Gould and Vrba 1982, 13).

Interestingly, at the time the paper was written, Gould did not explicitly connect exaptation with a hierarchical view of macroevolution. When Gould and Vrba initially submitted their manuscript to *Paleobiology*, it was nearly derailed by an extremely negative review (the paper also received ratings of "good" from Robert Raikow and "excellent" from David Hull). In rating the paper "poor" and "unacceptable" for publication, Joel Cracraft complained that the arguments about exaptation "reeked . . . of *non*-hierarchical thinking (i.e., a purely transformationist worldview permeates the ms, contrary statements notwithstanding)." Cracraft went on to argue that, in "restricting these concepts only to within-population variation" Gould and Vrba were committing "the same sin that Gould and Lewontin criticized." He concluded that "all the examples are precisely the kind of inappropriate hierarchical reasoning that Gould has discussed (and criticized) in recent papers" (Cracraft, "Comments on Gould and Vrba MS," 5 May 1981: PS-*P* "1981"). Fortunately for Gould, Schopf, who was in his final year as editor of the journal, thought the submission was "an extremely important paper," and ensured that the manuscript went out to a fourth reviewer, G. C. Williams, whose rating of "excellent" broke the impasse (Schopf to Gould, 1 June 1981: PS-*P* "1981"). In his reply to Schopf, Gould insisted that Cracraft had "misunderstood what we are doing," and maintained, "Our arguments neither support nor deny hierarchical thinking. They just aren't about this subject" (Gould to Schopf, 15 June 1981: PS-*P* "1981").

Whether or not Cracraft misunderstood the exaptation paper, hierarchical thinking became central in discussions of selection and macroevolution over the next several years. In 1984, Vrba and Eldredge published a long paper in *Paleobiology* in which they promoted a view of evolution centered around hierarchy: "Individuals, Hierarchies and Processes: Towards a More Complete Evolutionary Theory." The gist of the paper was that evolutionary theory would be incomplete until it recognized that different levels of taxonomic organization (e.g., from the individual up through the higher taxa) were arranged in a nested hierarchy, and that evolutionary change takes place through the process of "sorting" (e.g., variation at any level, produced either by adaptive or nonadaptive mechanisms) wherein effects at one level permeate and influence both upward and downward in the hierarchy. Species selection, Eldredge and Vrba argued, is one kind of sorting, which operates by differential birth and death rates at the focal level of the species (Vrba and Eldredge 1984, 164). Here they described the entities being selected and sorted in this kind of process as "emergent characters," features that manifest themselves at the hierarchical level where selection is taking place, and which are not reducible to the aggregate characters at lower levels of resolution (Vrba and Eldredge 1984, 165–66). In other words, species (and higher taxa) genuinely behave like individuals, since they have properties that cannot be reduced to the sum of their constituent parts, *and* these properties are heritable. In their conclusion, the authors stressed that their proposal was "not merely a matter of semantics or epistemology," but rather that it "concerns our explanations of the actual processes which are important in evolution," since "hierarchy is an essential feature of life" (Vrba and Eldredge 1984, 168–69).

The debate over sorting and hierarchy carried on for several years and, to a lesser degree, continues today. Cracraft, a defender of hierarchy in principle, criticized the Gould/Vrba/Eldredge assumption that species selection is a genuine process, arguing that at best it was a pattern produced by underlying microevolutionary mechanisms and thus could not constitute an independent theory of macroevolution (Cracraft 1982). This occasioned a rejoinder from Vrba, whose paper "What is Species Selection?" argued that species selection is a genuine causal process because it deals with "heritable, emergent character variation" that causes differences in rates of speciation and extinction among members of a monophyletic group (Vrba 1984, 322–23).

In 1986, Gould and Vrba published what Gould promised would be his final word on the subject—at least for the time being—in an essay in *Paleobiology* titled "The Hierarchical Expansion of Sorting and Selection: Sorting and Selection Cannot Be Equated." In his cover letter to Sepkoski, who had taken over coeditorship (along with Peter Crane) in 1984, Gould wrote, "I absolutely swear to you that you will not again see (at least from me) another theoretical paper on this topic. . . . This one completes the tale started with exaptation" (Gould to Sepkoski, 29 November 1984: PS-*P* "1986"). Like the earlier Eldredge and Vrba manuscript, this paper ran into trouble during the refereeing process. Sepkoski wrote to Gould, "I sent the manuscript to one person I judged to be 'in your camp' and another judged outside. . . . Not surprisingly, the reviews came back split: one rating the manuscript 'excellent,' the other 'fair.'" The negative review had come from Levinton, who faulted the paper for being repetitive, for erecting straw men, and for defining its concepts poorly. Levinton had particularly sharp words for Gould and Vrba's treatment of their opponents, reprimanding the authors for "slander[ing] many individuals as reductionists simply because they focused their efforts at the level of the individual," citing in particular their use of "Maynard Smith as a whipping boy twice, maybe even thrice" (Levinton, "Review of Manuscript by Vrba and Gould," 5 February 1985: PS-*P* "1986"). But even the more favorable review, written by Eldredge, noted that the manuscript was bloated and repetitive and urged significant revision before publication (Eldredge, "Review of 'All Sorting is Not Selection,'" 1 January 1985: PS-*P* "1986"). Rather than decline the manuscript, Sepkoski offered to let Gould revise and resubmit for review by a third, tie-breaking referee (Sepkoski to Gould, 12 March 1985: PS-*P* "1986"). Gould's response was uncharacteristically contrite: "I must have gotten religion or something—because I don't think I have ever truly revised and cut so substantially in the light of comments" (Gould to Sepkoski, 13 August 1985: PS-*P* "1986"). The manuscript was then sent to Raup, whose third review rated it "excellent," and it was published in early 1986.

The paper reiterated many of the proposals already made elsewhere by Gould, Eldredge, and Vrba but, as Gould took pains to explain to Sepkoski, it stressed that the hierarchical view implied distinct "levels of *causation* (with propagating effects across levels)," not simply accumulated effects of causation at a single level (i.e., the population) carried forward (Gould to Sepkoski, 13 August 1985: PS-*P* "1986"). This paper refined the earlier analogy between natural selection and species selection by clarifying that sorting and selection "are quite distinct and should be carefully separated" (Vrba and Gould 1986, 217). Natural selection, Gould and Vrba argued, is one kind of sorting, but it is not the only kind. Rather, sorting is a more general term for the patterns in which individuals—whether organisms or taxa—are differentially distributed in space and time. Some kinds of sorting involve heritable, nonrandom factors (e.g., natural selection), but others do not (genetic drift, stochastic macroevolution). *Selection* is an operative cause only at the focal level at which the individuals are recognized, but other kinds of sorting involve causes propagated upward or downward from one hierarchical level to another. As the authors put it, "A causal process of selection at a focal level is ontologically different from a process of sorting among focal individuals driven by events at a higher or lower level" (Vrba and Gould 1986, 219).

Gould and Vrba presented a number of instances of causation both "upward" and "downward" between different hierarchical levels via sorting (such as the "effect hypothesis" Vrba had proposed in an earlier paper), but one of their clearest examples of nonselective sorting was the constraining function of mass extinctions (Vrba 1980). Major extinction events, caused by environmental and perhaps abiotic changes, have effects that are random with respect to the focal level of the individual organism, since natural selection cannot prepare organisms for such rare and potentially catastrophic events. However, such events do "sort" individual species and higher taxa, and by creating new environmental conditions, these effects are propagated "downward," since these new conditions will become the basis for future selection at the level of the population. Towards the end of the essay, Gould and Vrba related this causal view of hierarchy back to the question of adaptation, arguing, "The scope of exaptation becomes vastly expanded under the hierarchical perspective, because *all upward or downward causation to new characters may lead to exaptation*." For example, mass extinctions "may generate a pervasive realignment of life's diversity" by creating "a largely fortuitous pool of exaptive potential" (Vrba and Gould 1986, 225). Here Gould fairly explicitly reversed his earlier claim that exaptation was unrelated to hierarchy. In this new formulation, exaptation is a paradigm example of the kind of "cross-level causation" that links sorting at one focal level to another.

We can also see the growing importance of historical contingency in Gould's developing view of evolution. Gould and Vrba observed, "If nature's hierarchy is factual, then it was built historically," and hierarchical causality "teach[es] us that we inhabit a world of enormous flexibility and contingency—a world built by irrevocable history" (Vrba and Gould 1986, 226). This is essentially the view Gould would popularize a few years later in his book *Wonderful Life*, and which in later years he would identify as the central pillar of his revision of Darwinian evolutionary theory. Gould's position was shaped by his diverse interests and investments in the paleobiological movement: his involvement with the MBL stochastic simulation project, his critique of adaptationism, his support for empirical studies of patterns in diversity and disparity, his fascination with analysis of the frequency and evolutionary importance of mass extinctions, and his championing of exaptation and hierarchy.

In a revealing personal letter, Gould wrote candidly to Sepkoski about the status of his developing views of the role of hierarchy in evolutionary theory:

It is different from other subjects much discussed in evolutionary theory. Take Sociobiology, for example (forget my known personal opinion about it, and just consider the structure of argument). It represents an extension of familiar Darwinian models to new areas. In this sense, it invites understanding and assent—at least no one will have trouble in comprehension. But hierarchy is different. It forces one to abandon some cherished ways of thinking about nature and to reconstruct it in a different light. People will misunderstand, fail to grasp simple things, state that the claims are trivial, well known or irrelevant—because they fail to grasp the core. Authors will be accused of obfuscation or unclarity—because it really is hard to say in a way that will force people to reassess fundamentally. The papers will be controversial. But I have a hunch—Lord knows it could be wrong in a gloriously big way—that when all the dust settles, the introduction of the hierarchical model will be

seen as a central reformulation that broke through old ways and freed a profession for examining new approaches (Gould to Sepkoski, 13 August 1985:
PS-*P* "1986").

One of the central new approaches Gould was referring to was the
study of patterns of mass extinction, which in 1985 was emerging as one
of the most visible and important areas of paleobiological contribution
to evolutionary biology. Gould believed that evolution was shaped by
two kinds of constraints, both internal and external. As he explained
to Sepkoski, "Hierarchy, as here discussed in its genealogical context,
is an 'internalist' theory about evolutionary dynamics. And we need to
formulate it properly if we are to tackle this internal dynamic with the
other great mover of life's patterns—the externals of geological history,
especially the mass extinctions, that so impact life's history . . . in other
words, all the data that you and your colleagues are treating in such new
and exciting ways. Hierarchy confronts the geological dynamic, and we
will not get it right until we reformulate both sides" (Gould to Sepkoski,
13 August 1985: PS-*P* "1986").

The vision of a purely nomothetic paleontology, in which the history
of life could be reduced to a series of equations, had been long abandoned at this point. But one of the central goals of paleobiology—the solution of Darwin's dilemma by "rereading the fossil record"—was proven
in the 1980s to be a dramatic success. Whatever the lasting impact of
theories like punctuated equilibria and species selection, the work of
paleobiologists on problems related to diversity and extinction over geologic time had shown, fairly definitively, that the fossil record *was* an important source of data about the patterns and processes of evolution.

As I have argued, several distinct strategies for reading this record
were deployed by paleobiologists throughout the discipline's history, and
not all met with equal success. The goal of providing a "literal" reading
of the record (as expressed, for example, in the 1972 Eldredge and Gould
paper) was shown to be naïve. As Raup and others emphasized during
the 1970s, the fossil record is simply too biased to sustain such an optimistic project. Nor was the opposite view, that the fossil record could
be read as an "idealized" document and reduced to a set of "gas laws"
wherein species behaved like particles, shown to be particularly viable.
As Raup, Gould, and others realized in the late 1970s and early 1980s,
there is simply too much particularity, contingency, and even "determinism" in the record to effectively erase historicity from the history of life.

But the third approach—which I have called "generalized" rereading—that involved seeking generalizations and producing models based on a careful tabulation and statistical analysis of the fossil record, was widely accepted as the basis for analytical paleobiology in the 1980s. This approach recognized the imperfections in the document it relied on, but found strategies for minimizing those imperfections so that reliable predictions could be made and tested. It was a mixture, as Gould put it, of "nomothetics and idiographics"; it produced models but also required hard data; it stretched the boundaries of traditional paleontology without entirely rejecting its basis; and it found a way to reconcile the tension between random influences and deterministic outcomes. It is no accident that this approach became the basis for paleobiology's future. The establishment of paleobiology had always depended on the ability for paleontologists to profitably deploy their most important resource: the fossil record itself.

Paleontology at the High Table?

One obvious question with which to conclude this book would be to ask whether paleontology, through the efforts of Steve Gould, Tom Schopf, Dave Raup, Jack Sepkoski, Niles Eldredge, Steven Stanley, and a host of others, genuinely succeeded in finding a place at the "high table" of evolutionary biology. We might resolve this, for example, by counting the number of citations to paleobiological papers in biology journals from the mid-1980s to the present, or by surveying geneticists and population biologists about their attitudes towards paleobiology. Indeed, some authors have attempted such a tally—both Michael Ruse and Michael Shermer found, for instance, that Gould's works had relatively little impact on citation indices past their first, heady days of introduction (Shermer 2002; Ruse 1999b). Other authors, mostly paleontologists, have lamented the extent to which the promise of the high table remains unfulfilled, or conversely have argued that it has been or will soon be achieved (Prothero 2009; McNamara 2002).

We might first reflect, however, on whether there has ever, in truth, been any single evolutionary high table, and on whether the success of paleobiology genuinely depended on the opinion of scientists outside the discipline of paleontology. Given the fruitful interactions we have examined—in particular, the cross-fertilization of paleobiology and theoretical ecology—there is little doubt that, between the 1960s and the 1990s, paleontologists had increasing and profitable contact with biological disciplines. But outside recognition was only one—and not necessarily the most important—of the goals of the paleobiological movement. From its earliest days in the hands of Othenio Abel and G. G. Simpson through its "golden age" in the 1950s and 1960s (to invoke Norman Newell's label) to its "revolutionary phase," the most central goal of those who self-identified as paleobiologists *was to change paleontology*. The major result of the paleobiological revolution, then, was not that it secured a place at anybody's table by the standards of any other discipline, but that it quite legitimately produced a subdiscipline of its own—paleobiology.

This might seem like a trivial or even circular conclusion: that "the success of paleobiology was that it established paleobiology." But if it is to be evaluated at all, the success of paleobiology should be measured first and foremost by the standards its proponents set for themselves. Paleobiology was not any single agenda (e.g., the stochastic view, the taxic approach, the hierarchical model, etc.), but was rather the collective enterprise of a diverse group of actors with overlapping motivations who shared programmatic aims. The goals of paleobiology, expressed variously but consistently from Simpson to Newell to Schopf to Gould, included: (1) making paleontology more theoretical and less descriptive; (2) introducing models and quantitative analysis into paleontological methodology; (3) importing ideas and techniques from other disciplines (especially biology) into paleontology; (4) establishing institutional foundations—journals, research groups, department appointments, funding opportunities, etc.—for paleobiology; and (5) emphasizing the evolutionary implications of the fossil record. External recognition was a hoped for *consequence* of achieving these goals, not the sole aim in itself (nor did recognition necessarily mean or require acceptance or consensus). I would argue that the evidence presented in this book clearly demonstrates that most of these goals were met. This book is about *how* this happened, not whether it happened, and that story tells us something more broadly about how scientists self-consciously construct and revise their own disciplinary identities. And while they may not all be original or surprising, I think several of those conclusions are especially important:

1. Revolutions in science. Revolutions *do* happen in science, and they are important, but they are processes, not events, and it is important how the term is defined. When describing the "revolution" in paleobiology, I have understood the term differently than some historians of science in the past. That is why I have more often used the term "revolutionary"; I have been more interested in the *act* of promoting radical change in a discipline than in the outcome of some successful paradigm shift. In this sense, what went on in paleobiology during the 1970s was not a revolution in the sense described by Thomas Kuhn in *The Structure of Scientific Revolutions*. There was no single dominant, overarching para-

digm, challenged by mounting anomalous evidence, and no new "world-view" that successfully replaced it. On the other hand, as Jan Golinski has pointed out, one of the most useful elements of Kuhn's analysis was his insistence that controversies between so-called paradigms involve rival subcultures (Golinski 2005, 20). Paleobiology is just such a subculture. The history of paleobiology is more a story about a conflict between rival networks of social organization and practice than a battle between competing logical claims about how the world "is." Paleobiologists reacted against a set of methodological prescriptions, and tried to find new institutional space for the kind of *practice* they favored. The paleobiological "revolution" was more like a political contest in which one group perceives itself to be disenfranchised and agitates for greater representation in government than a contest of lofty ideas. The 1970s was a period of revolution in paleontology because paleobiologists saw themselves, and described what they were doing, as revolutionary. To put it in terms of recent constructivist theory of knowledge, it is an actors' category, not a historical label.

2. The uses of history. One complicating factor in reconstructing the history of paleobiology was some paleobiologists' awareness of the potential historical significance of their own work. In some of the more strident examples, paleobiologists explicitly presented themselves as challenging a dominant paradigm and introducing a new worldview. That, in essence, that was the argument Eldredge and Gould made in 1972 when they presented punctuated equilibria as a challenge to the dominant paradigm of "uniformitarianism" (Eldredge and Gould 1972, 86). Eldredge and Gould were well aware of Kuhn's theory about scientific revolutions, which they cited directly, and they very self-consciously presented their theory as an example of paradigm change. Kuhnian revolutions were a hot topic in the early 1970s and Gould, particularly, latched onto the idea as a tool for agitating for his particular agenda. Gould inserted a discussion of paradigms into the second draft of the 1972 paper in order to use history—not just the facts of history, but the theory of history—to bolster a rhetorical claim for the purposes of methodological and institutional transformation.

Over his career Gould would mobilize history in the service of promoting his agenda in more and more sophisticated ways, ultimately using it to advance his scientific agenda on topics including evolutionary development (in *Ontogeny and Phylogeny*), human biological difference (*The Mismeasure of Man*), and the "hardening" of the modern synthesis

(in many of his essays during the 1980s assessing paleobiology's significance). The first several hundred pages of his magnum opus, *The Structure of Evolutionary Theory*, is a history of evolutionary thought: Gould
learned from Ernst Mayr's promotion of the modern synthesis that the
revolution is not complete until the history has been rewritten, and he
applied that lesson wherever he could. Gould was not alone in this effort:
Eldredge revised the history of punctuated equilibria in several popular books during the 1980s, Raup published an account of the extinction
controversy while the debate was still in full swing, and even Schopf edited a volume of Paleontological Society presidential addresses in which
his own long introduction attempted to highlight the development of a
"paleobiological" trend in modern paleontology (Eldredge 1985a, 1985b,
1995; Raup 1986b; Schopf 1980). By the mid-1980s, paleobiologists could
point to a kind of official history of their struggle: Darwin had deni-
grated the fossil record, Simpson came along to try to save paleontology,
Simpson's more radical ideas were pushed aside by the hardening modern synthesis, and paleobiology emerged to rectify this. Whether or not
that narrative was accurate, paleobiologists saw this history as an important element in the self-definition of their identity and common cause.

Another interesting way in which history was mobilized by paleobiologists had to do with evolutionary causality and the interpretation of
patterns—it was central to the way paleobiologists reread the fossil record. Paleontology has always been considered a "historical" science,
and indeed the fossil record is itself a kind of narrative. But during the
1970s some paleobiologists attempted to frame a different kind of historical account, free of what they saw as the unnecessary particularity
and determinism of individual events and causes. This effort, the MBL
model, was the source of some very fruitful analysis, but ultimately it
failed to take hold. The overwhelming conclusion reached by paleobiologists like Raup and Gould was that historicity cannot be removed from
history: individual events *do* matter, even if the aggregate patterns are
more interesting. Here was the ultimate success of the "nomothetic" approach to paleontology. According to its most broadly accepted definition (though not necessarily the one Gould employed), "idiographic" science looks for causes or descriptions of individual events (e.g., why did
this particular species become extinct at this time?). But it is not the case
that "nomothetic" explanations must be free from empirical grounding.
Jack Sepkoski's faunal analysis was genuinely nomothetic because it attempted to fit the actual fossil data into a generalized logistic model for

the history of diversification. But it was also empirical, since it attempted to make use of the best available fossil data to properly fit the curves. However, Sepkoski never tried to ascribe specific causes to particular events, as properly idiographic science does. Rather, the empirical data provided realistic values for modeling parameters, and the model itself was (ostensibly) generalizable to any time or group.

3. Self-promotion and fashioning. One place where we can take a lesson from Kuhn is in his early insistence that disciplinary identity is located in smaller social groups of scientists invested in particular kinds of practices. Paleobiologists were never a very large segment of the overall paleontological community, and the activist group in paleobiology's revolutionary phase was really just a handful of people. It is less striking that this small cabal was successful in promoting many of its aims than that its members were, as revealed not just in published works, but especially in private correspondence, so *self-conscious* about what they were doing. And while individual differences and disagreements emerged, it is also remarkable how clear and consistent the self-articulated goals of paleobiology were. These began to emerge with Simpson and Newell in the 1950s and 1960s, eventually becoming a kind of official platform during the 1970s. Nothing did more to codify this agenda than the founding of *Paleobiology*, which right from the start explicitly incorporated the basic themes—theoretical independence, relevance to biology, nomothetics instead of idiographics—in its promotional materials and editorial policies. The conscious self-fashioning of paleobiology also involved constructing a new image of who a paleontologist was. Previously, paleontology had been an almost exclusively field-oriented science, and its practitioners were collectors and describers. Graduate training involved learning stratigraphy and systematics, and a dissertation was almost always based on extensive field study of a particular taxonomic group, stratigraphic range, or paleoenvironment. The central proponents of paleobiology in the 1970s helped introduce a kind of new-model paleontologist who was not tied to the field and who did not focus on one empirical slice of the fossil record. Sepkoski and Raup pioneered a new approach in which paleobiology was pursued in an office, in front of a computer, with nary a specimen to be seen. This new image of the paleontologist has hardly been embraced by the entire profession, but it shows that paleobiologists were interested in redefining not just the subject matter of their discipline, but the very identity of the paleontologist.

4. Institutions, ideas, and technology. Another point the story of

paleobiology reinforces is the dependence of ideas on institutions that support them. The history of paleobiology offers us insight into some of the actual strategies deployed by scientists to institutionalize their ideas, and it emphasizes the practical requirements of discipline formation. Paleobiology is often thought of in terms of its major thinkers (Gould, Raup, Sepkoski, Stanley) and ideas (punctuated equilibria, species selection/hierarchy, mass extinction theory). But in the history we have examined, we see that institutional framework and agenda often comes *before* theories and ideas. That is, the desire for paleobiological reform came before the existence of any particular theoretical superstructure around which to organize the discipline. The theory of punctuated equilibria got so much attention, I would argue, not because it was widely accepted or particularly well liked even among paleontologists, but because of what it stood for. It was a theory from paleontology, using a uniquely paleobiological perspective, that purported to say something important about evolution. The opportunistic nature of theory selection is evident in the way Gould, who had no real experience with or research on the problem of speciation, immediately attached himself to the idea and folded it into his emerging agenda. The same could be said for his endorsements of the MBL model, periodicity, and exaptation. In each case the particular theory happened to coincide with institutional goals (space within the discipline, attention from biologists, promotion of friends and students), not the other way around.

Different individuals also played different roles in the institutional and conceptual development of paleobiology. Raup's contribution, for example, was heavily slanted towards conceptual innovation. He certainly had institutional roles, such as serving on the Paleontological Society council that recommended founding *Paleobiology*, coauthoring the textbook *Principles of Paleontology*, serving as PS president, co-organizing the Chicago macroevolution conference, and chairing the University of Chicago geophysical science department. But his major importance was as an innovator of new ideas such as theoretical morphology, the MBL model, and mass extinctions, and his primary commitment seems to have been to the generation of ideas. Schopf, on the other hand, was what the historian Joe Cain has called a "community architect" (*sensu* Cain 1994). As chapter 6 has shown, Schopf was the moving force behind the establishment of the journal *Paleobiology*, which was probably the single most important and visible ingredient in the success of paleobiology as a discipline. He also organized and edited the *Models*

in Paleobiology collection, and he had a major role in building the paleontology group at Chicago into an international powerhouse. Schopf also
had theoretical commitments, but those were secondary to his institutional accomplishments, and despite his passionate advocacy for his "stochastic view," he acknowledged that his primary role was as an institution
builder. Gould had his own unique role as well; he was the propagandizer,
the promoter, the silver-tongued pamphleteer of the paleobiological revolution. Of course, as in the other cases, Gould's role cannot be entirely
reduced to a single function. He was responsible for introducing a number of important ideas, and I hope this book will put to rest the erroneous
notion that he did not do any important science. But in the same way that
paleobiology needed Raup and Schopf, it needed Gould. His controver-
sialist attitude, popular appeal, and willingness to frame the "big questions" undeniably accounted for much of the movement's success.

The growth of paleobiology also relied heavily on technology. Many
of the research questions asked by paleontologists as far back as the 19th
century simply could not be answered before the advent of reasonably
powerful digital computers, and it is no accident that the rise of paleobiology to prominence closely mirrored the development of the computer
industry. By the early 1980s, analytical paleobiology was almost entirely
dependent on technology, and sophisticated computer-assisted statistical
analysis had become a standard part of the graduate training in departments where paleobiology thrived. This was more than merely opportunism or good fortune, however. From the start, key figures in paleobiology's revolutionary phase—Raup and Sepkoski, primarily—recognized
the promise of computers and actively developed a research program that
could take advantage of them, in many cases *before* the technology was
quite ready for the applications. In the 1970s, paleobiologists' computer
skills were essentially self-taught, and the programs and routines they
developed had to be written from scratch. By the early 1980s, however,
Raup and Sepkoski had widely shared their expertise, and researchers
could now use prepackaged programs to analyze data without knowing
much about how those programs worked. In this sense, computers were
"black-boxed" during the establishment of paleobiology.

In addition to being tools for analysis, computers—or, more precisely,
paper punch cards and later magnetic tape and floppy disks—became
the storage medium for the massive databases that paleobiologists like
Sepkoski were building. These databases were, in effect, collective en-
terprises: Sepkoski updated his *Compendium* annually with reports and

corrections received from colleagues, and the electronic version quickly made its way around the profession. This facilitated sharing and testing of the database, which served to reinforce its status and importance. Data sets also served as a kind of collective communal glue: once the databases became electronic, colleagues widely separated by geography could more easily share and collaborate, and once the Internet came into wide use in the 1990s, the databases became freely accessible for all. In short, computers and computerized data were more than just tools for paleobiologists; they were objects that facilitated and reinforced social interactions and connections between paleontologists, and consequently reinforced a particular kind of paleobiology.

One final question to ask is a historical and conceptual one. A central motivation behind paleobiological reform was the desire to challenge the orthodoxy of the modern synthesis. But was (and is) paleobiology genuinely non-Darwinian? It is not obvious that challenging the synthetic interpretation of evolution is the same as challenging Darwin. After all, the synthesis has significantly expanded on and modified the content of Darwin's evolutionary theory, and much has taken place in biology and related disciplines since 1859 that Darwin himself could never have predicted. But the synthesists always argued that their theory nonetheless stayed true to the logic of Darwinism—and in particular to the centrality of selection as the causal mechanism for evolutionary change. Did paleobiology depart from this logic?

There is certainly an argument to be made that at times, some paleobiologists flirted with explanations that, at the very least, stretched the traditional logic of Darwinism. One obvious case was the MBL simulation project, which in its most extreme interpretations radically downplayed the causal importance of selection in patterns of phylogeny. Another was Gould's critique of "adaptationism," which argued that not all features of organisms that are eventually recognized as adaptations arise directly in response to the functional needs of organisms. More broadly, the hierarchical view of macroevolution, which argues that selection operates in different ways on different levels, and that macroevolution is decoupled from microevolution, is sometimes interpreted as an attack on the central logic of Darwinism. It was on this basis, after all, that Gould once claimed that the modern synthesis "as a general proposition, is effectively dead." But one of Gould's main points was that the synthesists did not have a monopoly on interpreting Darwinism. In later writings, Gould argued for what he sometimes called a modified or "higher Darwinism," which preserved Darwin's core principle of descent with modification by natural selection while at the same time opening space for pluralistic interpretations of the processes and patterns of life's history. In his final, expansive treatment of evolutionary theory in 2002, Gould explained that his aim was "to expand and alter the premises of Darwinism, in order to build an enlarged and distinctive evolutionary theory that, while remaining in the tradition, and under the logic, of Darwinian argument, can also explain a wide range of macroevolutionary phenomena lying outside the explanatory power of extrapolated modes and mechanisms of microevolution" (Gould 2002, 1339).

For my own part, I conclude that the central macroevolutionary revision of evolutionary theory produced by paleobiologists in the 1970s and 1980s was not anti-Darwinian. Certain theoretical alternatives—such as Schopf's stochastic paleontology, and Raup's "bad luck not bad genes" explanation of extinction—came very close to crossing the line. But those approaches were eventually abandoned and were not a significant part of the macroevolutionary synthesis of the 1980s. In the end, as I have stressed in the last several chapters of the book, paleobiologists settled on an approach to reading the history of life that was "nomothetic," but which also took seriously the empirical fossil record and the significance of individual, contingent historical events. It should be emphasized that despite the radical language used by Gould and others, much of the most important work of recent paleobiology has confirmed Darwin's central insights and expanded our understanding of evolution along fairly traditional Darwinian lines. For example, one of Darwin's greatest anxieties concerned the absence of a fossil record earlier than the Cambrian (some 550 million years ago), when complex life seemed to simply appear, as if from nothing. Paleobiology has definitively put that fear to rest by exposing a remarkable fossil record of evolution back to the earliest microbial stages of life, extending the fossil record some seven times—or almost two billion years—further than had previously been known (Schopf 2009). Paleobiology has also helped to settle questions about how the earliest complex life diversified and "exploded" in the Cambrian oceans to produce the lineages that are antecedent to the modern phyla (Gould 1989), and time and again it has confirmed—in perfect accordance with Darwin's expectations—the fossil evidence of the major transitions in the evolutionary history of life (Shubin 2008).

Indeed, as Gould ultimately argued, "Answers must be sought in the particular and contingent prior histories of individual lineages, and not in general laws of nature that must affect all taxa in a coordinated and identical way" (Gould 2002, 1335). Paleobiology is no more a repudiation of Darwinism than is molecular genetics, or evo-devo, or any of the other countless developments in evolutionary biology that have come about since 1859. Modern evolutionary theory is remarkable both for the continuing validity of Darwin's original insights, and also for its adaptability to 150 years of continued investigation of evolutionary phenomena. Like all scientific theories, evolutionary theory adapts and evolves, and paleobiology has contributed to this ongoing process—and to a robust and pluralistic definition of what it means to be Darwinian.

Acknowledgments

A great many people provided advice, assistance, and support over the several years it took me to complete this project. First I would like to thank the scientists who graciously agreed to submit to taped interviews with me, and whose thoughts, recollections, and interpretations of what went on in paleobiology over the last 50 years have made this a richer story: Walter Alvarez, Francsico Ayala, Simon Conway Morris, Niles Eldredge, Adrian Friday, Tony Hallam, David Jablonski, Ken Joysey, Dave Raup, Martin Rudwick, Bill Schopf, Steve Stanley, Jim Valentine, and Leigh Van Valen. A number of other people discussed the project with me, read drafts, and/or provided materials. I am indebted to them for their assistance. Thanks to Gar Allen, Warren Allmon, Richard Bambach, Mark Borrello, Art Boucot, Faik Bouhrik, Paul Brinkman, Joe Cain, Fritz Davis, Dave Fastovsky, Michael Foote, Monica Gisolfi, Todd Grantham, Chris Haufe, Carla Hubbard, John Huss, Christine Janis, Arnie Miller, Isaac Miller, Kevin Padian, Roy Plotnick, Patricia Princehouse, Betty Smocovitis, and Derek Turner. While the intellectual influences for this project have been many and varied, this book could not have been written without the towering achievement of Martin Rudwick's scholarship on the history of paleontology.

Several people deserve special recognition for going above and beyond the call of duty in helping me with this project. Dave Raup not only hosted me for two days of fascinating conversation about paleontology, but generously and cheerfully read and commented on drafts of every chapter in this book. He has also been supremely patient with every follow-up e-mail I have sent him, and has been an invaluable source of information. Niles Eldredge has also gamely responded to all of my irritating historian's questions, and has provided me with personal copies of drafts, correspondence, and other historical documents not available in any archive. In addition to being my supportive stepmother, Christine Janis read the entire manuscript and shared the insights she gained as a paleontologist who experienced and participated in many of the developments described in this book. I am also grateful to the staff at the University of Chicago Press, who have nurtured this project from initial idea to finished product. Special thanks go to my editor, Christie Henry, who has been a staunch advocate and constant source of encouragement, and to my manuscript editor, Renaldo Migaldi, whose careful reading of the manuscript has unquestionably made this a better book.

Michael Ruse has had a formative and consistently supportive role throughout this project. It was he who first suggested that I might try my hand at writing a history of paleobiology. At the time, in 2004, I had no idea where to begin. Without Michael's experience, enthusiasm, and good humor, this book would not exist. Michael has been a constant anchor over the past six years, providing advice and intellectual encouragement whenever needed. On several occasions he has also provided a warm and constructive forum for discussing the ideas in this book in his informal workshops at Florida State University. I am grateful also to the participants at those meetings—students and faculty—for patiently listening to me talk about paleobiology as my ideas took shape. The entire Ruse family (Lizzie, Oliver, Emily, and Edward) has extended tremendous hospitality to me on these occasions, and those days of intense discussion followed by evenings of cookouts and never-ending beer have been among the happiest of my career.

I have also benefited from the assistance and support of a number of institutions. My thanks go especially to the staffs at the Smithsonian Institution Archive, the American Philosophical Society Library, the University of Illinois Department of Special Collections, and the Departments of Vertebrate Paleontology and Invertebrates at the American Museum of Natural History. While I was at the AMNH researching the careers of G. G. Simpson and Norman Newell, I had the chance to meet Gillian Newell, Norman's widow. Ms. Newell not only shared recollections of life with her husband but also allowed me access to his unprocessed papers, for which I am extremely grateful. My research was also supported by several grants, including a National Science Foundation Scholars Award (NSF grants SES 0523123 and 0715259). I am grateful to all of these institutions, and to my supportive colleagues at Oberlin College and the University of North Carolina Wilmington.

Finally, I thank my family for their love and support: Maureen Meter, Christine Janis, Ella Henning-Sepkoski, and Teri Chettiar. The inspiration for this project extends, in a sense, to my childhood: my father, the late Jack Sepkoski, was a prominent member of the generation of paleontologists who brought about the "paleobiological revolution" this book describes. Some of the formative experiences of my youth were accompanying my dad on summer paleontology field trips, beginning when I was only seven or eight. As I got older I became more interested in his work, and by the time I was in graduate school my dad and I had weekly phone conversations in which we discussed his field and mine. We even talked of someday jointly writing a book about the history of life, as a kind of popular account of his life's work. His death, in 1999 at age 50, put an end to those discussions. But, in many ways, pursuing this project and writing this book has been a continuation of our conversations. I am sure he would be embarrassed by the attention, and quite probably he would disagree with some of my interpretations and conclusions, but the book would undoubtedly have given us plenty to talk about. It is dedicated to his memory.

Abbreviations

Manuscript and Archival Sources

AMNH DVP [box], [folder]	American Museum of Natural History Department of Vertebrate Paleontology archives
AMNH Inv.	American Museum of Natural History Department of Invertebrates archives
AMNH Newell [box]	American Museum of Natural History, papers of Norman D. Newell
Eldredge pap.	Personal papers of Niles Eldredge
PS [box]	Paleontological Society archives, main records
PS-*P* [box]	Paleontological Society archives, *Paleobiology* editorial records
Schopf pap. [box], [folder]	Smithsonian Institution archives, papers of Thomas J. M. Schopf
Sepkoski pap. [folder]	American Philosophical Society, papers of J. John Sepkoski Jr.

Oral Interviews

Ayala interview	Interview with Francisco Ayala conducted by David Sepkoski, 26 October 2005

Eldredge interview   Interview with Niles Eldredge con-
                                      ducted by David Sepkoski, 19 January
                                      2006
Jablonski interview   Interview with David Jablonski con-
                                      ducted by David Sepkoski, 7 June 2008
Raup interview   Interview with David M. Raup con-
                                      ducted by David Sepkoski, 20 July 2005
Raup interview (Princehouse)   Interview with David M. Raup con-
                                      ducted by Patricia Princehouse, 11 May
                                      1998
Stanley interview   Interview with Steven M. Stanley con-
                                      ducted by David Sepkoski, 11 July 2005
Valentine interview   Interview with James Valentine con-
                                      ducted by David Sepkoski, 24 October
                                      2005

Works Cited

Abel, Othenio. 1912. *Grundzüge der Palaeobiologie der Wirbeltiere*. Stuttgart: E. Schweizerbart.

———. 1980. *Palaeobiologie und Stammesgeschichte*. New York: Arno Press.

Adler, Jerry, and John Carey. 1980. Is Man a Subtle Accident? *Newsweek*, 3 November, 95.

Alberch, Pere, Stephen Jay Gould, George F. Oster, and David B. Wake. 1979. Size and Shape in Ontogeny and Phylogeny. *Paleobiology* 5 (3): 296–317.

Allen, Garland E. 1975. *Life Science in the Twentieth Century*. New York: Wiley.

Alvarez, Luis W., Walter Alvarez, Frank Asaro, and Helen V. Michel. 1980. Extraterrestrial Cause for the Cretaceous-Tertiary Extinction. *Science* 208 (4448): 1095–1108.

Bambach, Richard K. 1977. Species Richness in Marine Benthic Habitats through the Phanerozoic. *Paleobiology* 3 (2): 152–67.

———. 2009. From Empirical Paleoecology to Evolutionary Paleobiology: A Personal Journey. In *The Paleobiological Revolution: Essays on the Growth of Modern Paleontology*, edited by D. Sepkoski and M. Ruse. Chicago: University of Chicago Press, 398–415.

Bassler, R. S. 1910. Adequacy of the Fossil Record. *Popular Science* 76.

Begley, Sharon, and John Carey. 1984. Evolution in Hard Places. *Newsweek*, 2 January, 60.

Bowler, Peter J. 1983. *The Eclipse of Darwinism: Anti-Darwinian Evolution Theories in the Decades around 1900*. Baltimore: Johns Hopkins University Press.

———. 1989. *The Mendelian Revolution: The Emergence of Hereditarian Concepts in Modern Science and Society*. Baltimore: Johns Hopkins University Press.

———. 1996. *Life's Splendid Drama: Evolutionary Biology and the Reconstruction of Life's Ancestry, 1860–1940*. Chicago: University of Chicago Press.

Brinkman, Paul D. 2010. *The Second Jurassic Dinosaur Rush: Museums and Paleontology in America at the Turn of the Twentieth Century* Chicago: University of Chicago Press.

Buckman, S. S. 1893. *Quarterly Journal of the Geological Society* 49.

Burma, Benjamin H. 1948. Studies in Quantitative Paleontology: I. Some Aspects of the Theory and Practice of Quantitative Invertebrate Paleontology. *Journal of Paleontology* 22 (6): 725–61.

———. 1949. Studies in Quantitative Paleontology II. Multivariate Analysis: A New Analytical Tool for Paleontology and Geology. *Journal of Paleontology* 23 (1):95–103.

Cain, Joseph A. 1993. Common Problems and Cooperative Solutions: Organizational Activity in Evolutionary Studies, 1936–1947. *Isis* 84:1–25.

———. 1994. Ernst Mayr as Community Architect : Launching the Society for the Study of Evolution and the Journal Evolution. *Biology and Philosophy* 9:387–427.

———. 2002. Epistemic and Community Transition in American Evolutionary Studies: The 'Committee on Common Problems of Genetics, Paleontology, and Systematics' (1942–1949). *Studies in History and Philosophy of Biological and Biomedical Sciences* 33:283–313.

Calvin, Samuel. 1910. Adequacy of the Fossil Record. *Popular Science* 76.

Carson, Hampton L. 1981. Macroevolution Conference (Letters). *Science* 211 (4484): 773.

Coleman, William. 1971. *Biology in the Nineteenth Century: Problems of Form, Function, and Transformation.* New York: Wiley.

Cooper, G. Arthur. 1958. The Science of Paleontology. *Journal of Paleontology* 35 (5): 1010–18.

Cowen, Robert C. 1984. "Death Star" and Dinosaur-Extinction Theories are Mostly Speculation. *Christian Science Monitor,* 10 May, 21.

Cox, C. Barry. 1975. Review of Models in Paleobiology. *Journal of Biogeography* 2:229–30.

Cracraft, Joel. 1982. A Nonequilibrium Theory for the Rate-Control of Speciation and Extinction and the Origin of Macroevolutionary Patterns. *Systematic Zoology* 31 (4): 348–65.

Cushman, Joseph A. 1938. The Future of Paleontology. *Bulletin of the Geological Society of America* 49: 359–66.

Darwin, Charles. 1964 [1859]. *On the Origin of Species.* Cambridge, MA: Harvard University Press.

———. 1987. *Charles Darwin's Notebooks, 1836–1844: Geology, Transmutation of Species, Metaphysical Enquiries.* Edited by P. J. G. Paul, H. Barrett, Sandra Herbert, David Kohn, and Sydney Smith. Ithaca, NY: Cornell University Press.

Davis, Marc, Piet Hut, and Richard A. Muller. 1984. Extinction of Species by Periodic Comet Showers. *Nature* 308:715–17.

Deevey, E. S. 1974. Review of *Models in Paleobiology*. *Limnology and Oceanography* 19:375–76.

Desmond, Adrian J. 1982. *Archetypes and Ancestors: Palaeontology in Victorian London, 1850–1875*. London: Blond & Briggs.

Dobzhansky, Theodosius. 1945. Genetics of Macro-evolution: A Review of *Tempo and Mode in Evolution*. *Journal of Heredity* 36:113–15.

Dobzhansky, Theodosius 1951. *Genetics and the Origin of Species*. 3d, rev. ed. New York: Columbia University Press.

Dritschilo, William. 2008. Bringing Statistical Methods to Community and Evolutionary Ecology: Daniel S. Simberloff. In *Rebels, Mavericks, and Heretics in Biology*, edited by O. Harman and M. Dietrich. New Haven: Yale University Press.

Dunbar, Carl O. 1959. A Half Century of Paleontology. *Journal of Paleontology* 33 (5): 909–14.

Durrant, Stephen D. 1954. Review of Major Features of Evolution. *Journal of Mammalogy* 35 (4): 600–601.

"E. R. C". 1889. Review of *Die Stamme des Thierreiches* by Von M. Neumayer. *Nature* 39:364–65.

Eldredge, Niles. 1971. The Allopatric Model and Phylogeny in Paleozoic Invertebrates. *Evolution* 25 (1): 156–67.

———. 1976. Differential Evolutionary Rates. *Paleobiology* 2 (2): 174–77.

———. 1979. Alternative Approaches to Evolutionary Theory. In *Models and Methodologies in Evolutionary Theory*, edited by J. H. Schwartz and H. B. Rollins. Pittsburgh: Carnegie Museum of Natural History.

———. 1985a. *Time Frames: The Rethinking of Darwinian Evolution and the Theory of Punctuated Equilibria*. New York: Simon and Schuster.

———. 1985b. *Unfinished Synthesis: Biological Hierarchies and Modern Evolutionary Thought*. New York: Oxford University Press.

———. 1995. *Reinventing Darwin: The Great Debate at the High Table of Evolutionary Theory*. New York: Wiley.

———. 2008. The Early "Evolution" of "Punctuated Equilibria." *Evolution, Education, and Outreach* 1:107–13.

Eldredge, Niles, and Stephen Jay Gould. 1972. Punctuated Equilibria: An Alternative to Phyletic Gradualism. In *Models in Paleobiology*. San Francisco: Freeman, Cooper & Co.

Filipchenko, Iurii A. 1927. *Variabilität und Variation*. Berlin: Gebrüder Borntraeger.

Fischer, Alfred G. 1960. Latitudinal Variations in Organic Diversity. *Evolution* 14 (1): 64–81.

Flessa, Karl W., and John Imbrie. 1973. Evolutionary Pulsations: Evidence from Phanerozoic Diversity Patterns. In *Implications of Continental Drift to the Earth Sciences, Vol. 1.* London: Academic Press.

Flessa, Karl W. 1986. Extinctions are In. *Paleobiology* 12 (3): 329–34.

Flessa, Karl W., and David Jablonski. 1983. Extinction is Here to Stay. *Paleobiology* 9 (4): 315–21.

Foote, Michael. 1999. J. John Sepkoski, Jr. (1948–1999). *Acta Palaeontologica Polonica* 44 (2): 235–36.

Fox, William T. 1970. Analysis and Simulation of Paleoecologic Communities through Time. In *Proceedings of the North American Paleontological Convention,* edited by E. L. Yochelson. Lawrence, KS: Allen Press.

Futuyma, Douglas J., Richard Lewontin, G. C. Mayer, J. Seger, and J. W. Stubblefield. 1981. Macroevolution Conference (Letters). *Science* 211 (4484):770.

Galton, Francis. 1901. Biometry. *Biometrika* 1:7–10.

Gasking, Elizabeth B. 1970. *The Rise of Experimental Biology.* New York: Random House.

Gause, G. F. 1934. *The Struggle for Existence.* Baltimore: Williams and Wilkins.

Ginenthal, Charles, Stephen Jay Gould, Immanuel Velikovsky, and Dale Ann Pearlman. 1996. *Stephen J. Gould and Immanuel Velikovsky: Essays in the Continuing Velikovsky Affair.* Forest Hills, NY: Ivy Press Books.

Glass, Bentley. 1945. Review of Tempo and Mode in Evolution. *Quarterly Review of Biology* 20:261–63.

Golding, Winnifred. 1952. Foreword. *Journal of Paleontology* 26 (3): 298.

Goldschmidt, Richard Benedict. 1940. *The Material Basis of Evolution.* New Haven: Yale University Press.

Golinski, Jan. 2005. *Making Natural Knowledge: Constructivism and the History of Science.* 2nd ed. Chicago: University of Chicago Press.

Gorman, James. 1980. The Tortoise or the Hare? *Discover* (October).

Gould, Stephen Jay. 1966. Allometry and size in ontogeny and phylogeny. *Biological Reviews of the Cambridge Philosophical Society* 41 (4): 587–640.

———. 1971. D'Arcy Thompson and the Science of Form. *New Literary History* 2 (2): 229–58.

———. 1976a. The Genomic Metronome as a Null Hypothesis. *Paleobiology* 2 (2): 177–79.

———. 1976b. Palaentology plus Ecology as Palaeobiology. In *Theoretical Ecology: Principles and Applications,* edited by R. M. May. Philadelphia: Saunders.

———. 1977. *Ontogeny and Phylogeny.* Cambridge, MA: Belknap Press of Harvard University Press.

———. 1978. Generality and Uniqueness in the History of Life: An Exploration with Random Models. *BioScience* 28 (4): 277–81.

———. 1980a. G. G. Simpson, Paleontology, and the Modern Synthesis. In *The*

Evolutionary Synthesis; Perspectives on the Unification of Biology., edited by E. Mayr and William B. Provine. Cambridge, MA: Harvard University Press.

———. 1980b. Is a New and General Theory of Evolution Emerging? *Paleobiology* 6 (1): 119–30.

———. 1980c. The Promise of Paleobiology as a Nomothetic, Evolutionary Discipline. *Paleobiology* 6 (1): 96–118.

———. 1980d. The Telltale Wishbone. In *The Panda's Thumb: More Reflections in Natural History.* New York: Norton.

———. 1981. But Not Wright Enough: Reply to Orzack. *Paleobiology* 7 (1): 131–34.

———. 1982. Darwinism and the Expansion of Evolutionary Theory. *Science* 216 (4544): 380–87.

———. 1983. The Hardening of the Modern Synthesis. In *Dimensions of Darwinism; Themes and Counterthemes in Twentieth-Century Evolutionary Biology*, edited by Marjorie Grene. Cambridge: Cambridge University Press.

———. 1984a. The Cosmic Dance of Siva. *Natural History* (August).

———. 1984b. The Life and Work of T. J. M. Schopf (1939–1984). *Paleobiology* 10 (2): 280–85.

———. 1985a. All the News That's Fit to Print and Some Opinions That Aren't. *Discover* 6 (11): 86–91.

———. 1985b. Challenges to Neo-Darwinism and Their Meaning for a Revised View of Human Consciousness. *The Tanner Lectures on Human Values*, http://www.tannerlectures.utah.edu/lectures/documents/gould85.pdf.

———. 1985c. The Paradox of the First Tier: An Agenda for Paleobiology. *Paleobiology* 11 (1): 2–12.

———. 1987. *Time's Arrow, Time's Cycle: Myth and Metaphor in the Discovery of Geological Time.* Cambridge, MA: Harvard University Press.

———. 1989a. Punctuated Equilibrium in Fact and Theory. *Journal of Social and Biological Structures* 12: 117–36.

———. 1989b. *Wonderful Life: The Burgess Shale and the Nature of History.* New York: W. W. Norton.

———. 2002. *The Structure of Evolutionary Theory.* Cambridge, MA: Belknap Press of Harvard University Press.

Gould, Stephen Jay, and Niles Eldredge. 1977. Punctuated Equilibria: The Tempo and Mode of Evolution Reconsidered. *Paleobiology* 3 (2): 115–51.

Gould, Stephen Jay, and Richard Lewontin. 1979. The Spandrels of San Marco and the Panglossian Paradigm: A Critique of the Adaptationist Programme. *Proceedings of the Royal Society of London, Series B, Biological Sciences* 205 (1161): 581–98.

Gould, Stephen Jay, David M. Raup, J. John Sepkoski Jr., Thomas J. M. Schopf, and Daniel S. Simberloff. 1977. The Shape of Evolution: A Comparison of Real and Random Clades. *Paleobiology* 3 (1): 23–40.

Gould, Stephen Jay, and Elisabeth S. Vrba. 1982. Exaptation: A Missing Term in the Science of Form. *Paleobiology* 8 (1): 4–15.

Graubard, Mark. 1945. A New Approach to Evolution: Review of Tempo and Mode in Evolution. *The Scientific Monthly* 61 (1): 76–77.

Hagen, Joel. 2003. The Statistical Frame of Mind in Systematic Biology from Quantitative Zoology to Biometry. *Journal of the History of Biology* 36 (2): 353–384.

Haldane, J. B. S., and J. Huxley. 1927. *Animal Biology.* Oxford: Clarendon Press.

Harland, W. B., ed. 1967. *The Fossil Record.* London: Geological Society.

Heaney, Lawrence R. 2000. Dynamic Disequilibrium: A Long-Term, Large-Scale Perspective on the Equilibrium Model of Island Biogeography. *Global Ecology and Biogeography* 9 (1): 59–74.

Henbest, Lloyd G. 1952. Significance of Evolutionary Explosions for Diastrophic Division of Earth History: Introduction to the Symposium. *Journal of Paleontology* 26 (3): 299–318.

Herbert, Sandra. 2005. *Charles Darwin, Geologist.* Ithaca, NY: Cornell University Press.

Howell, B. F. 1945. Paleontology in the Post-War World. *Bulletin of the Geological Society of America* 56:371–84.

Hubbard, J. A. E. B. 1973. Critique in Palaeontology: Review of Models in Paleobiology. *Nature* 243:208.

Hubbs, C. 1945. Review of Tempo and Mode in Evolution. *American Naturalist* 79:271–75.

Huggett, Richard J. 1997. *Catastrophism: Asteroids, Comets, and Other Dynamic Events in Earth History.* London: Verso.

Hull, David L. 1988. *Science as a Process: An Evolutionary Account of the Social and Conceptual Development of Science.* Chicago: University of Chicago Press.

Huss, John Edward. 2004. Experimental Reasoning in Non-experimental Science: Case Studies from Paleobiology. PhD dissertation, University of Chicago.

———. 2009. The Shape of Evolution: The MBL Model and Clade Shape. In *The Paleobiological Revolution: Essays on the Growth of Modern Paleontology,* edited by David Sepkoski and Michael Ruse. Chicago: University of Chicago Press, 326–45.

Hutchinson, G. Evelyn. 1944. Review of Tempo and Mode in Evolution. *American Journal of Science* 243:356–358.

———. 1959. Homage to Santa Rosalia; or, Why Are There So Many Kinds of Animals? *American Naturalist* 93:145–59.

———. 1965. *The Ecological Theater and the Evolutionary Play*. New Haven: Yale University Press.

Huxley, Julian. 1932. *Problems of Relative Growth*. London: MacVeagh.

———. 1942. *Evolution, the Modern Synthesis*. London: Harper & Bros.

———. 1945. Genetics and Major Evolutionary Change: Review of Tempo and Mode in Evolution. *Nature* 156:3–4.

Huxley, Thomas Henry. 1893. *Darwiniana: Essays*. London: Macmillan.

———. 1894. *Man's Place in Nature, and Other Anthropological Essays*. New York: D. Appleton.

Imbrie, John. 1956. Biometrical methods in the study of invertebrate fossils. *Bulletin of the American Museum of Natural History* 108:215–252.

———. 1964. Factor Analytic Model in Paleoecology. In *Approaches to Paleoecology*, edited by J. Imbrie and N. D. Newell. New York: Wiley, 407–22.

Jablonski, David. 1986. Background and Mass Extinctions: The Alternation of Macroevolutionary Regimes. *Science* 231 (4734): 129–33.

Jaffe, Mark. 2000. *The Gilded Dinosaur: The Fossil War between E. D. Cope and O. C. Marsh and the Rise of American Science*. New York: Crown.

Johnson, Ralph G. 1964. The Community Approach to Paleoecology. In *Approaches to Paleoecology*, edited by J. Imbrie and N. D. Newell. New York: Wiley, 107–34.

Kingsland, Sharon E. 1982. The Refractory Model: The Logistic Curve and the History of Population Ecology. *Quarterly Review of Biology* 57 (1): 29–52.

———. 1985. *Modeling Nature: Episodes in the History of Population Ecology*. Chicago: University of Chicago Press.

Knight, J. Brookes. 1947. Paleontologist or Geologist. *Bulletin of the Geological Society of America* 58: 281–86.

Kolata, Gina Bari. 1975. Paleobiology: Random Events over Geological Time. *Science* 189 (4203): 625–26, 660.

Krumbein, W. C. 1969. The Computer in Geological Perspective. In *Computer Applications in the Earth Sciences: An International Symposium*, edited by F. Merriam Daniel. New York: Plenum Press, 251–75.

Kuhn, Thomas S. 1961. The Function of Measurement in Modern Physical Science. *Isis* 52: 161–90.

Kukalova-Peck, Jarmila. 1973. *A Phylogenetic Tree of the Animal Kingdom (Including Orders and Higher Categories)*. Ottawa: National Museums of Canada Publications in Zoology.

Kutschera, U. 2007. Palaeobiology: The Origin and Evolution of a Scientific Discipline. *Trends in Ecology and Evolution* 22 (4): 172–73.

Lande, Russell. 1980. Microevolution in Relation to Macroevolution. Review of *Macroevolution: Pattern and Process*. *Paleobiology* 6 (2):233–238.

Laporte, Léo F. 2000. *George Gaylord Simpson: Paleontologist and Evolutionist*. New York: Columbia University Press.

Levinton, Jeffrey S. 1980. Review of *Macroevolution: Pattern and Process*. *Quarterly Review of Biology* 55:432–3.

Lewin, Roger. 1980. Evolutionary Theory Under Fire. *Science* 210 (4472): 883–87.

———. 1985a. Catastrophism Not Yet Dead. *Science* 229 (4714): 640.

———. 1985b. Pattern and Process in Life's History. *Science* 229 (4709): 151–53.

Lovtrup, Soren. 1981. Macroevolution and Punctuated Equilibria. *Systematic Zoology* 30 (4): 498–500.

Lyell, Charles Sir Deshayes G. P. 1830. *Principles of Geology; Being an Attempt to Explain the Former Changes of the Earth's Surface, by Reference to Causes Now in Operation*. London: J. Murray.

Lyons, Sherrie Lynne. 1999. *Thomas Henry Huxley: The Evolution of a Scientist*. Amherst, NY: Prometheus Books.

MacArthur, Robert H. 1957. On the Relative Abundance of Bird Species. *Proceedings of the National Academy of Sciences* 45:293–95.

———. 1960. On the Relative Abundance of Species. *American Naturalist* 94 (874): 25–36.

MacArthur, Robert H., and Edward O. Wilson. 1963. An Equilibrium Theory of Insular Zoogeography. *Evolution* 17 (4): 373–87.

———. 2001 [1967]. *The Theory of Island Biogeography* Princeton, NJ: Princeton University Press.

Maddox, John. 1984. Extinctions by Catastrophe? *Nature* 308 (6951): 685.

———. 1985. Periodic Extinctions Undermined. *Nature* 315 (6021): 627.

Matthew, W. D. 1923. Recent Progress and Trends in Vertebrate Paleontology. *Bulletin of the Geological Society of America* 34: 401–18.

Maynard Smith, John. 1984. Palaeontology at the High Table. *Nature* 309:401–2.

Mayr, Ernst. 1942. *Systematics and the Origin of Species from the Viewpoint of a Zoologist*. New York: Columbia University Press.

———. 1947. Ecological Factors in Speciation. *Evolution* 1 (4): 263–88.

———. 1954. Change of Genetic Environment and Speciation. In *Evolution as a Process*, edited by A. C. Hardy, E. B. Ford, and J. S. Huxley. London: Allen and Unwin.

———. 1963. *Animal Species and Evolution*. Cambridge, MA: Belknap Press of Harvard University Press.

———. 1980a. G. G. Simpson. In *The Evolutionary Synthesis: Perspectives on the Unification of Biology*, edited by Ernst Mayr and William B. Provine. Cambridge, MA: Harvard University Press.

———. 1980b. Some Thoughts on the History of the Evolutionary Synthesis. In *The Evolutionary Synthesis: Perspectives on the Unification of Biology*, edited by Ernst Mayr and William B. Provine. Cambridge, MA: Harvard University Press.

———. 1982a. *The Growth of Biological Thought: Diversity, Evolution, and Inheritance.* Cambridge, MA: Belknap Press.

———. 1982b. Speciation and Macroevolution. *Evolution* 36 (6): 1119–32.

———. 1989. Speciational Evolution or Punctuated Equilibria. 12 (2–3): 137–58.

Mayr, Ernst, and William B. Provine. 1980. *The Evolutionary Synthesis: Perspectives on the Unification of Biology,* edited by Ernst Mayr and William B. Provine. Cambridge, MA: Harvard University Press.

McGhee, George R. 1999. *Theoretical Morphology: The Concept and its Applications.* New York: Columbia University Press.

McNamara, Kenneth J. 2002. Paleobiology: Bridging the Gap. *Biology and Philosophy* 17:729–38.

Mello, James F. 1970. Paleontologic Data Storage and Retrieval. In *Proceedings of the North American Paleontological Convention,* edited by E. L. Yochelson. Lawrence, KS: Allen Press, 57–71.

Miller, Arnold I. 2000. Conversations about Phanerozoic Global Diversity. *Paleobiology* 26 (4, supplement): 53–73.

———. 2009. The Consensus that Changed the Paleobiological World. In *The Paleobiological Revolution: Essays on the Growth of Modern Paleontology,* edited by David Sepkoski and Michael Ruse. Chicago: University of Chicago Press, 364–82.

Moore, R.C., ed. 1969. *Treatise on Invertebrate Paleontology.* Part N. Geological Society of America and University of Kansas Press.

Morgan, Thomas Hunt. 1916. *A Critique of the Theory of Evolution.* Princeton, NJ: Princeton University Press.

Moseley, H. 1838. On the Geometrical Forms of Turbinated and Discoid Shells. *Philosophical Transactions of the Royal Society of London* 128:351–70.

Newell, Norman D. 1947. Infraspecific Categories in Invertebrate Paleontology. *Evolution* 1 (3): 163–71.

———. 1952. Periodicity in Invertebrate Evolution. *Journal of Paleontology* 26 (3): 371–85.

———. 1956a. Catastrophism and the Fossil Record. *Evolution* 10 (1): 97–101.

———. 1956b. Fossil Populations. In *The Species Concept in Palaeontology; A Symposium,* edited by P. C. Sylvester-Bradley. London: Systematics Association, 63–82.

———. 1959. Adequacy of the Fossil Record. *Journal of Paleontology* 33 (3): 488–99.

———. 1962. Paleontological Gaps and Geochronology. *Journal of Paleontology* 36 (3): 592–610.

———. 1963. Crises in the History of Life. *Scientific American* 208 (2): 76–92.

———. 1967. Revolutions in the History of Life. In *Uniformity and Simplicity.* Boulder, CO: Geological Society of America (GSA), 63–91.

Newell, Norman D., and Edwin H. Colbert. 1948. Paleontologist: Biologist or Geologist? *Journal of Paleontology* 22 (2): 264–67.

Newsweek. 1984. A Death-Star Theory is Born: Nemesis. 5 March.

New York Times. 1985a. Miscasting the Dinosaurs' Horoscope. 2 April, A26.

———. 1985b. Nemesis of Nemesis. 7 July, section 4, page 14.

Novacek, Michael J. 1980. Review of Macroevolution: Pattern and Process. *Systematic Zoology* 29 (2): 219–25.

Nyhart, Lynn K. 1995. *Biology takes Form: Animal Morphology and the German Universities, 1800–1900*. Chicago: University of Chicago Press.

———. 2009. *Modern Nature: The Rise of the Biological Perspective in Germany*. Chicago: University of Chicago Press.

Olson, Everett C. 1970. Current and Projected Impacts of Computers upon Concepts and Research in Paleontology. In *Proceedings of the North American Paleontological Convention*, edited by E. L. Yochelson. Lawrence, KS: Allen Press, 135–53.

Olson, Everett C., and Robert L. Miller. 1951. A Mathematical Model Applied to a Study of the Evolution of Species. *Evolution* 5 (4): 325–38.

Oreskes, Naomi. 1999. *The Rejection of Continental Drift: Theory and Method in American Earth Science*. New York: Oxford University Press.

Orzack, Steven Hecht. 1981. The Modern Synthesis is Partly Wright. *Paleobiology* 7 (1): 128–31.

Osborn, Henry Fairfield. 1914. Rectigradations and Allometrons in Relation to the Conceptions of the ""Mutations of Waagen," of Species, Genera, and Phyla. *Bulletin of the Geological Society of America* 25:411–16.

———. 1933. Biological Inductions from the Evolution of the Probioscidea. *Palaeontology* 19:159–63.

Pauly, Philip J. 1987. *Controlling Life: Jacques Loeb and the Engineering Ideal in Biology*. Oxford: Oxford University Press.

Pearl, Raymond, and Lowell J. Reed. 1920. On the Rate of Growth of the Population of the United States since 1790 and Its Mathematical Representation. *Proceedings of the National Academy of Sciences of the United States of America* 6 (6): 275–88.

Pearson, Carl. 1901. On Lines and Planes of Closest Fit to Systems of Points in Space. *Philosophical Magazine* 6 (2): 559–72.

Phillips, John. 1860. *Life on the Earth; Its Origin and Succession*. Cambridge and London: Macmillan and Co.

Porter, Theodore M. 1986. *The Rise of Statistical Thinking, 1820–1900*. Princeton, NJ: Princeton University Press.

———. 1995. *Trust in Numbers: The Pursuit of Objectivity in Science and Public Life*. Princeton, NJ: Princeton University Press.

Princehouse, Patricia M. 2003. Mutant Phoenix: Macroevolution in Twentieth-

Century Debates over Synthesis and Punctuated Evolution. PhD dissertation, Harvard University.

Prothero, Donald R. 2009. Stephen Jay Gould: Did He Bring Paleontology to the "High Table"? *Philosophy and Theory in Biology* 1:1–7.

Provine, William B. 1971. *The Origins of Theoretical Population Genetics*. Chicago: University of Chicago Press.

———. 1986. *Sewall Wright and Evolutionary Biology*. Chicago: University of Chicago Press.

Provine, William B. 1989. Founder Effects and Genetic Revolutions in Microevolution and Speciation: An Historical Perspective. In *Genetics, Speciation, and the Founder Principle*, edited by L. V. Giddings, Kenneth Y. Kaneshiro, and Wyatt W. Anderson. New York: Oxford University Press, 43–76.

Quinn, James F, David M Raup, J. John Sepkoski Jr., and Stephen M Stigler. 1983. Mass Extinctions in the Fossil Record: Discussion and Reply. *Science* 219 (4589): 1239–41.

Rainger, Ronald. 1982. The Understanding of the Fossil Past: Paleontology and Evolution Theory, 1850–1910. PhD thesis, Indiana University.

———. 1986. Just before Simpson: William Diller Matthew's Understanding of Evolution. *Proceedings of the American Philosophical Society* 130:453–474.

———. 1988. Vertebrate Paleontology as Biology: Henry Fairfield Osborn and the American Museum of Natural History. In *The American Development of Biology*, edited by Ronald Rainger. Philadelphia: University of Pennsylvania Press, 219–56.

———. 1991. *An Agenda for Antiquity: Henry Fairfield Osborn and Vertebrate Paleontology at the American Museum of Natural History, 1890–1935*. Tuscaloosa: University of Alabama Press.

———. 1993. Biology, Geology, or Neither, or Both: Vertebrate Paleontology at the University of Chicago, 1892–1950. *Perspectives on Science* 1 (3): 478–519.

———. 1997. Everett C. Olson and the Development of Vertebrate Paleoecology and Taphonomy. *Archives of Natural History* 24:383–96.

———. 2001. Subtle Agents for Change: The *Journal of Paleontology*, J. Marvin Weller, and Shifting Emphases in Invertebrate Paleontology, 1930–1965. *Journal of Paleontology* 75 (6): 1058–64.

Raup, David M. 1961. The Geometry of Coiling in Gastropods. *Proceedings of the National Academy of Sciences of the United States of America* 47:602–9.

———. 1962. Computer as Aid in Describing Form in Gastropod Shells. *Science* 138 (3537): 150–52.

———. 1966. Geometric Analysis of Shell Coiling: General Problems. *Journal of Paleontology* 40 (5): 1178–90.

———. 1967. Geometric Analysis of Shell Coiling: Coiling in Ammonoids. *Journal of Paleontology* 41 (1): 43–65.

———. 1970. Modeling and Simulation of Morphology by Computer. In *Proceedings of the North American Paleontological Convention*, edited by E. L. Yochelson. Lawrence, KS: Allen Press, 71–83.

———. 1972. Taxonomic Diversity during the Phanerozoic. *Science* 177 (4054): 1065–71.

———. 1975a. Taxonomic Diversity Estimation Using Rarefaction. *Paleobiology* 1 (4): 333–42.

———. 1975b. Taxonomic Survivorship Curves and Van Valen's Law. *Paleobiology* 1 (1): 82–96.

———. 1976a. Species Diversity in the Phanerozoic: A Tabulation. *Paleobiology* 2 (4): 279–88.

———. 1976b. Species Diversity in the Phanerozoic: An Interpretation. *Paleobiology* 2 (4): 289–97.

———. 1977. Probabilistic Models in Evolutionary Paleobiology. *American Scientist* 65 (1): 50–57.

———. 1978a. Approaches to the Extinction Problem. *Journal of Paleontology* 52 (3): 517–23.

———. 1978b. Cohort Analysis of Generic Survivorship. *Paleobiology* 4 (1): 1–15.

———. 1978c. Size of the Permo-Triassic Bottleneck and Its Evolutionary Implications. *Science* 206 (4415): 217–18.

———. 1979. Biases in the Fossil Record of Species and Genera. *Bulletin of the Carnegie Museum of Natural History* 13: 85–91.

———. 1981. Extinction: Bad Genes or Bad Luck? *Acta Geologica Hispanica* 16 (1–2): 25–33.

———. 1982. Large Body Impacts and Terrestrial Evolution Meeting, October 19–22, 1981. *Paleobiology* 8 (1): 1–3.

———. 1986a. Biological Extinction in Earth History. *Science* 231 (4745): 1528–33.

———. 1986b. *The Nemesis Affair: A Story of the Death of Dinosaurs and the Ways of Science.* New York: W. W. Norton & Co.

Raup, David M., Stephen Jay Gould, Thomas J. M. Schopf, and Daniel S. Simberloff. 1973. Stochastic Models of Phylogeny and the Evolution of Diversity. *Journal of Geology* 81 (5): 525–42.

Raup, David M., and Adolf Seilacher. 1969. Fossil Foraging Behavior: Computer Simulation. *Science* 166 (3908): 994–95.

Raup, David M., and J. John Sepkoski Jr. 1982. Mass Extinctions in the Marine Fossil Record. *Science* 215 (4539): 1501–3.

———. 1984. Periodicity of Extinctions in the Geologic Past. *Proceedings of the National Academy of Sciences of the United States of America* 81 (3): 801–5.

Raup, David M., and Steven M. Stanley. 1971. *Principles of Paleontology.* San Francisco: W. H. Freeman.

———. 1978. *Principles of Paleontology*. 2nd ed. San Francisco: W. H. Freeman.

Raup, David M., and Stephen Jay Gould. 1974. Stochastic Simulation and Evolution of Morphology: Towards a Nomothetic Paleontology. *Systematic Zoology* 23 (3): 305–22.

Raup, David M., and Arnold Michelson. 1965. Theoretical Morphology of the Coiled Shell. *Science* 147 (3663): 1294–95.

Reif, W. E. 1986. The Search for a Macroevolutionary Theory in German Paleontology. *Journal of the History of Biology* 19:79–130.

———. 1999. Deutschsprachige Paläontologie im Spannungsfeld zwischen Makroevolutionstheorie und Neo-Darwinismus (1920–1950). In *Die Entstehung der synthetischen Theorie. Beitruage zur Geschichte der Evolutionsbiologie in Deutschland 1930–1950*, edited by Eve-Marie Engels, Thomas Junker, and Michael Weingarten. Berlin: Verlag für Wissenschaft und Bildung, 151–88.

Reif, Wolf-Ernst, Thomas Junker, and Uwe Houfeld. 2000. The Synthetic Theory of Evolution: General Problems and the German Contribution to the Synthesis. *Theory in Biosciences* 119:41–91.

Rensberger, Boyce. 1980. Recent Studies Spark Revolution in Interpretation of Evolution. *New York Times*, 4 November.

———. 1984. Extinction Gaining Force in Theory of Evolution; Darwin's "Origin of the Species," 125 Years Old Today, Challenged by Biologists. *Washington Post*, 24 November, A1.

Rohlf, F. James, and Robert R. Sokal. 1962. The Description of Taxonomic Relationships by Factor Analysis. *Systematic Zoology* 11 (1): 1–16.

Roughgarden, Jonathan. 1973. Possibilities for Paleontology: Review of *Models in Paleobiology*. *Science* 179:1225.

Rowell, A. J. 1973. Review of *Models in Paleobiology*. *Systematic Biology* 22:94–95.

Rudwick, M. J. S. 2005a. *Bursting the Limits of Time: The Reconstruction of Geohistory in the Age of Revolution*. Chicago: University of Chicago Press.

———. 2005b. *Lyell and Darwin, Geologists: Studies in the Earth Sciences in the Age of Reform*. Burlington, VT: Ashgate.

Ruse, Michael. 1989. Is the Theory of Punctuated Equilibria a New Paradigm? In *The Darwinian Paradigm*, edited by Michael Ruse. London: Routledge.

———. 1999a. *Mystery of Mysteries: Is Evolution a Social Construction?* Cambridge, MA: Harvard University Press.

———. 1999b. This View of Stephen Jay Gould. *Natural History* 108 (9): 54–55.

Schopf, J. William. 2009. Emergence of Precambrian Paleobiology: A New Field of Science. In *The Paleobiological Revolution: Essays on the Growth of Modern Paleontology*, edited by David Sepkoski and Michael Ruse. Chicago: University of Chicago Press, 89–110.

Schopf, Thomas J. M. 1972a. An Approach to Understanding Evolutionary Relationships in the Phylum Ectoprocta. In *Papers on Marine Science: The*

Link Lecture Series, edited by A. L. Meyerson and C. S. Zois. Montclair, NJ: Montclair State College, 1–10.

———. 1972b. Varieties of Paleobiologic Experience. In *Models in Paleobiology.* San Francisco: Freeman, Cooper & Co.

———. 1973. Ergonomics of Polymorphism: Its Relation to the Colony as the Unit of Natural Selection in Species of the Phylum Ectoprocta. In *Animal Colonies; Development and Function through Time.* Stroudsburg, PA: Dowden, Hutchinson & Ross, 274–94.

———. 1974. Permo-Triassic Extinctions: Relation to Sea-Floor Spreading. *Journal of Geology* 82 (2): 129–43.

———. 1976a. Environmental versus Genetic Causes of Morphologic Variability in Bryozoan Colonies from the Deep Sea. *Paleobiology* 2 (2): 156–65.

———. 1976b. Ralph Gordon Johnson 1927–1976. *Paleobiology* 2 (4): 388–91.

———. 1979. Evolving Paleontological Views on Deterministic and Stochastic Approaches. *Paleobiology* 5 (3): 337–52.

———, ed. 1980. *Presidential Addresses of the Paleontological Society.* New York: Arno Press.

———. 1981. Punctuated Equilibrium and Evolutionary Stasis. *Paleobiology* 7 (2): 156–66.

———. 1982. A Critical Assessment of Punctuated Equilibria: I, Duration of Taxa. *Evolution* 36 (6): 1144–57.

Schopf, Thomas J. M., Antoni Hoffman, and Stephen Jay Gould. 1983. Punctuated Equilibrium and the Fossil Record: Discussion and Reply. *Science* 219 (4584): 438–40.

Schopf, Thomas J. M., David M. Raup, Stephen Jay Gould, and Daniel S. Simberloff. 1975. Genomic versus Morphologic Rates of Evolution: Influence of Morphologic Complexity. *Paleobiology* 1 (1): 63–70.

Schuchert, Charles. 1910. Biologic Principles of Paleogeography. *Popular Science* 76.

Sepkoski, David. 2005. Stephen Jay Gould, Jack Sepkoski, and the "Quantitative Revolution" in American Paleobiology. *Journal of the History of Biology* 38 (2): 209–37.

Sepkoski, David, and David M. Raup. 2009. An Interview with David M. Raup. In *The Paleobiological Revolution: Essays on the Growth of Modern Paleontology,* edited by David Sepkoski and Michael Ruse. Chicago: University of Chicago Press.

Sepkoski, J. John, Jr. 1975. Stratigraphic Biases in the Analysis of Taxonomic Survivorship. *Paleobiology* 1 (4): 343–55.

———. 1978. A Kinetic Model of Phanerozoic Taxonomic Diversity I. Analysis of Marine Orders. *Paleobiology* 4 (3):223–251.

———. 1979. A Kinetic Model of Phanerozoic Taxonomic Diversity II: Early Phanerozoic Families and Multiple Equilibria. *Paleobiology* 5 (3): 222–51.

——. 1981. A Factor Analytic Description of the Phanerozoic Marine Fossil Record. *Paleobiology* 7 (1): 36–53.

——. 1982. *A Compendium of Fossil Marine Families*. Milwaukee: Milwaukee Public Museum.

——. 1984. A Kinetic Model of Phanerozoic Taxonomic Diversity: III. Post-Paleozoic Families and Mass Extinctions. *Paleobiology* 10 (2): 246–67.

——. 1985. Some Implications of Mass Extinction for the Evolution of Complex Life. In *The Search for Extraterrestrial Life: Recent Developments*, edited by M. D. Papagiannis. Dordrecht: D. Reidel.

——. 1993. Ten Years in the Library: New Data Confirm Paleontological Patterns. *Paleobiology* 19 (1): 43–51.

——. 1994. What I Did with My Research Career; or, How Research on Biodiversity Yielded Data on Extinction. In *The Mass-Extinction Debates: How Science Works in a Crisis.*, edited by William Glen. Stanford, CA: Stanford University Press, 132–44.

Sepkoski, J. John, Jr., Richard K. Bambach, David M. Raup, and James W. Valentine. 1981. Phanerozoic Marine Diversity and the Fossil Record. *Nature* (London) 293 (5832): 435–37.

Sepkoski, J. John, Jr., David Jablonski, and Michael Foote. 2002. *A Compendium of Fossil Marine Animal Genera*. Ithaca, NY: Paleontological Research Institution.

Sepkoski, J. John, Jr., and David M. Raup. 1982. Macro-extinction. In *North American PaleontologicalCconvention III: Abstracts of Papers*. Tulsa, OK: Society of Economic Paleontologists and Mineralogists.

Sepkoski, J. John, Jr., and Michael A. Rex. 1974. Distribution of Freshwater Mussels: Coastal Rivers as Biogeographic Islands. *Systematic Zoology* 23 (2): 165–88.

Shermer, Michael B. 2002. This View of Science: Stephen Jay Gould as Historian of Science and Scientific Historian, Popular Scientist and Scientific Popularizer. *Social Studies of Science* 32 (4): 489–524.

Shubin, Neil. 2008. *Your Inner Fish: A Journey into the 3.5-Billion-Year History of the Human Body*. New York: Pantheon.

Simberloff, Daniel S. 1972. Models in Biogeography. In *Models in Paleobiology*, edited by Thomas J. M. Schopf. San Francisco: Freeman, Cooper & Co.

——. 1974. Permo-Triassic Extinctions: Effects of Area on Biotic Equilibrium. *Journal of Geology* 82:267–74.

Simpson, George Gaylord. 1926. Mesozoic Mammalia IV: The Multituberculates as Living Animals. *American Journal of Science* 11:228–50.

——. 1944. *Tempo and Mode in Evolution*. New York: Columbia University Press.

——. 1952. Periodicity in Vertebrate Evolution. *Journal of Paleontology* 26 (3): 359–70.

——. 1953. *The Major Features of Evolution*. New York: Columbia University Press.

——. 1960. The History of Life. In *Evolution after Darwin, Volume 1: The Evolution of Life*, edited by Sol Tax. Chicago: University of Chicago Press, 117–80.

——. 1964. Species Density of North American Recent Mammals. *Systematic Zoology* 13 (2): 57–73.

——. 1978. *Concession to the Improbable: An Unconventional Autobiography*. New Haven: Yale University Press.

Simpson, George Gaylord, and Anne Roe. 1939. *Quantitative Zoology: Numerical Concepts and Methods in the Study of Recent and Fossil Animals*. New York: McGraw-Hill.

Simpson, George Gaylord, Anne Roe, and Richard C. Lewontin. 1960. *Quantitative Zoology*. Rev. ed. New York: Harcourt.

Smocovitis, Vassiliki Betty. 1994. Disciplining Evolutionary Biology: Ernst Mayr and the Founding of the Society for the Study of Evolution and Evolution (1939–1950). *Evolution* 48 (1): 1–8.

——. 1996. *Unifying Biology: The Evolutionary Synthesis and Evolutionary Biology*. Princeton, NJ: Princeton University Press.

Sokal, Robert R., and F. James Rohlf. 1969. *Biometry: The Principles and Practice of Statistics in Biological Research*. San Francisco: W. H. Freeman.

Sokal, Robert R., and P. H. A. Sneath. 1963. *Principles of Numerical Taxonomy*. San Francisco: W. H. Freeman.

Stanley, Steven M. 1975. A Theory of Evolution above the Species Level. *Proceedings of the National Academy of Sciences of the United States of America* 72 (2): 646–50.

——. 1979. *Macroevolution: Pattern and Process*. San Francisco: W.H. Freeman.

Stanley, Steven M., Philip W. Signor III, Scott Lidgard, and Alan F Karr. 1981. Natural Clades Differ from "Random" Clades: Simulations and Analyses. *Paleobiology* 7 (1): 115–27.

Stebbins, G. Ledyard, and Francisco J. Ayala. 1981. Is a New Evolutionary Synthesis Necessary? *Science* 213 (4511): 967–71.

Stephenson, Lloyd William. 1942. Paleontology: An Appraisal. *Bulletin of the Geological Society of America* 53: 373–79.

Teichert, Curt. 1956. How Many Fossil Species? *Journal of Paleontology* 30 (4): 967–69.

Templeton, Alan R., and L. Val Giddings. 1981. Macroevolution Conference (Letters). *Science* 211 (4484): 770–73.

Thompson, D'Arcy Wentworth. 1992. *On Growth and Form: The Complete Revised Edition*. New York: Dover.

Time. 1985. Did Comets Kill the Dinosaurs? 6 May.

Tucker, Ledyard, and Robert MacCallum. 1993. *Exploratory Factor Analysis.* Accessed at http://www.unc.edu/~rcm/book/factornew.htm.

Valentine, James W. 1968. The Evolution of Ecological Units above the Population Level. *Journal of Paleontology* 42 (2): 253–67.

———. 1969. Patterns of Taxonomic and Ecological Structure of the Shelf Benthos during Phanerozoic Time. *Palaeontology* 12 (4): 684–709.

———. 1970. How Many Marine Invertebrate Fossil Species? A New Approximation. *Journal of Paleontology* 44 (3):410–15.

———. 1973a. *Evolutionary Paleoecology of the Marine Biosphere.* Englewood Cliffs, NJ: Prentice-Hall.

———. 1973b. Phanerozoic Taxonomic Diversity: A Test of Alternate Models. *Science* 180 (4090): 1078–79.

———. 2001. Scaling is Everything: Brief Comments on Evolutionary Paleoecology. In *Evolutionary Paleoecology: The Ecological Context of Macroevolutionary Change.,* edited by W. D. Allmon and D. J. Bottjer. New York: Columbia University Press, 9–13.

———. 2009. The Infusion of Biology into Paleontological Research. In *The Paleobiological Revolution: Essays on the Growth of Modern Paleontology,* edited by David Sepkoski and Michael Ruse. Chicago: University of Chicago Press, 385–97.

Van Valen, Leigh. 1973. A New Evolutionary Law. *Evolutionary Theory* 1:1–30.

———. 1980. One Man's View of Evolution. *BioScience* 30 (9): 620.

Velikovsky, Immanuel. 1955. *Earth in Upheaval.* Garden City, NY: Doubleday.

Vrba, Elisabeth S. 1980. Evolution, Species and Fossils: How Does Life Evolve? *South African Journal of Science* 76:61–84.

———. 1984. What is Species Selection? *Systematic Zoology* 33 (3): 318–28.

Vrba, Elisabeth S., and Niles Eldredge. 1984. Individuals, Hierarchies and Processes: Towards a More Complete Evolutionary Theory. *Paleobiology* 10 (2): 146–71.

Vrba, Elisabeth S., and Stephen Jay Gould. 1986. The Hierarchical Expansion of Sorting and Selection: Sorting and Selection Cannot Be Equated. *Paleobiology* 12 (2): 217–28.

Wade, Nicholas. 1984. From Nemesis to Noah. *New York Times,* 4 May, A30.

Wallace, David R. 1999. *The Bonehunters' Revenge: Dinosaurs, Greed and the Greatest Scientific Feud of the Gilded Age.* Boston: Houghton Mifflin.

Weller, J. Marvin. 1947. Relations of the Invertebrate Paleontologist to Geology. *Journal of Paleontology* 21 (6): 570–75.

———. 1948. Paleontologist: Biologist and Geologist. *Journal of Paleontology* 22 (2): 268–69.

Wilford, John Noble. 1983. Study Hints Extinctions Strike in Set Intervals. *New York Times*, 11 December, section 1, page 70.

———. 1984. Search for "Nemesis" Intensifies Debate. *New York Times*, 18 December, section C, page 1.

Wilson, Edward O. 1961. The Nature of the Taxon Cycle in the Melanesian Ant Fauna. *The American Naturalist* 95 (882): 169–93.

Wilson, Edward O., and William H. Bossert. 1971. *A Primer of Population Biology*. Stamford, CT: Sinauer Associates.

Wilson, Edward O., and G. Evelyn Hutchinson. 1989. Robert Helmer MacArthur. *Biographical Memoirs* 58:318–27.

Wolfram, Stephen. 2002. *A New Kind of Science*. Wolfram, IL: Wolfram Media.

Woodruff, David S. 1980. Evolution: The Paleobiological View. *Science* 208 (4445): 716–17.

Wright, Sewall. 1945. A Critical Review. *Ecology* 26:415–19.

York, Derek. 1985. Pattern of Mass Extinctions Not Just Chance, Theorists Say. *Globe and Mail*, 29 July.

Index

Italicized page numbers indicate illustrations. Page numbers followed by the letter t indicate tables.